Handbook of technical textiles

Second edition

The Textile Institute and Woodhead Publishing

The Textile Institute is a unique organisation in textiles, clothing, and footwear. Incorporated in England by a Royal Charter granted in 1925, the Institute has individual and corporate members in over 90 countries. The aim of the Institute is to facilitate learning, recognise achievement, reward excellence, and disseminate information within the global textiles, clothing, and footwear industries.

Historically, The Textile Institute has published books of interest to its members and the textile industry. To maintain this policy, the Institute has entered into partnership with Woodhead Publishing Limited to ensure that Institute members and the textile industry continue to have access to high calibre titles on textile science and technology.

Most Woodhead titles on textiles are now published in collaboration with The Textile Institute. Through this arrangement, the Institute provides an Editorial Board which advises Woodhead on appropriate titles for future publication and suggests possible editors and authors for these books. Each book published under this arrangement carries the Institute's logo.

Woodhead books published in collaboration with The Textile Institute are offered to Textile Institute members at a substantial discount. These books, together with those published by The Textile Institute that are still in print, are offered on the Elsevier website at http://store.elsevier.com/. Textile Institute books still in print are also available directly from the Institute's website at www.textileinstitutebooks.com.

A list of Woodhead books on textile science and technology, most of which have been published in collaboration with The Textile Institute, can be found towards the end of the contents pages.

Related titles

Smart textiles for protection
ISBN (9780857090560)

Fibrous and composite materials for civil engineering applications
ISBN (9781845695583)

Textile advances in the automotive industry
ISBN (9781845693312)

Woodhead Publishing Series in Textiles: Number 169

Handbook of technical textiles

Second edition

Volume 1: Technical Textile Processes

Edited by

A Richard Horrocks, Subhash C. Anand

The Textile Institute

ELSEVIER

AMSTERDAM • BOSTON • CAMBRIDGE • HEIDELBERG
LONDON • NEW YORK • OXFORD • PARIS • SAN DIEGO
SAN FRANCISCO • SINGAPORE • SYDNEY • TOKYO
Woodhead Publishing is an imprint of Elsevier

WP
WOODHEAD
PUBLISHING

Published by Woodhead Publishing in association with The Textile Institute
Woodhead Publishing is an imprint of Elsevier
80 High Street, Sawston, Cambridge, CB22 3HJ, UK
225 Wyman Street, Waltham, MA 02451, USA
Langford Lane, Kidlington, OX5 1GB, UK

Notices
Knowledge and best practice in this field are constantly changing. As new research and experience broaden our understanding, changes in research methods, professional practices, or medical treatment may become necessary.

Practitioners and researchers must always rely on their own experience and knowledge in evaluating and using any information, methods, compounds, or experiments described herein. In using such information or methods they should be mindful of their own safety and the safety of others, including parties for whom they have a professional responsibility.

To the fullest extent of the law, neither the Publisher nor the authors, contributors, or editors, assume any liability for any injury and/or damage to persons or property as a matter of products liability, negligence or otherwise, or from any use or operation of any methods, products, instructions, or ideas contained in the material herein.

ISBN: 978-1-78242-458-1 (print)
ISBN: 978-1-78242-481-9 (online)

British Library Cataloguing-in-Publication Data
A catalogue record for this book is available from the British Library

Library of Congress Control Number: 2015937281

For information on all Woodhead Publishing publications
visit our website at http://store.elsevier.com/

Working together
to grow libraries in
developing countries

www.elsevier.com • www.bookaid.org

Contents

Contributors

S.C. Anand Institute of Materials Research and Innovation, The University of Bolton, Bolton, UK

X. Chen University of Manchester, Manchester, UK

R. Conway Textiles Division, School of Materials, University of Manchester, Manchester, United Kingdom

K.L. Gandhi Consultant, Boothstown, Worsley, Manchester, UK

R.H. Gong University of Manchester, Manchester, UK

J.W.S. Hearle University of Manchester, Manchester, United Kingdom

I. Holme International Dyer, World Textile Information Network, Leeds, United Kingdom

B.J. McCarthy University of Manchester, Manchester, United Kingdom

M. Miraftab Institute for Materials Research and Innovation, University of Bolton, Bolton, United Kingdom

S.J. Russell Nonwovens Research Group, School of Design, University of Leeds, Leeds, United Kingdom

P.A. Smith Nonwovens Research Group, School of Design, University of Leeds, Leeds, United Kingdom

W.S. Sondhelm Consultant, Didsbury, Manchester, UK

L.W. Taylor University of Manchester, Manchester, UK

L.-J. Tsai Department of Textiles and Clothing, Fu Jen University, New Taipei City, Taiwan

Preface

The first edition of the *Handbook of Technical Textiles* was published in 2000. It has proven to be such a successful text that the authors have produced this updated and expanded version, not only because of its success, but also because the subject has seen considerable development over the last 15 years. To reflect the significant expansion in the whole area of technical textiles, this second edition has been divided into two parts. Volume 1 features *Technical Textile Processes* and Volume 2 focuses on *Technical Textile Applications*.

Almost 20 years ago, technical textiles were reported to be the fastest growing sector of the textile industrial sector. In 1997, they accounted for almost 19% (10 million tons) of the total world fibre consumption for all textile uses, totalling 53 million tons. Since then, the diversity of technical textiles as a defined sector has increased. Its borders have increasingly become less sharp, and overlaps with what might be termed the conventional textile sector have occurred. Thus it is difficult to identify a meaningful current global set of figures in terms of fibre usage or financial value. However, in the EU, such figures are often assumed to be more easily defined, with an estimated 30% of total textile industry value estimated as turnover, excluding clothing. The financial value has been put as low as 30 billion and as high as 50 billion Euros, and so even in the EU, there still remains great uncertainty in such figures.

Global figures published for 2007 suggested a value of US$126 billion, with a growth rate four times greater than conventional textiles. As the estimated global value in 1997 was more than US$60 billion, the total value, if not volume, has doubled within a 10-year period. It is most likely that even with the world economic downturn of 2008 and beyond, growth rates, while reducing slightly, will still have remained positive. This signifies the continuing buoyancy and excitement of the sector. This is confirmed by a recent estimate for the global value in 2012 of US$134 billion with a predicted growth to US$160 billion by 2018 (Transparency™ Market Research, August 2015), thus indicating an average growth rate of 3.3%.

The uniqueness and challenge of technical textiles lies in the need to understand and apply the principles of textile science and technology to provide solutions, mainly to technological problems, but also often to engineering problems, which are defined by the respective application area. With the emphasis on measurable textile performance in a particular field of application, this requires the technologist to have not only an intricate knowledge of fibres and textile science and technology. But they also must have an understanding of the demands of the particular application and the scientists, technologists, and engineers who service it. Thus, for example, the producer of geotextiles requires an intricate knowledge across the civil engineering and built environmental areas and related challenges within an often complex supply chain from producer to end-user. Similarly, the medical textile producer must have some knowledge of the requirements of consultant, medical practitioner, and nurse. This handbook assists in providing a bridge between producer and end-user.

In Volume 1, the main principles involved in the selection of raw materials and their conversion into yarns and fabrics, having various structures, are considered in detail. This discussion is followed by the essential dyeing, finishing, and coating technologies required for technical textiles. In this edition, in addition to updating the chapters previously presented in the first edition, there are chapters on the production and processing of both three-dimensional and one-dimensional textile structures. These structures include rope, cord, twine, webbing, and nets; the additional chapters provide a more complete coverage of all technical textile structures than hitherto.

Each of the chapters has been specially prepared and edited by internationally recognised experts to cover current developments, as well as future trends in the principles and manufacturing processes in their respective fields.

Woodhead Publishing Series in Textiles

An overview of the technical textiles sector

B.J. McCarthy
University of Manchester, Manchester, United Kingdom

Chapter Outline

1.1   Introduction

The economic scope and importance of technical textiles extend far beyond the textile industry itself and has an impact on just about every sphere of human economic and social activity. In spite of the economic downturn triggered by the financial crisis of 2008, the global technical textiles sector is growing at approximately 4% annually – a major driver of innovation and invention.

Technical textiles represent the fastest growing sector in the textile and clothing industries. It has never been seen as a single coherent industry sector and market segment. The barriers between traditional definitions of textiles and technical textiles are continually changing. *Technical textiles* can be seen as niche, high-added-value products. Manufacturers in the sector have strong basic textile skills in manipulating fibres, fabrics, and finishing techniques as well as an understanding of how these parameters interact and perform in different combinations and environments.

It would be useful to begin with a broad overview of textiles generally. Initially, *textile* was originally defined as a woven fabric, but that term (and the plural *textiles*) is now also applied to fibres, filaments and yarns, natural and manufactured, and most products for which these materials are a principal raw material.

Textiles can be woven or knitted, braided, or produced as layers of nonwoven materials. Textiles are normally made from polymers – natural or synthetic – molecules that consist of covalently bonded, repeating units (McCarthy, 2013).

These polymers can derive from natural sources (e.g. wool and cotton) or from synthetic sources (e.g. polyester and polyamide). Textiles can be converted into clothing: normally garments, articles of dress that cover the body and/or limbs. The term *apparel* can also be used to refer to personal outfits, clothing, and attire.

Handbook of Technical Textiles. http://dx.doi.org/10.1016/B978-1-78242-458-1.00001-7

Natural fibres such as cotton, flax, jute, and sisal have been used for centuries in applications ranging from tents and tarpaulins to ropes, sailcloth, and sacking. There is evidence of woven fabrics being used in Roman times and before to stabilise marshy ground for road building – early examples of what would now be termed *geotextiles* and *geogrids* (Byrne, 2000).

TexData reports that according to Bremen Cotton Exchange, the world's total fibre consumption will rise rapidly in the next 5–10 years while cotton production is potentially close to its global production limit. One reason for this textile stimulus is the huge growth of technical textiles and nonwovens, another the increase of the world human population leading to the continually increasing demand for textiles and clothing (www.texdata.com).

Synthetic fibres constitute the largest share of fibre production with some 58% of global volume. Global volumes are predicted to increase from 40 to 56 million tonnes by 2020. This would mean a growth of 40% within 6 years. Cotton will continue to be the most popular fibre in the apparel industry in the near future (Chalupsky, 2014). For the first time ever, the worldwide fibre demand in 2013 exceeded 90 million tonnes or 4.4% more than in 2012.

The global per capita use of textile materials for clothing, home textiles, carpets, and technical textiles was estimated at 12.7 kg in 2014.

Synthetic fibres in 2014 continued their growth rate and increased 5.7% to 54.4 million tonnes, mainly because of the growth of polyester and polyamide production, whereas the volume growth of cellulosic fibres amounted to 10.4%, or nearly 6 million tonnes.

In contrast, cotton demand in 2014 increased slightly by 0.9% to 23.6 million tonnes. The worldwide stocks of cotton in storage exceeded the 20 million tonnes mark for the first time ever (mainly because of stockpiling in China), which covers 85% of demand.

Wool continued to expand in the fourth of successive years in 2014 and presented a growth of 1.2% to 1.1 million tonnes (Anon, 2014).

1.2   Definition of technical textiles

The current definition of technical textiles accepted by the internationally authoritative Textile Terms and Definitions (published by The Textile Institute) is "textile materials and products manufactured primarily for their technical and performance properties rather than their aesthetic or decorative characteristics" (www.ttandd.org).

The European Union (EU) definitions read that technical textiles are defined as technical fibres, materials, and support materials meeting technical rather than aesthetic criteria, although in some markets (e.g. work wear or sportswear) both types or criteria are specified.

The sector of technical textiles that continues to register positive economic and employment trends in the EU is an example of a traditional sector able to reinvest itself on a new business model fully suited to the needs of the new industrial revolution (smarter, more inclusive, and more sustainable) (Butaud-Stubbs, 2013).

Technical textiles can be made into clothing – but they normally offer enhanced protection (e.g. against heat, cold, flame, chemical, biological and nuclear agents, detection, that is, camouflage) and even ballistic threats.

Byrne (2000) maintains that no two published sources, industry bodies or statistical organisations ever seem to adopt precisely same approach when describing and categorising specific products and applications as technical textiles. Technical textiles have never been a single coherent industry sector and market segment. He states that it is not surprising that any attempt to define too closely and too rigidly the scope and content of technical textiles and their markets is doomed to failure.

For example, a simple cotton T-shirt would not be regarded as a technical textile. But what about a cotton T-shirt printed with a thermo-chromic image – which changes colour when the user wears it? Or a T-shirt with embedded electro-luminescent yarns that light up when attached to a power supply?

Textile materials and technologies are key innovations that can provide solutions to a broad range of social challenges. Technical textiles are enablers in a wide range of industrial sectors by offering:

- Alternative materials: lightweight, flexible, soft, multifunctional, and durable.
- New technologies: flexible, continuous, versatile.
- Functional components: reliable, multifunctional, cost-effective, user-friendly parts of large technology systems.
- Material substitution: replacement of traditional materials such as steel and cement by more environmentally sustainable materials.

Technical textiles can offer a broad range of properties: lightness, resistance, reinforcement, filtration, fire retardancy, conductivity, insulation, flexibility, antimicrobial, nano, absorption, and so on.

Technical textiles are part of a wider field that David Rigby Associates terms the "engineering of flexible materials' including foams, films, powders, resins, and plastics". The term *advanced flexible materials* has also been used (DRA, 2010). For many years, the term *industrial textiles* was widely used to encompass all textile products other than those intended for apparel, household, and furnishing end uses.

The technical textiles supply chain can be complex and extended, beginning with the manufacture of synthetic polymers for technical specialist fibres, specific coating or specialty membranes through to the converter and mill fabricators who incorporate technical textiles into finished products or use them as a component or subunit of a larger component. As an example, it has been estimated that only 10% of technical textiles embedded in a modern automobile are visible.

According to Euratex, the technical textiles industry in the EU represents roughly 30% of the total turnover in textiles (excluding clothing) – this would equate to a total annual turnover in the EU of some 30 billion euros. It could represent a higher percentage market share in some member states – e.g. technical textiles production is estimated at 50% of total textile production in Germany, 45% of total production in Austria and approximately 40% in France. The sector covers some 15,000 companies and 300,000 employees. Certain analysts consider that other parts of the EU industries should be added including a part of the textile machinery industry as well as the "textile" part of the manufacturing activities of other sectors such as tyres or the revetment of roads or buildings with geotextiles. This is why the size/annual turnover of the EU technical textiles industry as a whole could in reality be even larger – with revised estimates suggesting it could total some 50 billion euros per annum (www.euratex.eu).

In 2014 the United Kingdom technical textile sector is estimated to represent a £1.2 billion turnover.

The leading international trade exhibition for technical textiles – Techtextil (http://techtextil.messefrankfurt.com/frankfurt/en/besucher/willkommen.html) – defines 11 main application areas:

- Agrotech
- Buildtech
- Clothtech
- Hometech
- Indutech
- Medtech
- Mobiltech
- Oeko-tex
- Packtech
- Protech
- Sportech

The Techtextil product groups (with functional apparel textiles introduced in 2013) represent the entire spectrum of technical textiles and nonwovens. The full product groups are listed as follows:

- R&D, planning, and consulting – Universities, research institutes, and so on.
- Technology, processes, and accessories – Production processes, finishing technologies, CAD/CAM, CMT, and such.
- Fibres and yarns – Synthetic fibres and yarns, glass fibres, metal and ceramic yarns, natural fibres, and so on.
- Woven fabrics, laid fabrics, braiding, and knitted fabrics – Woven fabrics, laid webs, braidings, nets, belts, ropes, and such.
- Nonwovens.
- Coated textiles – Coated textiles, laminated textiles, tent fabrics, sacking, awnings, and so on.
- Composites – Reinforcing textiles, prepregs, structural components, membrane systems, and such.
- Bondtec – Finishing processes, adhesive and moulding materials, materials pretreatment, and so on.
- Functional apparel textiles – New materials, new finishing techniques, smart and/or wearable textiles, multifunctionalisation, nanotechnology, and so on.
- Associations.
- Publishers.

The Innovations in Textiles Web site (www.innovationintextiles.com) also separates the technical textiles market sectors into 11 segments:

- Sports and outdoor
- Protection
- Medicine, health, and hygiene
- Transportation and aerospace
- Clothing and footwear
- Sustainability
- Interiors
- Construction and architecture
- Industry

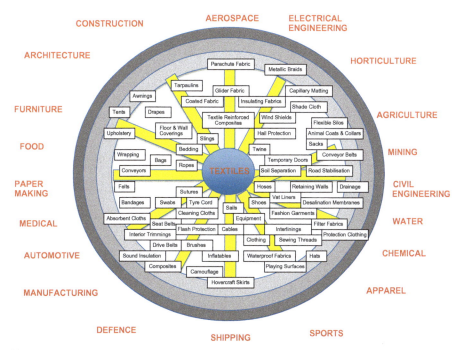

Figure 1.1 The diversity of technical textiles.

- Packaging
- Agriculture, horticulture, and forestry

The search continues for an all-embracing term that will link such terms as *technical*, *industrial*, *performance*, *functional*, *engineered,* and *high-tech textiles*.

An illustration of the range of technical textiles and end-use sectors is given in Figure 1.1.

1.3 Emergence and milestones

Although the development of technical and industrial applications for textiles can be traced back many years, a number of more recent milestones have marked the emergence of technical textiles as we now know them. Primarily, these have centred on new materials, new processes, and new applications. The major fibre types are now reviewed.

1.3.1 Natural fibres

Until the emergence of synthetic fibres, natural fibres dominated the textile sector. The major cellulosic fibres are cotton (pre-eminent), flax, jute, sisal, and hemp. The total amount of cotton produced worldwide in 1990 was 87.2 million 480-pound bales. Production peaked in 2011 at 126,590 bales and for 2014, it was 115,921 bales (http://www.statista.com/statistics/259392/cotton-production-worldwide-since-1990/).

Average global cotton prices have also varied, peaking at US cents 155.7 per 1 pound of cotton and standing at 94.04 cents in 2014 (http://www.statista.com/statistics/259431/global-cotton-price-since-1990/).

As the world's leading animal fibre, wool is produced in about 100 countries on half a million farms. Major producers are Australia, Argentina, China, India, the Islamic Republic of Iran, New Zealand, Russia, South Africa, United Kingdom, and Uruguay. Depending on the country and region, wool producers range from small farmers (e.g. Peru) to large-scale commercial grazing operations (e.g. Australia).

Annual wool production is around 2.1 million tonnes. Australia produces one-fifth, and each of China, New Zealand, Iran, Argentina, and the United Kingdom produces more than 50,000 tonnes. An estimated 50% of wool, both raw and partially processed, is exported to major textile centres in other countries to be spun and woven. China is the number one importer of raw wool (310,000 tonnes in 2007), followed by Italy. The annual retail value of sales of wool products is $80 billion a year (http://www.iwto.org/wool/the-natural-fibre/) (all dollars are in US currency unless otherwise noted). Wool is valued for its insulation and flame retardancy properties and is extensively used in high-temperature and protective clothing applications (e.g. firefighters' clothing).

Silk fabric was first developed in ancient China and later spread around the world via the "Silk Road" and became popular among the super rich and high society. Today silk is an affordable luxury for the middle class in Europe and the United States and continues to hold its own in Asia as traditional ceremonial wear (http://www.fibre-2fashion.com/industry-article/38/3793/the-global-silk-industry1.asp).

According to the International Sericultural Commission, global silk production rose from 120,396 tonnes in 2008 to 152,868 tonnes in 2012. China is by far the largest silk producer followed by India. Silk plays a technical role in the production of medical textiles (e.g. surgical suture thread) and tissue scaffolds (http://inserco.org/en/statistics). Considerable research work is being conducted on the potential exploitation of spider silk.

The significant changes over the last 20 years can be summarised in Table 1.1.

1.3.2 Viscose rayon

Brussels-based International Bureau for the Standardization of Man-Made Fibers (BISFA) (http://www.bisfa.org/) defines *viscose* as being "a cellulose fibre obtained by the viscose process". It is known as rayon fibre in the United States. Although several cellulosic fibres were made experimentally during the nineteenth century, not until 1905 was viscose, which has become the most popular cellulosic fibre, produced.

Table 1.1 **World production of cotton, wool, and man-made fibres**

Year	Cotton (%)	Wool (%)	Man-made (%)
1992	46	5	49
2002	35	2	63
2012	31	1	68

http://www.cirfs.org/KeyStatistics.aspx.

Viscose fibres are made from cellulose – wood pulp. The cellulose is ground and treated with caustic soda. After an aging waiting period, the ripening process during which depolymerisation occurs, carbon disulphide is added. This forms a yellow crumb known as *cellulose xanthate,* which is easily dissolved in more caustic soda to give a viscous yellow solution. This solution is pumped through a spinneret, which can contain thousands of holes, into a dilute sulphuric acid bath where the cellulose is regenerated as fine filaments as the xanthate decomposes (http://www.cirfs.org/manmadefibres/fibrerange/Viscose.aspx).

In recent years, the United States, Europe, and Japan have successively withdrawn from the viscose fibre industry because of labour costs, environmental protection, and other factors while the Asia-Pacific Development Zone, including China, has entered a period of accelerated development of viscose fibre with capacity and production increasing considerably.

In 2013, the global viscose fibre output exceeded 4.9 million tonnes, up more than 13% over the previous year, with China contributing more than 65%.

The Austria-based Lenzing Group and the India-based Aditya Birla Group, two viscose fibre giants worldwide, gained capacity of more than 800,000 tonnes/annum each in 2013. Their Chinese counterparts also obtained increased capacity.

Viscose fibre includes viscose filament yarn (VFY) and viscose staple fibre (VSF); VSF capacity/output accounts for about 90% in China. In 2013, China's VSF capacity approximated 3.45 million tonnes, 51.8% of which came from the top five enterprises, namely Fulida Group, Sanyou Chemical, Aoyang Technology, CHTC HELON, and Shandong Yamei, showing a high industry concentration.

As a large consumer of viscose fibre around the globe, China's output of viscose fibre is anticipated to keep a growth rate of 10% or so in the coming years, which beyond doubt stimulates a rise in the demand for dissolving pulp. After an antidumping tariff is levied by China on the imported dissolving pulp in 2014, the output of home-made dissolving pulp in China is expected to grow steadily, and it is expected to be 1 million tonnes in 2016.

1.3.3 Polyamide and polyester

Globally, polyester is by far the greatest synthetic fibre. It is the workhorse for many, many applications because of its generally good properties at the lowest price. Apparel accounts for a large share of use of polyester fibres or as blends. Industrial use, such as tyre cord fabrics and unspun uses, such as furniture fillings and nonwovens, are both expanding rapidly (http://www.cirfs.org/ManmadeFibres/Fibrerange/Polyester.aspx).

BISFA (http://www.bisfa.org/) *polyamide fibre* as being "a fibre composed of linear macromolecules having in the chain recurring amide linkages, at least 85% of which are joined to aliphatic or cycloaliphatic units". Many polyamide fibres are made but only two, described below, are made in significant quantities. The first fibres made from polyamide polymers were produced in 1938 in the United States and Germany. In the United States, the raw materials, which were used to produce the polymer, were adipic acid and hexamethylene diamine. Because both chemicals contain six carbon atoms, the new polymer was named polyamide 6.6. In Germany caprolactam was polymerised to produce a different fibre known as polyamide 6.

In weaving, the main end use of polyamide is for outerwear and technical fabrics, such airbags and cap piles for tires. In knitting, stockings, hosiery, tights, and outerwear are important outlets for polyamide. Carpets and ropes are also important sectors.

1.3.4   Polyolefins

Two polyolefin polymers are used to make synthetic fibres, polypropylene, and polyethylene; polypropylene is by far the more important. The BISFA (http://www.bisfa.org/) definition for polyethylene fibres is "fibre composed of linear macromolecules of un-substituted saturated aliphatic hydrocarbons" and for polypropylene fibres "composed of linear macromolecules made up of saturated aliphatic carbon units in which one carbon atom in two carries a methyl side group…". Polyethylene was first produced in the United Kingdom in 1933 by polymerising ethylene under pressure. In 1938 in Germany, polyethylene was made by polymerising ethylene in an emulsion. Polypropylene was commercialised in 1956 by polymerising propylene with catalysts. Both of these polyolefins that are spun into synthetic fibres on a large scale are very important in plastic moulding and for making plastic sheeting.

1.3.5   Aramids

Aramid is a contraction of *aromatic* and *polyamide.* BISFA (http://www.bisfa.org/) defines it as "fibre composed of linear macromolecules made up of aromatic groups joined by amide or imide linkages". The two types of aramid are meta-aramid (m-aramid) and para-aramid (p-aramid).

The p-aramid fibres have a very high strength, five times stronger than steel and lose little strength during repeated abrasion, flexing and stretching. It has an excellent dimensional stability. The m-aramid fibres are used for their excellent heat resistance. Some of the main end uses for m-aramids are protective clothing, hot gas filtration, and electrical insulation. *p*-Aramids are used to replace asbestos in brake and clutch linings, as tire reinforcement, and in composites such as materials for aircraft, boats, high-performance cars, and sports equipment. Members of police forces and armed forces wear antiballistic aramid apparel.

1.3.6   Glass and ceramic

There are many inorganic fibres, including glass, carbon, metal, and ceramic. They are used especially in the industrial fibre sector. Glass is the most important inorganic fibre.

Several types of glass fibre are produced. They have in common high moduli, high rot resistance, low moisture uptake, are brittle and have low breaking extensions. Glass is used extensively for insulation in the form of a felt and for reinforcing plastics to make boats, caravans, automobile parts, and so on. It is used less often for flame-resistant curtains and décor fabrics.

Ceramic fibres are small-dimension filament or thread composed of a ceramic material, usually alumina and silica, used in lightweight units for electrical, thermal, and sound insulation; filtration at high temperatures; packing and reinforcing other ceramic materials.

1.3.7 Carbon

Carbon fibre is a reinforced fibre used in the fields of aerospace, wind power, automotive, sports, and leisure. In 2013, global industrial applications, such as wind power and automobile, accounted for some 60% of the market followed by aerospace, which is steadily increasing. Global carbon fibre capacity was estimated at some 120,000 tonnes (based mainly in Japan and the United States). The top five global companies – Toray, Teijin, Zoltek, Mitsubishi Rayon, and Formosa Fabrics – accounted for some 55% of total capacity with Toray as the dominant company (Anon Knitting, 2014).

Boeing and the BMW Group have signed a collaborative agreement to participate in joint research on carbon fibre recycling and share knowledge about carbon fibre materials and manufacturing.

Boeing's 787 Dreamliner is made up of 50% carbon fibre material. BMW has introduced two vehicles with passenger compartments made of carbon fibre in 2013. Recycling composite material at point of use and the end of product life is critical to both companies.

1.3.8 Metal

Metal fibres are also included. Metallic wires can be processed into products such as cables, woven or knitted screens, and meshes and structures for tires. They can be manufactured from steel, aluminium (e.g. aluminised nylon), silver, and gold. Metals can also be coated on fibres and yarns to make them conductive.

1.4 Textile processes

Figure 1.2 provides an overview of the complex processes utilised to produce technical textiles.

Plaiting and knotting are used for manufacturing ropes (e.g. rope ladders, scramble nets, gym ropes, and decorative rope) and nets (e.g. mosquito nets). Rope is being used in barrier/garden decking, tree felling, scaffolding, marquee, gymnasium (battling ropes), timber, fencing, marine, construction, transportation, brewery and equestrian applications, and in manufacturing, for example, paper mills.

Manufacturers' sales of cordage, rope, twine, and netting in the United Kingdom in 2013 amounted to £51 million (http://www.statista.com/statistics/290667/textiles-manufacturers-sales-in-the-united-kingdom-uk/).

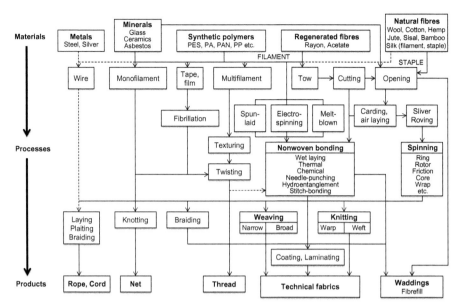

Figure 1.2 Textile processing routes.

1.5   Applications and end-use sectors

The leading international trade exhibition for technical textiles – Techtextil (http://techtextil.messefrankfurt.com/frankfurt/en/besucher/willkommen.html) – defines 12 main application areas:

- Agrotech – Textile products for agriculture, forestry, horticulture, and landscape gardening.
- Buildtech – Textile products for membrane construction, lightweight and solid structures, earthworks, hydraulic engineering, and road construction.
- Clothtech – Innovations in shoe and clothing manufacture.
- Geotech – Products in road construction, civil engineering, and dam and waste site construction.
- Hometech – Innovations in the manufacture of furniture, upholstery, floor coverings, and carpets.
- Indutech – Products for mechanical engineering and for chemical and electrical industries.
- Medtech – Innovations in medical and hygiene products.
- Mobiltech – Textiles for ship and aerospace construction as well as automotive, railway, and space travel.
- Oeko-tex – Products for environmental protection, waste disposal, and recycling.
- Packtech – Innovations in packaging, covering, and transportation.
- Protech – Innovations in personal and property protection.
- Sportech – Innovations in the sport and leisure world.

The main Techtextil event in Frankfurt, Germany, in 2013 attracted 1330 exhibitors from 48 countries and more than 27,400 visitors. Regional exhibitions are now held in India, China, the Middle East, and North America. Clearly, these products are not limited to the terms *technical* or *industrial*. Terms such as *performance textiles*, *functional textiles, engineered textiles,* and *high-tech textiles* are also used in some contexts.

1.6   The impact of globalisation

The geographical distribution of production in the textile, clothing, and footwear (TCF) industries has changed dramatically in the past 25 years, resulting in sizable employment losses in Europe and North America and important gains in Asia and other parts of the developing world.

Worldwide, the development of technical textiles production is best illustrated by reviewing fibre consumption. Technical textiles consumed worldwide about 22 billion tonnes of fibres in 2010, representing 27.5% of a total consumption of 80 billion tonnes for all textile and clothing applications. Europe accounted for about 15% of the global consumption of technical textiles. This section briefly reviews the situation in Europe and the United Kingdom, India, China, Brazil, the United States, and Pakistan as key examples.

1.6.1   Europe and the United Kingdom

Of the EU member states, the top five that export technical textiles (Germany, Italy, France, the United Kingdom, and Belgium) represent 60% of the total exports to the world. Moreover, the member states whose technical textiles represent the highest share of exporting their textiles (excluding clothing) are Finland, Denmark, Sweden, the Czech Republic, and Hungary.

Two technologies primarily drive growth in Europe:

–   Nonwovens with a growth rate of 60% over the past decade.
–   Composites with a growth rate of 75% over the past decade.

Of crucial importance is the leading position of the EU in textile machinery manufacturing with 75% of the global market – especially Italy, Germany, and France. This is a major driver of innovation.

Recent research (2014) in Germany confirmed that the technical textiles companies belonging to this cross-sectorial branch and material supplier to several industrial segments have a high innovative capacity and realise more than 25% of their turnover from innovative products, ranking third after automotive and electronics industries (http://developpement.euratex.org/content/1st-euratex-convention-istanbul-16th-november-2012-booth-availability-companies-tentative-pr).

The UK technical textile sector is estimated to represent a £1.2 billion turnover. However, the aging workforce is expected to significantly challenge the key technical capability of UK textile companies as they face global growth in demand. The sector is significantly disadvantaged in recruiting able young people into the industry as the result of a widespread lack of understanding of opportunities within the sector and a continued belief that the majority of young people are best served by further and higher education, causing the consequent decline in industry-led apprenticeship and technical program entrants.

1.6.2 India

The size of the Indian textile and apparel industry is estimated to be INR 530,000 crores (around $86 billion). It contributes 14% to industrial production, and its share of Indian exports stands at 12%. India is one of the few countries with a complete and integrated textile value chain having production at each level of textile manufacturing with an overall annual growth of 8.9%.

With per capita consumption of only 0.4 kg/person in India, the technical textiles sector began perhaps a decade ago at a very low level. The sector has developed with local substitution of some traditional textiles such as carpet felts and filters. The main focus is on automobile, filter, and industrial felt needle-punched products. The technical textile share is estimated to be 11.5–12% of all textile production. The output of the technical textile/nonwoven sector has doubled in the last 5 years and is expected to be worth $15 billion by 2015 (an average compound annual growth rate – CAGR – of 12%) (Tyagi, 2014). The industry itself is expected to grow at a rate of 20% annually to reach $36 billion by 2016–2017.

Technical textile is an important part of the overall textile sector in India. It not only grew at an annual rate of 11% during 2006–2011 but is also estimated to expand at a rate of 20% to reach $36 billion by 2016–2017 (Figure 1.3).

The technical textiles market in India, which is currently estimated at $14 billion, is likely to reach a level of $32 billion by 2023 by shifting to nonwoven technical textiles and forging global partnerships with similar industries. The textile industry is one of the largest of India's export sectors, contributing nearly 13.25% of the country's total export basket at $41.57 billion (about INR 265,100 crores) in fiscal year 2013–2014.

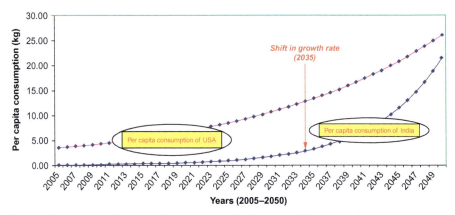

Figure 1.3 Actual and Projected Comparison of India versus USA per capita nonwoven consumption from 2005 to 2050 (Turaga et al., 2014).

In 2014, India overtook Germany and Italy to move into the number two position in textiles in terms of market value – with China remaining in first place. The textile industry has two broad segments, namely the unorganised sector of the handloom, handicraft, sericulture, power loom and the organised sector of spinning, apparel, garmenting, and made-ups (https://www.thedollarbusiness.com/another-promising-year-awaits-indias-textiles-exports/).

A study conducted by PHD Chamber of Commerce reported in 2014 that the size of the Indian textiles and apparel industry would rapidly expand to $226 billion by 2023 (http://www.phdcci.in/).

Currently, the Indian textiles and apparel industry is estimated to be worth $99 billion, which includes both domestic consumption and exports. It is projected to grow by 8.6% of CAGR to reach $226 billion in 2020.

"Given the scope of the technical textiles in emerging economies, the Indian Government and industry need to build a close partnership and roll out their joint vision for development and promotion of the sector, setting targets for 2025", according to Ajai Shankar, member secretary of National Manufacturing Competitiveness Council (http://www.nmcc.nic.in/).

The study cites vertically integrated supply chain and diverse range of products as the main factors expected to contribute to the growth of the industry.

In India's textiles and apparel exports, textiles represent 60% compared with that of apparel, which is 40%. The textiles and apparel sector contributes 5.2% to the country's gross domestic product (GDP).

1.6.3 China

According to Chinese Nonwovens & Industrial Textiles Association estimates (http://www.cnita.org.cn/en/), China completed 11.3 million tonnes of industrial textile fibre processing in 2013, showing annual growth of 11.9% and accounting for a 23.3% share of total global textile fibre processing.

Included in this, the total output of nonwovens reached 3.84 million tonnes, a growth rate of 11.7% since 2012. According to data from the Chinese National Bureau of Statistics (http://www.stats.gov.cn/english/), total nonwovens fabric production from large and small and medium size enterprises reached 2,573,000 tonnes, growing 12.4% since 2012.

In 2013, China's industrial textiles exports gradually increased: From zero growth in 2012, the industry has completed $19.27 billion worth of various types of industrial textiles exports, an annual increase of 8.4%, or three percentage points lower than the entire textile industry export growth rate.

Plastic-coated fabrics, medical textiles, nonwovens, canvas textiles, bags, and fibre glass are six categories of export products whose value accounted for nearly 80% of China's entire industry exports. Among them, nonwoven fabrics export growth rate reached 21.2%; the other five categories achieved only single-digit growth.

The dominant Chinese textile industry is now being affected by rising labour costs and the Chinese government's decision to tackle growing environmental and pollution concerns. Countries such as Vietnam, Cambodia, and Bangladesh are emerging as serious textiles manufacturers but often with Chinese capital backing. Latin America will also feature and will show significant future growth.

The China Nonwovens & Industrial Textiles Association (CNITA; http://www.cnita.org.cn/en/) predicts that China's imports of technical textiles will grow at a compound annual growth rate of 7% from $3 billion in 2010 to $4.2 billion per annum in 2015.

China's technical textiles industry has toppled creaking barriers to trade and employment, forging a global labour market and rapid industrialisation. About 620 million people globally have been lifted from poverty by moving from farm to factory, and China's GDP per capital rose from 3% of the level in advanced economies in 1980 to 20% by 2010, according to the McKinsey Global Institute (http://www.mckinsey.com/insights/mgi/).

By shifting labour-intensive parts of production to countries such as China and keeping the higher-value aspects at home, Chinese companies lowered their costs and raised their returns on capital.

1.6.4   Brazil

The Brazilian Textile industry is composed of 3000 companies employing 275,000 people producing 1.9 million tonnes of textiles annually. The total income of the industry was $22 billion of which $1.115 billion was for exports in 2013. The Brazilian apparel sector comprises some 30,000 companies employing 1.4 million people and produces 1.2 million tonnes annually. It invoiced $43 billion in 2013 with $150 million going for export.

Brazil is the largest producer and consumer of technical textiles and nonwovens in South America, producing 600,000 tonnes and consuming 690,000 tonnes. The number of technical textile companies has increased from 194 (2008) to 210 (2012).

Brazil produces more than $2 billion of geotextiles per annum and 500,000 bulletproof vests (Febratex News 12–1 August 2014, p. 13; http://www.febratex.com.br/inicio).

1.6.5   The United States

The US textile industry is one of the more important employers in the manufacturing sector with more than 230,000 workers, representing 2% of the country's manufacturing workforce. The United States is a globally competitive manufacturer of textiles, including textile raw materials, yarns, fabrics, apparel, home furnishings, and other textile-finished products. Key strengths are in cotton, synthetic fibres, and a wide variety of yarns and fabrics, including those for apparel and industrial end uses.

Textile industry workers are highly skilled, and the industry is technologically advanced with investments of more than $1.2 billion annually in total capital expenditures. In recent years, US textile companies have focused on retooling their businesses, finding more effective work processes, investing in niche products and markets, and controlling costs.

The industry is globally competitive, ranking fourth in global export value behind China, India, and Germany. US textile exports increased by 12% between 2010 and 2012 to $17.1 billion. More than 65% of US textile exports go to free trade agreement partner countries.

The United States exported some $314 billion of specialty and industrial fabrics in 2010. By 2014, this had grown to $410 billion. The major export markets are Mexico, Canada, Japan, China, and Hong Kong.

1.6.6 Pakistan

Pakistan is seeking to increase its textile exports over the next 9 years, aiming to double figures by 2023.

This decision comes after Pakistan signed an agreement with Tadjikistan and Kirghizistan at the beginning of December 2014 to work on enhancing the country's energy supply. Pakistan had been experiencing power blackouts, causing problems in the textile industry. With the improvements underway, Pakistan hopes to increase 2013s €466 million profit further; $26 billion profit in textile exports are expected by 2023.

Plans have been underway to improve the textile industry for some time. In 2013 the government sought to increase the country's profit from exports by $1 billion. This was successful, as exports to the European Union in the first 9 months of 2014 reached $900 million, a welcomed increase and 2013 profits increased 29% from 2012 (http://www.retailgazette.co.uk/articles/31010-textile-exports-hope-to-sew-the-way-for-pakistans-financial-future).

1.7   The future

Analysts forecast the global technical textiles market to grow at a CAGR of 3.71% over the period 2014–2019 (http://www.whatech.com/market-research-reports/press-release/industrial/35624-explore-global-technical-textiles-market-that-is-expected-to-grow-at-a-cagr-of-3-71-to-2019).

The report "Global Technical Textiles Market 2015–2019" analysed the major regions of China and India. One trend is the emergence of key markets. According to the report, the high demand for nonwoven technical textiles is a major factor driving market growth. The report identified lack of sufficient resources for small and medium-sized enterprises (SMEs) as a major challenge to the industry growth.

Technical textiles including nonwovens comprise about 25% of total global fibre consumption. In the United States and other developed nations, the consumption is up to 75%. Research into new products and new applications is ongoing. The needs and potential are enormous in areas such as water purification, medical devices, and responsive fibrous systems.

Gherzi (http://www.gherzi.com/) sees five main growth drivers for the application of nonwovens:

- The global growth of the population and its demands for hygiene, medical applications, consumer goods, and so on.
- The industrialisation of emerging markets with more automobiles, geotextiles, protective apparel, and so forth.

- The further industrialisation of the agriculture with covers for seed, mulch mats, drainage, shading, and others.
- The displacement of fabrics in the areas of filtration, protective apparel, roofing membranes, geotextiles, and so on.
- A growing environmental awareness, which drives the need for more products for filtration, oil absorbents, landfill-backings, and such.

Technical textiles appear to have no limits, yet a producer who does not want to work with upstream and downstream supply chains will be forced to operate the entire production chain. However, in most cases, this is not possible, especially in the supply of fibres and yarns. What is the alternative? The producer needs reliable partners or suppliers – a sustainable supply chain – to provide the exact raw material to produce the requested high-tech products. For example, producing fire-resistant protective apparel requires many specialists, including:

- Fibre and yarn producers
- Weavers
- Finishers
- Apparel manufacturers
- Retailers

Significant growth is expected in the future in the areas of medical textiles and sportswear/performance clothing. The following sections suggest the future for nonwovens, dyestuffs, e-textiles, personal protective equipment, and textile machinery again as key examples.

1.7.1 Nonwovens

The global nonwoven fabrics market is expected to reach \$42.1 billion by 2020. Global demand for nonwovens is forecast to grow by an average of 5.4% per annum between 2012 and 2017, to 9.1 million tonnes. In developed countries, however, demand is forecast to grow by only 2.4% annually although this represents an improvement over the decline that occurred between 2007 and 2012. In developing countries, demand is forecast to grow by 7.2% annually.

Filter media represent the fastest growing application for nonwovens. The filtration industry has been greatly affected by changes in environmental legislation as governments worldwide face pressure to reduce levels of pollution.

Transportation remains the biggest market for nonwoven filter media in using filters to reduce contamination from fuels and oils. However, in the EU, there is growing demand for nonwoven filter media in heating, ventilation and air conditioning (HVAC) applications to provide higher indoor air quality, especially in view of an EU requirement that all new public buildings and all other new buildings be "nearly zero energy" by the end of 2018 and 2020, respectively. Furthermore, nonwovens are forecast to play an increasingly important role in the production of clean water from wastewater as global water consumption increases.

1.7.2   Dyestuffs

The global dyestuff (black colour) market for textile fibres was valued at $1.10 billion in 2013 and is projected to reach $1.88 billion by 2020, growing at a CAGR of 8.0% from 2014 to 2020.

Textile dyestuffs primarily include basic, acid, disperse, direct, reactive, sulphur, and vat dyes. They are used to produce coloured textile fibres. Acid, basic, and disperse dyes are primarily used in the production of black coloured textile fibres.

Furthermore, several types of dyes such as direct, reactive, and vat are used for dyeing textile fibres for specific applications. Textile dyestuffs are used to dye various types of textile fibres such as polyester, polyamide, acrylic, and olefin. These dyed textile fibres are used in several end-user industries including home textiles, apparel, agricultural textiles, automotive textiles, and protective clothing.

Rising demand for technical textiles is predicted to drive demand for dyestuff in the manufacture of textile fibres in the next few years. Furthermore, demand for dyestuff in the production of textile fibres is anticipated to increase significantly in the near future because of the steady growth in the global apparel market. However, volatility in raw material prices and strict environmental regulations regarding the usage of synthetic dyestuff can hamper growth of the dyestuff market for textile fibres in the next few years.

1.7.3   E-textiles

The value of the global wearable technology ecosystem was estimated to be more than $4 billion as of 2012 and is expected to reach more than $14 billion by 2018, growing at a CAGR of more than 18% from 2013 to 2018.

The total addressable market (TAM) for wearable technology is estimated to be more than $14 billion, as of 2012, and the current level of penetration for wearable technology was estimated to be roughly 18% and is expected to accelerate (increasing the rate of penetration every year) over the next 5 years, reaching roughly 46% penetration level in the TAM by 2018.

Wearables are electronics that can be worn on the body, either as an accessory or as part of material used in clothing. One of the major features of wearable technology is its ability to connect to the Internet, enabling data to be exchanged between a network and the device.

The global wearable electronic textiles market is expected to grow faster than that of the overall wearable electronics market with increasing demand and consumer adoption for wearable electronic smart textiles, along with expected commercialisation of more advanced wearable electronic e-textiles by 2016.

The European market for wearable technology will grow 42.1% annually over the next 5 years to a total value of $2545.51 million in 2019, according to a new report.

The market is characterised by new technology and dominated by few players, which means the price of products is high. But the applications for these wearable devices cover wide sectors including health and well-being, infotainment, medicine, sport and leisure, industry, and military.

Transparency Market Research (http://www.transparencymarketresearch.com/) reported that the fitness and wellness segment accounted for the largest share of wearable technology in 2012 and is expected to maintain its position throughout the forecast period. Germany was found to be the largest market (32% in 2012).

A 2014 webinar conducted by market research firm IDTechEx (http://www.idtechex.com/) divided the wearable electronics industry into two segments: apparel and textiles and devices. Apparel and textiles can be woven, washable, stretchable, tightly rolled, folded, printed, transparent, or invisible and can provide integral energy harvesting, whereas those with devices tend to be rigid or bendable with largely conventional electronics provided by batteries and usually offer no energy harvesting.

1.7.4 Personal protective equipment (PPE)

Protective clothing is primarily used to protect personnel or workers against any hazard caused by working condition inside or outside. Protective clothes, which have various designs, are manufactured and designed according to the need of companies and their operations and factors such as chemical handhelds and regulatory reforms. The global protective clothing is expected to grow at a high CAGR over the next 6 years. The growing demand for chemicals throughout the globe is expected to be the primary driver of this market during the forecast period.

According to the estimates of U.S. Occupational Safety and Health Administration (OSHA; https://www.osha.gov/), global prevalence of protective clothing is expected to grow in industries such as pharmaceutical, research laboratories, and health care from 2013 to 2020. Other drivers, which lead to the growth of protective clothing market, include industrial development in the emerging economies, importance of safety in industries, and the emergence of the blue collar workforce. The application of protective clothing is mainly to repel the chemicals from causing any harm in the body of the workforce; it also includes apparels that are flame resistant. The market for protective clothing has a huge need as various accidents are taking place in the industries over the globe; the use of protective clothing will minimise those accidents and lead to working conditions with fewer accidents. The factor that restricts the increase in protective clothing market includes the availability of imprecise numerical safety data from industries.

1.7.5 Textile machinery

Today, Switzerland is the leading manufacturer of textile machinery supplying the global marketplace (with an estimated global market share of 33%) and the major players themselves (http://www.swissmem.ch/txm) who consider its outstanding quality and innovative potential. Swiss textile machinery holdings employ more than 28,000 people in more than 100 companies and branch units around the world.

World demand for textile machinery depends indirectly on the demand in sectors including housing and automobiles. As textile machinery is used for manufacturing home carpets and upholstery for furniture and automobiles, changing fashion trends affect the demand for textile equipment.

Production and manufacturing process of textile machinery is cyclical in nature and depends on equipment purchases, which slows during a recession and accelerates when the economy is sound. Textile mills tend to invest in machinery when interest rates prevailing in the market are considerably low.

The rapid pace of technological innovations taking place in the textile machinery market has resulted in the production of more efficient machines at low prices. Traditional textile machinery relies on cheap labour; hence, although high volumes of textiles are sold, they are of poor quality. Demand for sophisticated machines that produce high-quality cloth with fewer defects is increasing.

The international textile yarn machinery market experienced another poor year in 2013, primarily because of a decrease in shipments of short staple spindles, long staple spindles, open-end rotors, and double heater false-twist draw texturing spindles to the textile industry in China; there had been no shipments in 2012. Supplies of single heater false-twist draw texturing machinery remained minimal.

On the other hand, there were increases in shipments of some textile machinery to the textile industries in Egypt, India, Indonesia, Spain, Turkey, the United States, and Vietnam. However, shipments of short staple spindles rose, and shipments of short staple spindles, open-end rotors, and double heater false-twist draw texturing machinery remained higher than during the 10-year period ending in 2013 (http://www.knittingindustry.com/world-markets-for-textile-machinery-part-1-yarn-manufacture/#sthash.3ntYPqQg.dpuf).

1.7.6 The future of technical textiles

The future of technical textiles includes a much wider economic sphere of activity than only the direct manufacturing and processing of textiles. The industry's suppliers include raw materials producers (both natural and synthetic), machinery and equipment manufacturers, information and management technology providers, R&D services, testing and certification bodies, consultants, and education and training organisations. Its customers and key suppliers include almost every conceivable downstream industry and field of economic activity – architects, engineers, designers, and other advisors employed by those industries.

A growing number of specialist and generalist publications have undertaken the task of disseminating and communicating information to these organisations and individuals:

Textile Intelligence – https://www.textilesintelligence.com/.
Innovation in Textiles – http://www.innovationintextiles.com/.
Fibre2fashion – http://www.fibre2fashion.com/.
Technical-textiles.net – http://www.technical-textiles.net/.
TechTextil – http://techtextil.messefrankfurt.com/frankfurt/en/besucher/willkommen.html.

Technical and performance textiles represent an emerging generation of products combining the latest developments in advanced flexible materials with advances in computing and communications technology, biomaterials, nanotechnology, and novel process technologies.

References

Anon, 2014. The Fibre Year 2014 – World Survey on Textiles and Nonwovens (Issue 14), May.

Anon Knitting, 2014. Wind power and automobiles account for 60% of carbon fibre use. Knitting International (Issue 5), p. 15.

Butaud-Stubbs, E., 2013. http://edz.bib.uni-mannheim.de/edz/doku/wsa/2012/ces-2012-1966-en.pdf.

Byrne, C., 2000. Technical textiles market – an overview. In: Horrocks, A.R., Anand, S.C. (Eds.), Handbook of Technical Textiles. Woodhead Publishing Limited, UK.

Chalupsky, M., 2014. http://www.techtextil-blog.com/en/texfact-of-the-week-kunstfasern-auf-dem-vormarsch/#more-1862.

DRA, 2010. Technical Textiles and Nonwovens: World Market Forecasts to 2010. www.davidrigbyassociates.com.

McCarthy, B.J., 2013. Preface. In: McCarthy, B.J. (Ed.), Polymeric Protective Technical Textiles. Smithers Rapra, Shawbury, p. iii.

Turaga, U., Singh, V. and Ramkumar, S. Technical Textiles: Creating Value for Spunbond Nonwovens Presented at the Southern Gujarat Chamber of Commerce & Industry, Surat, India – March 24 2014.

Tyagi, M., 2014. Where next for India's textile industry? Textile Month International (Issue 2), pp. 34–35.

Technical fibres: Recent advances

2

M. Miraftab
Institute for Materials Research and Innovation, University of Bolton, Bolton,
United Kingdom

Chapter Outline

2.1   Introduction

In the first published edition of *Handbook of Technical Textiles* in 2000, under the "Technical fibres" heading, fibres were classified into conventional and nonconventional materials. Under these general categories, natural, regenerated, and synthetic fibres were first discussed followed by high-strength and high-modulus as well as high chemical- and combustion-resistant fibres based on the general chemical groupings of "organics" and "inorganics". Ultrafine and novelty fibres, mainly referring to microfibres and more curiosity-driven fibres with often exotic properties, were also discussed. Subsequent sections highlight fibres in each of the above categories, such as civil engineering, automotive and aeronautic sectors, medicine and hygiene, and defence. Finally, general assessment of the fibre growth market and futuristic outlooks were given [1].

However, since there has been a paradigm change in the way that fibres are viewed and used to the extent that today a variety of opportunities has emerged in fibre manufacture, utilisation, and applications, which embrace nearly all engineering, medical, and technical application areas.

Fibres today have a larger role than ever beyond their traditional use as mere fundamental units within textile materials. This increased role is partly because of better appreciation of their potential benefits in combined and composite settings and partly, or perhaps more importantly, because of their descending size association to improve performance characteristics, flexibility, and larger surfaces of engagement. Once the millimetre was the commonly understood smallest unit of measurement for fibrous materials in terms of length and even diameter for monofilaments, but within the last 30 years, this generality has shifted to micrometre scales until today when the nanometre scale has become the dominant unit of measurement and exposition for such materials.

As the surface area to volume ratio increases, fibres' behavioural and functional abilities change too. Generally, since increased fibre fineness leads to increased specific surface area, higher absorption of liquids and shorter drying periods can be expected. Such fine fibres would also dye or require impregnation to a given shade or amount with less dye or impregnator concentration and therefore help to reduce costs. They are more flexible as a result (since flexural moduli relate inversely to the fourth power of fibre diameter) and have better drape and therefore feel smoother to the touch and are more easily mouldable. Molecular orientation and stronger intermolecular forces within these fibres make them stronger, too, giving them an inherent advantage compared to their coarser counterparts.

Furthermore, the demand for functional and smart fibres has pushed the application boundaries to newer and more exciting horizons so that building or modifying the structure of a single fibre and its technical manipulation has increasingly become the target of change and innovation.

This chapter builds on information in a publication issued 15 years ago [1] and reviews only the latest advances in fibre production, processing, functionality, and potential applications in various fields of engineering and medical applications.

2.2 Nanofibres

Nanofibres are defined as fibres having diameters of less than 100 nm. To produce nanofibres, a number of processing techniques including drawing [2], template synthesis [3,4], phase separation [5], self-assembly [6], and electrospinning [7,8] have been used. Amongst these, electrospinning has proven to be the simplest, most convenient, and relatively cheapest method of producing nanofibres. The origins of electrospinning or, as it was initially known electrospraying, goes back to the sixteenth century when Gilbert [9] showed that when an electrically charged piece of amber was placed in close vicinity to a water droplet, a cone would form and small droplets would escape from the tip of the cone. Not much happened for almost three centuries until 1900 when J.F. Cooley filed a patent on the electrospinning process [10] and was granted exclusivity. Since then, various versions of the processing techniques including melted versus solvent electrospinning were developed to allow commercialisation [11,12]. In 1938, the first truly commercially electrospun nanofibres were made from cellulose acetate and used in gas masks in the former USSR. Reportedly, by 1960s, the production of filtration material based on this technology had risen to 20 million square metres per year [10]. By the late 1990s, it was shown that a whole range of organic polymers could be electrospun and since then, the technology has been revitalised and has gained worldwide popularity. Today, many synthetic and natural polymers have been electrospun including polylactic acid (PLA), polyurethane (PU), polycaprolactone (PCL), polylactic-*co*-glycolic acid, polyethylene-*co*-vinylacetate, and polylactide-*co*-caprolactone (PLLA-CL). Amongst natural polymers, collagen, chitosan, hyaluronic acid, and silk fibrion have been successfully electrospun into nanofibres, thus opening up a whole range of application potentials for these fibres [10,13].

In electrospinning, nanofibres can be produced by applying a high voltage (up to 30 kV) to a polymer solution/melt. The solution is typically drip fed from the tip of a needle, and the electrical potential applied provides a charge to the polymer. At the tip of the needle, an interesting phenomenon occurs; negative ions at the end of some molecules within the solution align themselves in the inner wall of the nozzle, forcing their positively charged ends to the centre. Concurrently, the polarised neutral species in the solution also align themselves in the centre of the solution, adding to the net positive charge directly facing the negative or the earthed collector. The solution is consequently forced out towards the collector; however, the resistance offered by the solution's surface tension causes the combined effect to form a conelike profile known as the Tailor Cone [14]. Under these circumstances, the resistance because of surface tension holding the charged ions and the neutral species gradually diminishes as the result of the thinning of the surface tension layer as highlighted by the tapering black and white line shown in Figure 2.1. Eventually, the solution is allowed to escape from the apex of the cone where surface tension has become minimal. As the solution escapes the overall surface, tension at the tip of the cone is largely unaffected (i.e. the structure does not collapse).

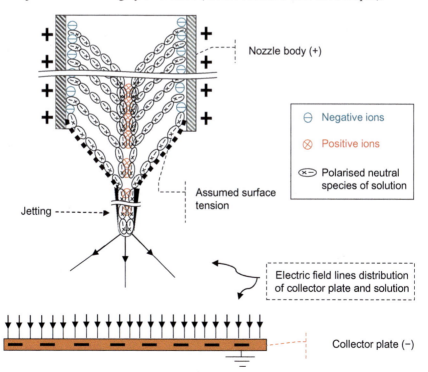

Figure 2.1 Schematic of collated ions at nozzle tip when subjected to external electric field [15].

Once released from the effects of surface tension, the aligned chains of polarised ions tend to repel one another because of repulsive forces, which are greater than the attractive forces of the collector, leading to an umbrella-like explosion (or splay). The electrostatic forces between the splayed fibres and the collector become increasingly dominant, causing a violent whipping action as the fibres head towards the collector. This action also contributes to drawing of the fibres and, increasing their surface areas, which also allows the solvent to evaporate. The net result is the formation of random deposition of the nanofibres on the collector surface. Much research has gone into trying to control or orient electrospun fibres, including anything from parallel plate arrangement of the collectors [16] to the rotation of the collector at variable speeds [17,18]. Figure 2.2a and b shows examples of the continuous production of oriented nanofibres produced parallel to the horizontal axis and at 45° to the horizontal axis [19].

When electrospinning from the melt, there exists a much better possibility of controlling the generated nanofibres and their subsequent ease of handling. In fact, they do not form a splay as they do in the solvent but conform to a highly ordered line of nanothreads that behave like a pen with the full ability to manoeuvre on demand. This gives melted spun nanofibres a tremendous application potential in all kinds of precision engineering applications as well as giving an alternative method to 3D printing [20]. Figure 2.3 is an example of the construction versatility of melted electrospun

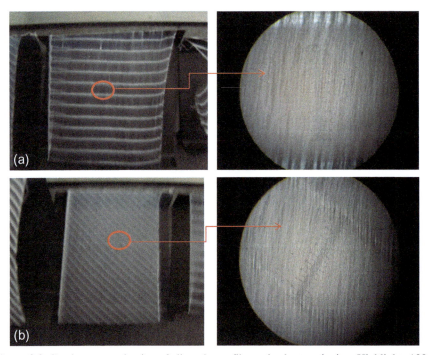

Figure 2.2 Continuous production of aligned nanofibres via electrospinning. Highlights 100× magnified [19]. (a) Vertically aligned between horizontal nodes and (b) aligned at pre-set angle between adjacent nodes.

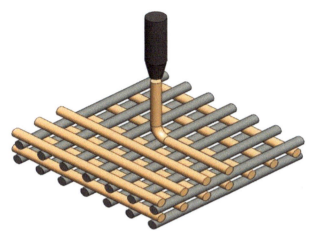

Figure 2.3 Controlled laying down of the filament in melt electrospinning.

nanofibres. Melted spun nanofibres, however, are not as fine as those produced via solvent spinning where solvent evaporation contributes to fibre fineness, but attempts are ongoing to achieve this in melted spun nanofibres too.

Nanofibres are today considered for many technical applications; for example, in medicine, they include anything from tissue engineering (2D and 3D scaffolds), drug delivery, and implants to wound healing and wound dressings. Nanofibres in engineering applications include anything from filtration, photovoltaic cells, membrane fuel cells, dye-sensitised solar cells to carrier materials for various catalysts and photocatalytic air/water purifiers. In apparel form, potential applications cover clothing, shoes, sportswear, diapers, and napkins. Table 2.1 shows some areas of more specialised application for nanofibres [21].

2.3   Auxetic fibres

The concept of auxetic behaviour was first brought to attention in 1987 by R.S. Lakes' publication, as cited by Lisiecki et al. [22], based on a study of foam structures with negative Poisson's ratio. Since then, many other researchers have tried to exploit this exclusive behaviour [23–25].

Poisson's ratio describes the quantitative relationship of lateral contraction to elongation in uniaxial extension of a homogenous isotropic material (i.e. $\upsilon = \Delta\acute{L}/\Delta L$, where υ is Poisson's ratio, $\Delta\acute{L}$ is the change in length in Y direction, and ΔL is the change in length in X direction). Whenever υ is ≥ 0, the material is said to have positive Poisson's ratio, and when this value is negative, the material is said to have negative Poisson's ratio (i.e. a material is said to have an auxetic behaviour when it becomes fatter as it is stretched). Figure 2.4 illustrates this exclusive behaviour as compared with conventional materials.

Table 2.1　Range of application areas for nanofibres

Area of application	Specific role	Application and mechanism of operation
Medicine	Drug delivery	Active pharmaceutical ingredients can be incorporated into the nanofibres for fast release of drugs, resulting in bioavailability improvement, especially for poorly soluble drugs
	Wound healing	Enhanced surface area of contact provides better healing when using organic and inorganic antimicrobials/drugs
	Tissue engineering	Scaffolds are suitable for implantation with different types of cells, enabling tissue replacement prepared from a patient's cells. They also allow for incorporation of different bioactive materials (e.g. growth factor)
	Barrier textiles	Suitable applications of sandwich type structures between which a nanofibre mat sits include surgical gowns, drapes, disposable face masks, and the like
Environmental protection	Filtration	Electrospun nanofibre mat provides dramatic increases in filtration efficiency at relatively small decreases in permeability
	Absorbers	Metal ion absorption, when functionalised or surface coated, nanofibre mats with enhanced surface area improve adsorption capability
	Capacitors	Electrospun nanofibre mats as catalyst carrier, given the extremely large surface a huge number of active sites, is provided, thus enhancing the catalytic capability
	Sensors	Electrospun nanofibres in combination with pyrenemethanol, for example, is an ideal sensor for detecting metal ions
Energy capture	Storage and entrapments	Nanofibrous materials have significantly higher energy conversion and storage efficiency than their bulk counterparts
		Nanofibrous are ideal media for solar and fuel cell constructions with high output efficiencies. Also as mechanical energy harvesters; like energy scavengers, supercapacitors, and parts of battery assemblies
Other applications	Miscellaneous	Electrospun nanofibres have shown excellent capabilities in absorbing sound and noise pollution as field-effect transistors and nanocomposite scaffolds

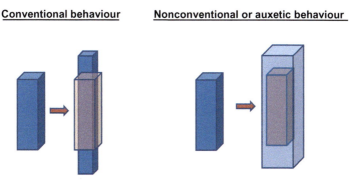

Figure 2.4 Comparative behaviour of conventional versus nonconventional under stretch.

At the structural level, auxetic materials work by virtue of their inherent design in which the hinge-like structures resemble a bow that flexes outwards when stretched. Auxetic materials can be made/formed at the macroscopic level. The bow-like structural elements of an auxetic material open out in the manner shown in Figure 2.5a when stretched; hence, the folded configuration expands unlike the contracting

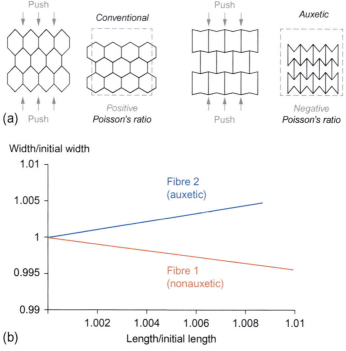

Figure 2.5 Conventional honeycomb structure versus auxetic bow structure displaying differences in width and length variation (a) schematically and (b) graphically.

effect of honeycomb structures in conventional materials. This effect is also represented graphically in Figure 2.5b.

This characteristic behaviour results in a number of positive attributes, including increased shear stiffness, improved fracture toughness, and increased indentation resistance, all of which are very important desired mechanical and engineering factors. From an application point of view, such materials can be used in the biomedical industry to address a number of requirements from surgical implants to suture/ligament locking devices, dilators in blood vessels, and controllable heart valves. When made into foams, auxetic filters carry a lower burden of pressure drop anomalies and can be cleaned more efficiently by the stretching process (i.e. opening up of pores) [26].

Use of auxetic materials becomes even more interesting when they are produced as fibres [27,28] in single or multiple filaments or processed into woven, knitted, or braided structures. As fibres, they can enhance composite structural properties by means of locking the assembly under stress, whereas fibres in conventional fibre-reinforced composites will decrease in diameter and thus tend to separate from the resin matrix under tension. In fibrous or fabric form, they can be used in safety garments including body armour, helmets, and the like, in mechanical lungs and drug delivery systems, and, more conventionally, in upholstery fabrics with built-in abrasion resistance and entrapments for various additives, including fire retardants. These are achieved by the enhanced indentation features, ease of curvature and drape, and the ability to open up and release additives when stretched, respectively.

Auxetic behaviour can also be achieved purely by carefully selected knit designs using conventional yarns [29]. Attempts have also been made to produce fine auxetic fibres, but their mechanical characteristics are a long way from ideal in terms of strength and ease of process and are as yet unsuitable for most practical applications [27,28,30].

2.4   Piezoelectric fibres

Piezoelectricity or electricity resulting from pressure is the electric charge that is generated when certain materials are subjected to mechanical stress. Such materials include crystals, certain ceramics, some biological matters (e.g. bones, proteins, and DNA). This phenomenon was first discovered by Jacques and Pierre Curie in 1880 [31]. The piezoelectric effect is completely reversible, which means when stress is applied, electricity is generated and when electricity is applied, stress is generated (Figure 2.6).

This unique behaviour renders these materials ideal for a number of useful applications, including microbalances, scanning probe microscopes, and ultrafine focussing as well many other highly sensitive ones. In the 1940s, a new group of synthetic materials, known as ferroelectrics [32], was discovered; they displayed piezoelectric constants many times larger than natural materials. Applications for such devices have been extensively explored and today appear in most gadgets used around the home, at work, and in cars. The following are examples of synthetic materials with piezoelectric properties:

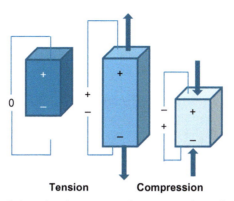

Tension   Compression

Figure 2.6 Schematic of piezoelectric response when pressure is applied.

Ceramics	*Polymers*
Lead zirconite	Poly(vinylidene fluoride)
Lead titanate	Poly(vinylidene chloride)
Barium titanate	Polyacrylonitrile
Inorganic salts	*Copolymers*
Sodium potassium tartrate	Vinylidene fluoride-trifluoroethylene
Ammonium tartrate	Vinylidene cyanide-vinyl acetate
Lithium sulphate monohydrate	

Polymers like poly(vinylidene fluoride), commonly referred to as PVDF, are popular sensing/actuator materials because of their unique piezoelectric property. In the β-crystalline phase, PVDF displays piezoelectricity several times greater than many natural minerals such as quartz. In these materials, the attraction and repulsion of the intertwined long chain of molecules create the piezoelectric effect rather than the crystal structure of the material, which is the case with ceramics. Traditionally, production of PVDF piezoelectric β-phase has been limited to drawn films followed by so-called poling (this process involves aligning all individual dipole moments so that they all point in the same general direction), and because of their flexibility, they have secured a number of key application areas. However, bicomponent fibres with PVDF sheath and a high-density polyethylene/ linear low-density polyethylene, HDPE/LLDPE + 10 wt% carbon black, have been produced via melt spinning whereby β-phase transformation is achieved at high draw ratios. These fibres have a number of application potentials in textile sensors, actuators, and energy absorption. A piezoelectric fibre has high flexibility, is lightweight, and is relatively simple to produce. These fibres can subsequently be woven or knitted into textile-based garments or serve as the main textile garments [33,34].

2.5   Photovoltaic fibres

Because traditional energy sources such as coal, oil, and gas diminish or, more relevantly, are blamed for their role in environmental pollution and its aftereffects, inherently clean and plentiful solar energy is increasingly seen as a suitable alternative. To this end, much research has gone into developing solar cells that directly convert sunlight energy into electrical energy. Their efficiency, however, is not very high, and very much depends on solar intensity, the surface area used to capture the sun's incident radiation, and the types of semiconductors used [35]. Other than silicon-based solar cells, which are expensive to produce, today polymer-based organic films are considered more appropriate as cells because of their ease of operation and their cost effectiveness. To take this a step further, production of photovoltaic fibres suitably made to be used in clothing and wearable products are being considered to generate sufficient electrical energy to run personal devices such as mobile phones, IPads, and laptops [36,37]. Figure 2.7 is an example of such a fibre made from polypropylene core material.

In these fibres, the polypropylene core is coated with a highly conductive compound (anode) that is surrounded by a photoactive material and subsequently with a very thin, highly conductive layer to allow the sun rays to pass through (cathode). Within the right environment, these fibres are said to be able to generate 300–360 mV, a current of 0.11–0.27 mA/cm^2, and a power conversion efficiency of 0.010–0.021% depending on the types of materials used. Although these power output values are fairly modest, they show promise for future wearable photovoltaic textiles. Attempts to combine piezoelectric active fibres with photovoltaic coating are being made to generate hybrid functionality (i.e. to be power generating when exposed to movement as well light) [38].

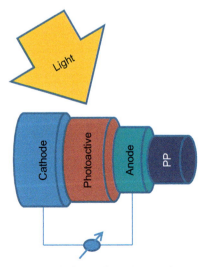

Figure 2.7 Schematic of a photovoltaic fibre with polypropylene core fibre.

2.6   Shape memory fibres

That shape memory alloys remember their permanent shape status and return to it after a temporary shape change by application of heat or other stimuli is now well established [39,40]. However, the advent of shape memory polymers is relatively recent and has opened many other possibilities [41,42]. For a start, shape memory polymers have very high plastic deformations (i.e. up to 400%), low density, low cost, and are potentially biocompatible and could even be biodegradable. They also have a wide range of application temperatures, which could be neatly tailored to address specific demands. Despite their earlier discovery, shape memory polymers have been seriously considered only since the 1980s, and much work has since concentrated on a number of polymers including poly(methyl methacrylate) or PMMA, poly(norbornene) or PN and polystyrene copolymers [42]. The key to shape memory operation is in their molecular network structure containing at least two different phases. When the trigger is temperature, the phase showing the highest thermal transition is the temperature that must be surpassed to initiate the permanent shape configuration. However, for a temporary shape modification, the temperature must be raised above the lower thermal transition phase. The switching segments are those with the ability to soften beyond the transition temperature and adopt the temporary shape. These transitions could either be the glass transition temperature or the melting temperature. Figure 2.8 shows the principle of the switching mechanism and its effect on the polymer.

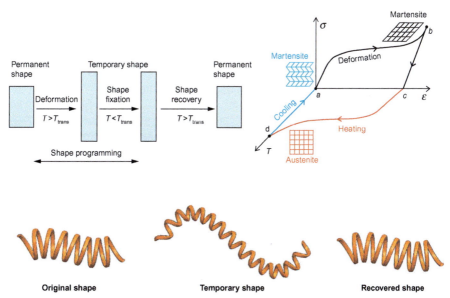

Figure 2.8 Principles of shape memory transition from original to temporary and recovered state.

Their application in everyday use textiles includes long sleeve shirts where the sleeves shorten with increasing temperature (i.e. making the wearer cooler) or reduce the need to iron clothes after washing when a breeze of dry air from dryers or the like could trigger the shape memory effect and remove creases. More precisely engineered clothing could include body suits intended for sailors and deep-sea divers when body temperature could be maintained with self-adjustable shape memory weaves/functions that would modify the thermally insulating or conducting properties. By including shape memory alloys in clothing, many visually attractive and fashionable features could also be included [43]. Attempts have also been made to produce bicomponent fibres with shape memory effects whose individual fibres in the given structure curl up based on their different thermal expansions, thereby regulating the passage of air and thus controlling the temperature. In medical applications, upon application of the trigger mechanism, shape memory sutures would tighten the stitches where inserting sutures in hard-to-reach places cannot provide adequate tension to close cuts and wounds [44]. More work in the medical application area has focussed on shape memory nanofibres, exploring their potentials in tissue engineering, drug delivery/release, wound dressings, and many other non-medical applications [45,46].

2.7   Animal- and plant-based fibres for medical and technical applications

Traditionally, all fibres used in clothing, apparel, and furnishing goods were made from a range of well-known natural fibres, including wool, silk, cotton, jute, and hemp. However, with the advent of synthetic fibres, the way that fibres were used either as replacements for natural fibres or in blends was revolutionary. These additions meant that fibres would increasingly be designed and/or blended to achieve particular functional and/or performance requirements. Cotton/polyester blend shirts, for example, are now synonymous with hard wearing, quick drying, and easy (or no) ironing. In more recent years, however, natural fibres of different origin and functionality have come to the forefront of fibre technology, helping to resolve issues of incompatibility, environmental friendliness, innate antimicrobial resistance, and natural healing, particularly in medical applications.

Cotton is a highly absorbent fibre, soft and pleasant to the touch. It has been routinely used in many countries as an aid to wound dressings either as loose fibres or in the form of gauze. However, cotton is not a biocompatible fibre and could potentially produce more problems than it solves. The human body sees cotton as an alien material and therefore tries to reject it if it is not physically removed. Cotton trapped in a healing wound is not only traumatic to remove for the patient but also can act as a platform for bacteria and infection within a wound, thus prolonging recovery [47].

More body friendly and naturally antimicrobial fibres include bamboo, alginate, and chitosan, which are today increasingly being considered in areas of health and hygiene. Bamboo fibres are extracted from bamboo plants. Despite having a long history, bamboo has only recently gained popularity as a "green fibre" because of its fast growth and impact of little environmental damage [48]. It uses less water than cotton and can be used for a variety of applications, including food and shelter in the form of buildings from the bark and its branches. Bamboo also contains bamboo "kun", an antimicrobial agent that gives the plant natural resistance to pests and fungi infestation, a property that has been used to push bamboo fibres as an inherently antimicrobial material. However, this claim is often questioned because the innate antimicrobial property is said to be process dependent [49].

Alginate, originating from seaweed, is an anionic polysaccharide that can absorb up to 200–300 times its own weight in water. It consists of two monomer units of L-guluronic (G) acid and D-mannuronic (M) acid. As a soluble salt (i.e. sodium alginate), alginate can be converted to insoluble fibre or film when treated with a calcium-rich coagulation bath. Alginate fibres made into wound dressings are today extensively used to treat low or medium exuding wounds [50]. They work principally by allowing ion exchanges at the wound surface between calcium and sodium ions that are present in the dressing and the exudate, respectively, thus allowing low to moderate exudate absorption. Furthermore, the soluble nature of sodium alginate eliminates the need for its complete removal (i.e. it can be harmlessly absorbed by the body). The health industry has benefited from this fibre enormously within the last 30 years or so, and it has been produced under different brand names in many countries (see Table 2.2).

Chitosan, which is derived from chitin by a chemical deacetylation process, is found in the hard shell of crustaceans such as crabs, lobsters, and shrimp

Table 2.2 Alginate dressings marketed under different brand names

Types of alginate dressings with and without antimicrobials	Manufacturers
Katostat	ConvaTec Ltd
Algisite M	Smith and Nephew
Algosteril	Les Laboratoires
Comfeel Alginate	Coloplast Ltd.
Sorbsan	Bertek Pharmaceuticals
Sorbalgon	Hartman-Conco Medicare
Algicell	Derma Sciences Products
Curasorb/Curasorb Plus and Curasorb Zn	Kendall Healthcare
Melgisorb	Molnlycke Health Care
Tegagen	3M Healthcare
Trionic	Trionic Sverige AB

along with minerals and proteins. Once separated from impurities, the chitin is deacetylated to enhance solubility and processing routines. Chitosan is of immense interest because of its natural antimicrobial tendencies, that is, carrying a positive charge, which enables it to attract negatively charged bacteria/microorganisms and then kills them. It is naturally haemostatic and is wildly believed to have good healing properties [51]. The latter are largely believed to result from the breakdown of chitosan via lysozyme, an enzyme present in the exudate medium, which is ultimately responsible for promoting macrophage activities and rebuilding damaged cellular structures in the body. It is biocompatible, and no significant allergic effects have been reported when using it. It is made into variety of forms including capsules, films, and fibres. Chitosan fibres are rather brittle and hence are difficult to process. Furthermore, unlike alginate fibres, chitosan fibres have limited absorption capacity. These deficiencies in function have limited their widespread availability as fibres.

To take advantage of alginate and chitosan properties while eliminating their shortcomings, a unique hybrid fibre based on these two components has been developed at the University of Bolton. This fibre offers very high water absorption (i.e. >30 g/g), very good gelling ability, a property much desired by clinicians and nurses, and the natural ability to heal while being naturally antimicrobial and haemostatic [52,53].

Collagen, the most abundant protein in the human body, acts as structural support for most tissues in the body. There are up to 30 varieties of collagens, but most of them in the human body are of type I, II, or III. Reconstituted type I collagen fibres have increasingly become more popular because of their high strength and resilience, often matching fibres found in human skin, blood vessels, teeth, and tendons. Reports of their use in repairing tendons in rabbits and anterior cruciate ligaments are widely available in the literature [54,55]. To match natural collagen elasticity, a property that is often lost in the regeneration of collagen, resilin – a highly elastic protein extract from arthropods – is added to overcome this shortcoming [56]. *In vitro* and *in vivo* studies of cell growths on these fibres have also shown good growth pattern and adaptation [57]. To date, various attempts have been made to make textile-like fibres from regenerated type I collagen albeit with limited success. US Patent No. 5,378,469 [58] describes continuous production of collagen fibres, but the longest length produced is about 5 m (18 ft). Likewise, US Patent No. 5,997,896 [59] describes a production technique by which segmented collagen fibres are made by a so-called injectable technique. Others including Jeffrey et al., Kew et al., and Shepherd et al. [60–62] have attempted to produce continuous collagen fibres with limited success. The challenge is reportedly the result of isolable long fibres that do not allow continuous filament production [63]. Recent attempts at the University of Bolton have employed a unique technique that produces reconstituted collagen fibres with no interruptions, thus opening many possibilities for the application of these fibres beyond tendon regeneration [64].

Figure 2.9 illustrates monomer units of materials referred to in this section.

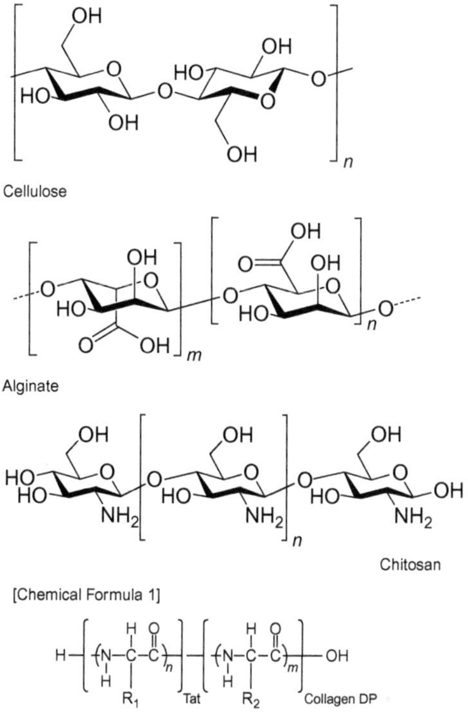

Figure 2.9 Monomer units of cotton, alginate, chitosan, and collagen.

2.8 Synthetic- (or chemical-) based fibres for medical and environmental applications

Synthetic or chemical fibres have played a major part in broadening the application potential of fibres with provisions to alter and tailor make properties. This freedom in structural manipulation has resulted in a range of products encompassing low to high tenacity and modulus and variations in extensibilities. More specialised synthetic fibres have surpassed the standard heat, combustion resistance, and specific strength of earlier generations of such fibres and have today become prominent in most high-tech applications including tyres, airbags, parachutes, harnesses, and a wide range of composites. While most of these fibres were reviewed in the first edition of this volume [1], some of these fibres are described elsewhere in the current edition with reference to their application (see Volume 2). The inertness and compatibility of many of these fibres have also made them a natural choice for biomedical applications and implants for which resilience, durability, and long-term performance under various forces are important.

However, the extensive use and indestructible nature of these petrochemical-derived, hydrocarbon-based synthetic fibres have increasingly become controversial and even blamed for the pollution of land, water, and air. In the UK alone, more than 3 million tonnes of plastic-based materials, including textiles, are used every year, but only 7%, or 210,000 tonnes, are recycled [65]. The rest end up in landfill sites where the leaching out of chemicals to the land, pollution of underground water, and spread of bacteria via feeding birds have become targets of much restriction and regulatory control. One way to overcome or minimise this problem is through synthetic biodegradability sourced from renewable resources rather than from finite fossil fuel, the normal source of synthetic fibres, which has become a major research issue that has consequentially led to very interesting developments.

2.9 Degradable and nondegradable synthetics

Biological degradation of natural and some synthetic polymers occurs when polymer depolymerisation takes place as the result of enzyme secretions by some microorganisms. These enzymes either oxidise or hydrolyse the polymer by attacking the polymer chain's main functional chain and end groups so that polymer chains with either oxygen or nitrogen or both with low levels of crystallinity are most susceptible. Natural fibres fitting most of these criteria are therefore highly biodegradable whereas synthetic fibres such as polypropylene and polyethylene are least likely to be biologically attacked and then degrade. Aromatic polyesters and polyamides, despite having oxygen and nitrogen in their respective molecular chains, also resist biodegradation largely because of their molecular chain rigidity and high levels of crystallinity. Aliphatic polyesters, however, are highly susceptible to biodegradation despite being thermoplastic because of the hydrolytic sensitivity of the ester groups present and relatively low crystallinities, although in many ways they are identical to the partly aromatic polyesters such as poly(ethylene terephthalate) in terms of their general textile properties. Furthermore, they are derived from renewable resources with virtually no finite limitations. For example, poly(lactic acid) or PLA is a thermoplastic aliphatic polyester derived from corn starch or sugar cane and poly(L-lactic acid) (PLLA), another derivative of PLA, has an enhanced glass transition and melting temperature. Poly(glycolic acid) (PGA) is another biodegradable, thermoplastic polymer and in fact is the simplest linear, aliphatic polyester with a melting point of around 230 °C. Poly(caprolactone) or PCL is also a biodegradable polyester with a low melting point of around 60 °C. Poly(2-hydroxybutyrate) (PHB) is derived from microorganisms acting on nutrients with good ultraviolent (UV) resistance, biocompatibility, tensile strength, and reasonably high melting point. These fibres/polymers are today extensively used as biomaterials/biomedical devices in implants (e.g. sutures and tissue scaffolds as well as consumable packaging). See Table 2.3 for further details on some of these polymers.

These biodegradable polyesters are derived from polymerisation of glycolic acid (PGA), lactic acid (PLA), 2-hydroxybutyric acid (PHB), and caprolactone (PCL). Physical or chemical blends of such polymers with traditional synthetics have also been tried and proven to be effective in encouraging degradation (e.g. aliphatic–aromatic copolyesters or AAC) [66].

Table 2.3 Origins and associated properties of biodegradable polymers

Polymers	T_g (°C)	Melting point (°C)	Origin	Commercial products	Monomer unit
Poly(lactic acid), PLA	~55	~175	Corn, tapioca roots, sugar cane	Food packaging, cups, etc.	(structure)
Poly(L-lactic acid), PLLA	60–65	173–178	Corn, tapioca roots, sugar cane	Packaging, films, fibres	(structure)
Poly(glycolic acid), PGA	35–40	225–230	Polycondensation of glycolic acid	Absorbable sutures, Dexon	(structure)
Poly(caprolactone), PCL	~ −60	~60	Ring opening polymerisation of ε-caprolactone	Sutures, tissue engineering, capsules	(structure)
Poly(2-hydroxybutyrate), PHB	2	175	Biodegradable polyester (microorganisms/ fermentation)	Medicine, tissue engineering	(structure)

2.10   Conclusion and future prospects

The increasing desire for smart applications and demands for multi-functionality by consumers has initiated an enormous drive for the development and exploitation of fibres that would address these challenges. As a result, an array of fibres, some of which are referred in this chapter, has been developed and gradually introduced to the marketplace. Much of what has been achieved has resulted directly from careful study and prudent observation of functional features of the natural habitat. From microbes to mammals and to plants, almost anything on the planet is rich in clues and information on how evolution has enabled species to combat danger, resist bacterial infiltration, and rebuild and manage unforeseen circumstances.

Just as there are many evolving species, countless opportunities remain undiscovered and unexplored, so there are ample opportunities for further research and uses in which fibres, as fundamental building blocks, will no doubt continue to play a key role in the intricacy and complexity of demands. Just as electronic devices have become smaller and smaller, functional fibres with intrinsic and built-in abilities will likewise play a major role in increasingly lower dimensions to meet all challenges.

While considering the extent of possibilities, future designers must consider the burden of synthetic wastes and their detrimental consequences on the environment and then adopt credible methodologies and/or separating technologies to allow recycling and reuse of such materials.

Bibliography

Fibre technology is an ongoing development and a scientific refinement process regarding which much remains to be explored and ultimately employed. This chapter has made only a brief reference to some of these fibres and their potential applications. Readers wishing to explore these concepts further are recommended to read a good selection of the references given in this chapter and refer to specialised textbooks for more in-depth studies.

References

[1] Miraftab M. Technical fibres. In: Horrocks AR, Anand SC, editors. Handbook of technical textiles. Cambridge, UK: Woodhead Publishing; 2000. p. 24–41.
[2] Ondarçuhu T, Joachim C. Drawing a single nanofibre over hundreds of microns. EPL (Europhysics Letters) 1998;42(2):215–20.
[3] Feng L, Li S, Li H, Zhai J, Song Y, Jiang L, et al. Super-hydrophobic surface of aligned polyacrylonitrile nanofibers. Angew Chem Int Ed 2002;41(7):1221–3.
[4] Martin CR. Membrane-based synthesis of nanomaterials. Chem Mater 1996;8:1739–46.
[5] Ma PX, Zhang R. Synthetic nano-scale fibrous extracellular matrix. J Biomed Mater Res 1999;46:60–72.

[6] Whitesides GM, Grzybowski B. Self-assembly at all scales. Science 2002;295:2418–21.
[7] Deitzel JM, Kleinmeyer J, Hirvonen JK. Controlled deposition of electrospun poly(ethylene oxide) fibers, BeckTNC. Polymer 2001;42:8163–70.
[8] Huang Z-M, Zhang Y-Z, Kotaki M, Ramakrishna S. A review on polymer nanofibers by electrospinning and their applications in nanocomposites. Compos Sci Technol 2003;63:2223–53.
[9] Gilbert W. De Magnete, Magneticisque Corporibus, et de Magno Magnete Tellure (On the Magnet and Magnetic Bodies, and on That Great Magnet the Earth). London: Peter Short; 1628.
[10] Tucker N, Stanger JJ, Staiger MP, Razzaq H, Hofman K. The history of the science and technology of electrospinning from 1600 to 1995. J Eng Fibers Fabr 2012;7(2):63.
[11] Mota C, Puppi D, Gazzarri M, Bartolo Paulo and Chiellini Federica, Melt electrospinning writing of three-dimensional star poly(ε-caprolactone) scaffolds, Polym Int 2013;62(6):2013–900.
[12] Luo CJ, Nangrejo M, Edirisinghe M. A novel method of selecting solvents for polymer electrospinning. Polymer 2010;51(7):1654–62.
[13] Vasita R, Katti DS. Nanofibres and their applications in tissue engineering. Int J Nanomed 2006;1(1):15–30.
[14] Yarin AL, Koombhongse S, Reneker DH. Tailor cone and jetting from liquid droplets in electrospinning of nanofibres. J Appl Phys 2001;90(9):4836–46.
[15] Hamzeh S, Miraftab M, Yoosefinejad A. Study of electrospun nanofibre formation process and their electrostatic analysis. J Ind Text 2014;44(1):147–58.
[16] Formhals A. Method and apparatus for spinning, US Patent 2160962, 1939.
[17] Theron A, Zussman E, Yarin AL. Electrostatic field-assisted alignment of electrospun nanofibres. Nanotechnology 2001;12:384–90.
[18] Sundaray B, Subramanian V, Natarajan TS, Xiang RZ, Chang CC, Fann WS. Electrospinning of continuous aligned polymer fibres. Appl Phys Lett 2004;84:1222–4.
[19] Hamzeh S. Private reports. PhD draft thesis, University of Bolton; 2013.
[20] Lee M, Kim H-Y. Toward nanoscale three-dimensional printing: nanowalls built of electrospun nanofibers. Am Chem Soc 2014;30:1210–4.
[21] Tong L. Nanofibers—production, properties and functional applications, ISBN 978-953-307-420-7, November, 2011.
[22] Lisiecki J, Błażejewicz T, Kłysz S. Flexible auxetic foams—fabrication, properties and possible application areas. Res Works Air Force Inst Technol 2010;27(1).
[23] Darja R, Tatjana R, Alenka P-C. Auxetic textiles. Acta Chim. Slov. 2013;60:715–23.
[24] Liu Y, Hu H. A review on auxetic structures and polymeric materials. Sci Res Essays 2010;5(10):1052–63.
[25] Javadi AA, Faramarzi A, Farmani R. Design and optimization of microstructure of auxetic materials. Eng Computation 2012;29(3):248.
[26] Yang W, Li Z-M, Shi W, Xie B-H, Yang M-B. Review on auxetic materials. J Mater Sci 2004;39:3269–79.
[27] Alderson KL, Simkins VR. Auxetic filamentry materials, UK Patent 11/018,062, University of Bolton, December 2004.
[28] Alderson KL. Progress in auxetic fibres and textiles. Actual Chimique 2012;(360–361):73–7.
[29] Glazzard M, Breedon P. Weft-knitted auxetic textile design. Phys Status Solidi B 2014;251(2):267–72.
[30] Uzun M, Patel I. Tribological properties of auxetic and conventional polypropylene weft knitted fabrics. Arch Mater Sci Eng 2010;44(2):120–5.
[31] Busch Vishniac Ilene J. Electromechanical sensors and actuators. New York: Springer; 1999, ISBN: 0-387-98495 X.

[32] Varghese J, Whatmore RW, Holmes JD. Ferroelectric nanoparticles, wires and tubes: synthesis, characterisation and applications. J Mater Chem C 2013;1:2618–38.

[33] Hadimani R, Vatansever Bayramol D, Soin N, Shah T, Qia Limin, Shi Shaox, et al. Continuous production of piezoelectric PVDF fibre for e-textile applications. Smart Mater Struct 2013;22(7):075017.

[34] Hamimani (Ravi) ML, Siores E, Vatansever D, Prekas K. Hybrid energy conversion device, Patent PCT/GB2011/051829, University of Bolton, UK, 2011.

[35] Gu F, Huang W, Wang S, Cheng X, Hu Y, Li C. Improved photoelectric conversion efficiency from titanium oxide-coupled tin oxide nanoparticles formed in flame. J Power Sources 2014;268:922–7.

[36] Paraskevopoulos I, Tsekleves E. Simulation of photovoltaics for defence applications: power generation assessment and investigation of the available integration areas of photovoltaic devices on a virtual infantryman. In: SIMULTECH 2012—Proceedings of the 2nd International Conference on Simulation and Modeling Methodologies, Technologies and Applications; 2012. p. 384–96.

[37] Lee S, Lee Y, Park J, Choi D. Stitchable organic photovoltaic cells with textile electrodes. Nano Energy 2014;9:88–93.

[38] Hadimani (Ravi) ML, Siores E, Vatansever D, Prekas K. Hybrid energy conversion device, US 20130257156 A1, PCT/GB2011/051829, Bolton, UK, 2013.

[39] Jani JM, Learya M, Aleksandar S, Gibson MA. A review of shape memory alloy research, applications, and opportunities. Mater Des 2014;56:1078–113.

[40] Tadaki T, Otsuka K, Shimizu K. Shape memory alloys. Annu Rev Mater Sci 1988;18:25–45.

[41] Jinlian H. Shape memory polymers and textiles. Cambridge: Woodhead Publishing; 2007.

[42] Liu C, Qinb H, Mather PT. Review of progress in shape-memory polymers. J Mater Chem 2007;17:1543–58.

[43] Lendlein A. Textiles come to life. Mater Today 2008;11(3):59.

[44] Small W, IV, Singhal P, Wilson TS, Maitland DJ. Biomedical applications of thermally activated shape memory polymers. J Mater Chem 2010;20(18):3356–66.

[45] Huang WM. Thermo-moisture responsive polyurethane shape memory polymer for biomedical devices. Open Med Devices J 2010;2:11–9.

[46] Hearon K, Singhal P, Horn J, Small W, IV, Olsovsky C, Maitland KC, et al. Porous shape-memory polymers. Polym Rev 2013;53(1):41–75.

[47] Bajpai V, Bajpai S, Jha MK, Dey A, Ghosh S. Microbial adherence of textile materials: a review. J Environ Res Dev 2011;5(3).

[48] Cui YM, Zhang JH. Natural environmentally friendly green fiber—the application of bamboo fiber fabric in design of brand clothing. In: 3rd international conference on energy and environmental protection, ICEEP 2014; Xi'an; China; 26 April 2014 through 28 April 2014; Advanced Materials Research, 955–959. 2014. p. 108–11.

[49] Hardin IR, Wilson SS, Dhandapani R, Dhende V. An assessment of the validity of claims for "Bamboo" fibers. In: American Association of Textile Chemists and Colorists International Conference; 2009.

[50] Broussard KC, Powers JG. Wound dressings: selecting the most appropriate type. Am J Clin Dermatol 2013;14(6):449–59.

[51] Kong L, Ziegler GR, Bhosale R. Fibers spun from polysaccharides. Handbook of carbohydrate polymers: development, properties and applications 2011;1–43.

[52] Miraftab M, Smart GM. Composite fibre of alginate and chitosan, Patent US2009099353 (A1)—2009-04-16, Bolton, GB.

[53] Miraftab M, Iwu C, Okoro C, Smart G. Inherently antimicrobial alchite fibres developed for wound care applications. Med Healthcare Text 2010;76–83.

[54] James R, Kesturu G, Balian G, Chhabra AB. Tendon: biology, biomechanics, repair, growth factors, and evolving treatment options, review article, ASSH Published by Elsevier, Inc. Vol. 33A, January 2008.

[55] Medlen JC. Collagen ligament and tendon regeneration method and material, Patent WO 1985000511 A1, 1985.

[56] Pikkarainen J, Kulonen E. Relations of various collagens, elastin, resilin and fibroin. Comp Biochem Physiol B 1972;41(4):705–12.

[57] Sanami M, Sweeney I, Shtein Z, Meirovich S, Sorushanova A, Mullen AM, Miraftab M, Shoseyov O, O'Dowd C, Pandit A, Zeugolis DI. The influence of poly(ethylene glycol) ether tetrasuccinimidyl glutarate on the structural, physical, and biological properties of collagen fibers: multibranched PEG crosslinked collagen fibers. J Biomed Mater Res B Appl Biomater 2015.

[58] Collagen threads, Patent No. 5,378,469, January 1995; issued date October 7, 1991; filing date 07/772,529.

[59] Carr A, Jr. et al. Reconstituted collagen fiber segment compositions and methods of preparation thereof, Patent No. US 5,997,896 1999.

[60] Caves JM, Cui W, Wen J, Kumar VA, Haller CA, Chaikof EL. Elastin-like protein matrix reinforced with collagen microfibers for soft tissue repair. Biomaterials 2011;32:5371e5379.

[61] Kew SJ, Gwynne JH, Enea D. Regeneration and repair of tendon and ligament tissue using collagen fibre biomaterial. Acta Biomater 2011;7(9):3237–47.

[62] Lew CW, Aufdemorte TB, McGuff HS, Alderson GL. Method and apparatus for delivery of a medicament in the oral cavity, US Patent 5181505 A, 1993.

[63] Meyer M, Baltzer H, Schwikal K. Collagen fibres by thermoplastic and wet spinning. Mater Sci Eng C 2010;30:1266.

[64] Miraftab M, Sanami M. Patent pending, University of Bolton, 2014.

[65] http://www-g.eng.cam.ac.uk/impee/topics/RecyclePlastics/files/Recycling%20 Plastic%20v3%20PDF.pdf.

[66] Suggs LJ, Moore SA, Mikos AG. Synthetic biodegradable polymers for medical applications. Physical Properties of Polymers Handbook, second ed., 2007;939–50.

Technical yarns

3

R.H. Gong, X. Chen
University of Manchester, Manchester, UK

Chapter Outline

3.1   Introduction

Technical yarns are produced for the manufacture of technical textiles. They must meet the specific functional requirements of the intended end use. This may be achieved by using special yarn production techniques or selecting special fibre blends, or a combination of both. This chapter describes the yarn production technologies that are applicable to technical yarns and discusses the structures and properties of the yarns that may be produced using these technologies.

3.2   Staple fibre yarns

3.2.1   Ring spinning

Ring spinning is currently the most widely used yarn production method. Initially developed in America in the 1830s, its popularity has survived the emergence of much faster spinning technologies. In addition to the superior yarn quality produced, ring spinning is extremely versatile. It is capable of producing yarns of wide ranges of count and twist from a great variety of fibre materials. It is also used for doubling and twisting multifold and cabled yarns.

Fibre materials must be properly prepared before they can be used on the ring-spinning machine. The preparation processes depend on the fibre material. Figures 3.1 and 3.2 illustrate the typical process routes for cotton and wool. The ultimate objectives of the many preparation processes are to produce a feed material that is clean, even, homogenous, and free from fibre entanglement for the final spinning process. The fibres must also be in the preferred orientation.

On the ring-spinning machine, the feed material is attenuated to the required linear density by a drafting system, typically a roller drafting system with three lines of rollers. The drafted fibre strand is then twisted by the ring spindle illustrated in Figure 3.3.

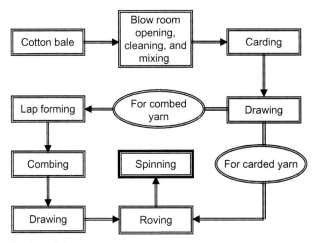

Figure 3.1 Production of ring-spun cotton yarn.

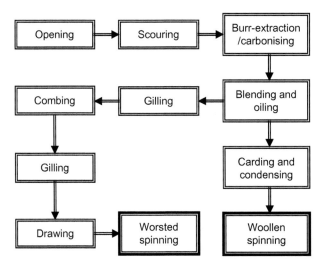

Figure 3.2 Production of wool yarn.

The yarn leaving the front rollers is threaded through a yarn guide (the lappet), which is located directly above the spindle axis. The yarn then passes under the C-shaped traveller onto the bobbin, which is mounted on the spindle and rotates with it. When the bobbin rotates, the tension of the yarn pulls the traveller around the ring. The traveller rotational speed, the spindle rotational speed, and the yarn delivery speed follow the following equation:

$$N_t = N_s - \frac{V_d}{\pi D_b} \tag{3.1}$$

where N_t is the traveller rotational speed (rpm); N_s the spindle rotational speed (rpm); V_d the yarn delivery speed (m/min); and D_b the bobbin diameter (m).

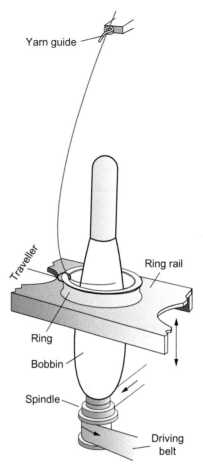

Yarn guide

Traveller

Ring rail

Ring

Bobbin

Spindle

Driving belt

Figure 3.3 The ring spindle.

During production, the bobbin rail moves up and down to spread the yarn along the length of the bobbin so that a proper package can be built. The movement of the ring rail is quite complicated; its purpose is to build a package that is stable and easy to unwind and that contains the maximum amount of yarn. As the yarn is wound on the bobbin, the bobbin diameter increases steadily during production. The spindle speed and the yarn delivery speed are normally kept constant; it is therefore obvious from Equation (3.1) that the traveller speed increases during production.

Each rotation of the traveller inserts one twist in the yarn, so the twist inserted in a unit length of yarn can be calculated by:

$$t = \frac{N_t}{V_d} \tag{3.2}$$

where t is the yarn twist (turns/m).

Because the traveller speed is not constant, the twist in the yarn also varies. However, because this variation is usually very small, it is commonly ignored, and the twist is simply calculated from the spindle speed:

$$t = \frac{N_s}{V_d} \tag{3.3}$$

As Equation (3.3) indicates, for a given yarn twist, the higher the spindle speed is, the higher is the yarn delivery speed. A spindle speed of up to 25,000 rpm is possible, although speeds of between 15,000 and 20,000 rpm are more usually used. The spindle speed is restricted by the traveller speed that has an upper limit of around 40 m/s. When the traveller speed is too high, the friction between it and the ring will generate too much heat, which accelerates the wear of the traveller and the ring and may also cause yarn damage. The yarn between the yarn guide and the traveller rotates with the traveller and balloons out as the result of centrifugal force. The tension in the yarn increases with the rotational speed of the yarn balloon. When the spindle speed is too high, the high yarn tension increases the yarn breakage. The traveller speed and the yarn tension are the two most critical factors that restrict the productivity of the ring-spinning system. The increasing power cost incurred by rotating the yarn package at higher speeds can also limit the economic viability of high spindle speeds. For the same traveller linear speed, using a smaller ring enables a higher traveller rotational speed and increases delivery speed. A smaller ring also reduces yarn tension because the yarn balloon is also smaller. However, a smaller ring leads to a smaller bobbin, which results in more frequent doffing.

Ring-spun yarns have a regular twist structure and, because of the good fibre control during roller drafting, the fibres in the yarn are well straightened and aligned. Ring-spun yarns therefore have excellent tensile properties, which are often important for technical applications.

The ring-spinning system can be used for spinning cover yarns when a core yarn, spun, or filament is covered by staple fibres. This can provide yarns with a combination of technical properties. For example, a high-strength yarn with good comfort characteristics may be spun from a high-strength filament core with natural fibre covering.

Other technical yarns, such as flame-retardant and antistatic yarns, can also be made by incorporating flame-retardant and electricity-conductive fibres.

The main limitation of the ring-spinning system is the low productivity. The other limitations are the high drafting and spinning tensions involved. These high tensions can become a serious problem for spinning from fibres such as alginate fibres that have low strength.

3.2.2 Rotor spinning

The productivity limitation of the ring-spinning system was recognised long before the commercial introduction of rotor spinning in 1967. In the ring spinning, the twist insertion rate depends on the rotational speed of the yarn package. This is so because of the continuity of the fibre flow during spinning. Numerous attempts were made before the end of the nineteenth century, particularly since the 1950s, to introduce a break into the fibre flow so that only the yarn end needs to be rotated to insert twist. Very high twisting speeds can thus be achieved. In addition, separating twisting from package winding provides much more flexibility in the form and size of the yarn package built on the spinning machine. This increases the efficacy of both the spinning machine and the subsequent processes. Rotor spinning was the first such new technology to become commercially successful and is the second most widely used yarn production method after ring spinning.

The principles of rotor spinning are illustrated in Figure 3.4. The fibre material is fed into an opening unit by a feed roller in conjunction with a feed shoe. The feed material is usually a drawn sliver. An opening roller is located inside the opening unit. The opening roller is covered with carding wire, usually sawtooth type metallic wire. The surface speed of the opening roller is in the region of 25–30 m/s, approximately 2000 times the feed roller surface speed. This high-speed ratio enables the fibres to be opened up into a very thin and open fibre flow. The fibres are taken off the opening roller by an air stream with a speed about twice that of the opening roller. The fibres are carried by the air stream through the fibre transportation tube into the spinning rotor. The airspeed in the transportation tube increases because of the narrowing cross-section of the tube as the air reaches the exit point inside the rotor. This ensures that the fibres are kept aligned along the airflow direction and as straight as possible. The exit angle of the fibres from the transportation tube is tangent to the rotor wall, and the surface speed of the rotor is faster than the exit airspeed, so the fibres emerging from the transportation tube are pulled into the rotor, keeping the fibres aligned in the direction of the fibre flow. The centrifugal force generated by the rotor forces the fibres into the rotor groove. Because of the high surface speed of the rotor, only a very

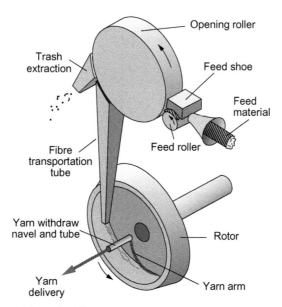

Figure 3.4 Rotor-spinning principle.

thin layer of fibres, usually one or two fibres in the cross-section, is deposited in the rotor when it goes past the fibre exit point of the transportation tube. Many such layers of fibres are needed to make up the yarn. This doubling up of the fibres in the rotor is called *back-doubling*.

The tail of the yarn arm inside the rotor is thrown against the rotor groove because of the centrifugal force. The yarn arm rotates with the rotor, and each rotation of the yarn arm inserts one twist in the yarn. As the yarn is withdrawn continuously through the navel and tube, the contact point of the yarn arm with the rotor groove must move around the rotor. Because the yarn arm is also rotating axially, the fibres in the rotor groove are twisted into the yarn. The machine twist of the yarn can be calculated by

$$t = \frac{N_y}{V_d} \qquad (3.4)$$

where t is the yarn twist (turns/m); N_y the rotational speed of the yarn arm (rpm); and V_d the yarn delivery speed (m/min).

The following relationship exists between the yarn arm speed, the rotor speed, and the yarn delivery speed:

$$\left(N_y - N_r\right)\pi D = V_d \qquad (3.5)$$

where D is the diameter of rotor groove.

The relative speed between the yarn arm and the rotor is normally very small in comparison with the rotor speed, and the machine twist of the yarn is commonly calculated by

$$t = \frac{N_r}{V_d} \qquad (3.6)$$

The doubling-back ratio β is equal to the ratio of the rotor speed to the relative speed between the yarn arm and the rotor:

$$\beta = \frac{N_r}{N_y - N_r} = \frac{N_r}{V_d}\pi D = t\pi D \qquad (3.7)$$

Because there is no need to rotate the yarn package for the insertion of twist, rotor spinning can attain much higher twisting speeds than ring spinning. The rotor speed can technically reach 200,000 rpm. The roving process needed in ring spinning is eliminated in rotor spinning, further reducing the production cost. The package can be much larger with fewer knots in the product and with a more suitable form for subsequent processes.

Because the yarn is formed in an enclosed space inside the rotor, trash particles remaining in the fibres can accumulate in the rotor groove. This leads to a gradual deterioration of yarn quality and in severe cases, yarn breakage. The cleanness of fibres is more critical for rotor spinning than for ring spinning. In order to improve the cleanness of the fibres, a trash extraction device is used at the opening roller.

As the twist in the yarn runs into the fibre band in the rotor groove, inner layers of the yarn tend to have higher levels of twist than outer layers. Fibres landing on the rotating fibre band close to the yarn tail or directly on the rotating yarn arm when the yarn arm passes the exit of the transportation tube tend to wrap around the yarn instead of being twisted into the yarn. These wrapping fibres are characteristic of rotor-spun yarns. Unlike ring-spun yarns, rotor-spun yarns cannot be fully untwisted because of the twist differential between fibres in different radial yarn layers and the wrapping fibres. For the same reasons, measuring the twist level of rotor yarns is difficult because twist-measuring methods are usually based on the yarn untwisting. To measure rotor yarn twist more accurately, the wrapping fibres and the fibres near the yarn surface need to be removed.

Rotor-spun yarns usually have lower strength than corresponding ring-spun yarns because of the poorer fibre disposition in the yarn. This is the result of using an opening roller to open up the fibres, transporting them by airflow and the low yarn tension during yarn formation. The wrapping fibres also lead to a rougher yarn surface. Rotor yarns have better short-term evenness than ring-spun yarns because of the back-doubling action.

The main advantage of rotor spinning is the high production rate. However, because of the lower yarn strength, rotor spinning is limited to medium to course yarn counts. It is also limited to spinning short staple fibre yarns because longer fibres require larger rotors that reduce production speed.

3.2.3 Friction spinning

Friction spinning is an open-end spinning technique. Instead of a rotor, two friction rollers are used to collect the opened-up fibres and twist them into the yarn. The principle is shown in Figure 3.5.

The fibres are fed in sliver form and opened by a carding roller. The opened fibres are blown off the carding roller by an air current and transported to the nip area of two

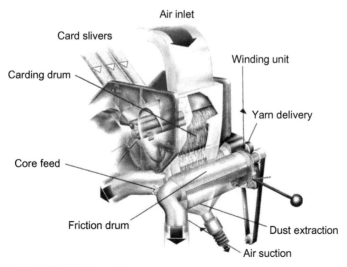

Figure 3.5 The DREF 2 friction spinner.

perforated friction drums. The fibres are drawn onto the surfaces of the friction drums by air suction. The two friction drums rotate in the same direction and because of the friction between the fibre strand and the two drum surfaces, twist is inserted into the fibre strand. The yarn is withdrawn in the direction parallel to the friction drum axis and delivered to a package-forming unit. The friction drum diameter is much larger than the yarn diameter. The diameter ratio can be as high as 200. A high twisting speed can thus be achieved by using a relatively low speed for the friction drums. Because of the slippage between the drum surface and the yarn end, the yarn takes up only 15–40% of the drum rotation. Nevertheless, a high production speed of up to 300 m/min can be achieved. For a finer yarn, the twist insertion rate is higher with the same drum speed, so the delivery speed can be practically independent of yarn count.

Because the yarn is withdrawn from the side of the machine, fibres fed from the machine end away from the yarn delivery tend to form the yarn core whereas fibres fed from the machine end closer to the yarn delivery tend to form the sheath. This characteristic can be conveniently used to produce core–sheath yarn structures. Extra core components, filaments or drafted staple fibres, can be fed from the side of the machine while the fibres fed from the top of the machine, the normal input, form the sheath.

Unlike ring- or rotor-spinning machines that are produced by many manufacturers around the world, friction-spinning machines were made only by Dr. Ernst Fehrer AG (DREF) of Austria, although production has now ceased. The diagram shown in Figure 3.5 is the DREF 2 machine that was later renamed as DREF 2000. The company also produced the DREF 3 (later DREF 3000) machine that has an extra drafting unit on the side of the machine for feeding drafted staple fibres as a core component.

The fibre configuration in friction-spun yarns is very poor. When the fibres come to the friction drum surface, they must decelerate sharply from a high velocity to almost stationary. This causes fibre bending and disorientation. Because of the very low tension in the yarn formation zone, fibre binding in the yarn is also poor. As a result, the yarn has a very low tensile strength and only coarse yarns, 100 tex and above, are usually produced.

The main application of friction spinning is for the production of industrial yarns and for spinning from recycled fibres. It can be used to produce yarns from aramid and glass fibres and with various core components including wires. The yarns can be used for tents, protective fabrics, backing material, belts, insulation, and filter materials. For core-spun yarns, the core gains no twist during yarn formation. This allows the core strength to be fully used in the final yarn. The core yarn placement can also be accurately controlled. In ring spinning, the core fibres are twisted together with the sheath fibres. This leads to lower strength contribution of the core fibres. Because of fibre migration, the core components also may migrate to the yarn surface.

3.2.4   Wrap spinning

Wrap spinning is a yarn formation process in which a twistless staple fibre strand is wrapped by a continuous binder. The process is carried out on a hollow spindle machine as illustrated in Figure 3.6. The hollow spindle was invented by DSO Textile in Bulgaria. The first wrap-spinning machine was introduced in the 1979 ITMA (This is a major textile machinery exhibition organised by ITMA - International Textile machinery Association.).

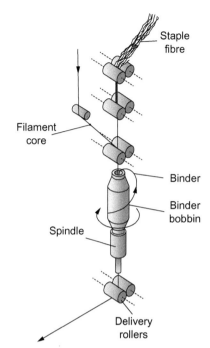

Figure 3.6 The wrap-spinning principle.

The staple roving is drafted on a roller drafting system similar to those used on ring frames and is passed through a rotating hollow spindle that carries a binder bobbin. The rotation of the hollow spindle and the bobbin wraps the binder around the staple strand.

To prevent the drafted staple strand from falling apart before it is wrapped by the binder, a false twist is usually generated in the staple strand by the spindle. To introduce the false twist, the staple strand is not threaded through the hollow spindle directly but is wrapped around either a twist regulator at the bottom of the spindle or part of the spindle top. The false twist also allows the stable strand to be compacted before the binder is wrapped around it. This improves the yarn strength.

Two hollow spindles can be arranged, one above the other, to wrap the staple strand with two binders in opposite directions. This produces special-effect yarns with a more stable structure. Real twist may also be added to the yarn by passing the wrapped yarn onto a ring spindle, usually arranged directly underneath the hollow spindle.

Core yarns, mostly filaments, can be added to the feed. This can be used to provide extra yarn strength or other special yarn features. An example is to use this method to spin alginate yarns. Alginate fibres are very weak and cause excessive breakages during spinning without the extra support of core filaments.

A variety of binders can be used to complement the staple core or to introduce special yarn features. For example, a carbon-coated nylon filament yarn can be used to produce yarns for antistatic fabrics. Soluble binders can be used for making yarns for medical applications.

Wrap spinning is highly productive and suitable for a wide range of yarn counts. Yarn delivery speeds of up to 300 m/min is possible. Because the binder is normally very fine, each binder bobbin can last many hours, enabling the production of large yarn packages without piecing. Because the staple core is composed of parallel fibres with no twist, the yarn has a high bulk, good cover, and very low hairiness.

The main limitation of wrap spinning is that it is suitable only for the production of multicomponent yarns. The binder can be expensive, increasing the yarn cost.

3.2.5   Air-jet spinning

The air-jet spinning technology was first introduced by Du Pont in 1963, but its Murata system has been commercially successful only since 1980. Du Pont used only one jet, which produced a low strength yarn. The Murata system has two opposing air jets, which improved the yarn strength. The twin-jet Murata jet spinner is illustrated in Figure 3.7. Staple fibres are drafted using a roller drafting system with three or four pairs of rollers. The fibres are then threaded through the twin-jet assembly. The second jet (N_2) has a higher twisting torque than the first jet (N_1). Immediately after leaving the front drafting rollers, the fibres in the core of the yarn are twisted in the twist direction of N_2. The fibres on the edges of the drafted ribbon are twisted by the weaker N_1 and wrap around the core fibres in the opposite direction. Because the jet system is located between the front drafting rollers and the yarn-delivery rollers, neither of which rotates around the axis of the yarn, the twist inserted by the jets is not real twist and after the yarn passes through the jet system, the core fibres become twistless. The yarn strength is imparted by the wrapping of the edge fibres. Because of the small jet dimensions, very high rotational jet speeds are possible. Although twist efficiency is only 6–12% because of the yarn's twist resistance, delivery speeds of up to 300 m/min is possible. A recent development by Murata is the replacement of the second jet with

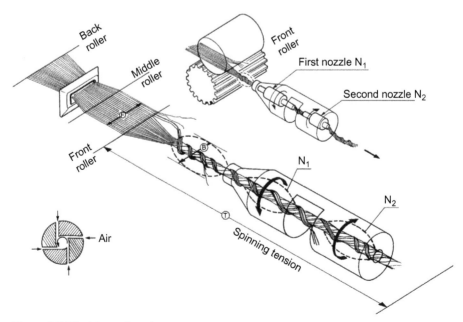

Figure 3.7 The Murata jet spinner.

a pair of roller twisters. The principle of yarn formation is similar to that of the twin-jet system. The new machine, the roller-jet spinner, is capable of delivery speeds of up to 400 m/min. However, the yarn has a harder handle.

Air-jet spinning is used mainly for spinning from short staple fibres, especially cotton and polyester blends. The Vortex spinner, the latest addition to the Murata jet spinner range, was shown at ITMA 99 for spinning from 100% cotton.

The air-jet system can be used to produce core–sheath yarn structures by feeding the core and sheath fibres at different stages of the drafting system. Fibres fed in from the back of the drafting system tend to spread wider under the roller pressure and form the sheath of the yarn whereas fibres fed in nearer to the front tend to form the yarn core. Filament core can also be introduced at the front drafting roller. Two spinning positions can be combined to produce a two-strand yarn that is then twisted using the usual twisting machinery.

Because air-jet yarns have no real twist, they tend to have higher bulk than ring and rotor yarns and better absorbency. They are more resistant to pilling and have little untwisting tendency. Because the yarn strength is imparted by the wrapping fibres, not the twisting of the complete fibre strand, air-jet yarns have lower tensile strength than ring and rotor yarns. The system is suitable only for medium- to fine-count yarns because the effectiveness of wrapping decreases with the yarn thickness. The rigid yarn core of parallel fibres makes the yarn stiffer than ring and rotor yarns.

Vortex spinning is increasing in popularity because of its suitability for producing 100% cotton yarns. Except for Murata, major textile machine manufacturers such as Rieter have started to manufacture vortex spinning machines.

3.2.6   Twistless spinning

Numerous techniques have been developed to produce staple yarns without twisting so that the limitations imposed by twisting devices, notably the ring traveller system, can be avoided and production speed can be increased. Because of the unconventional yarn characteristics, these techniques have not gained widespread acceptance commercially, but they do offer an alternative and could be used for producing special products economically.

Most of these twistless methods use adhesives to hold the drafted staple fibre strand together. They can produce fine-count yarns at a high speed. The adhesives may be removed after the fabric is made. The fibres are then bound by the interfibre forces imposed by fabric constraints. This type of yarn has high covering power because of the untwisted yarn structure. However, these processes primarily involve additional chemicals and require high-power consumption. The yarns can be used only for fabrics that offer good interfibre forces.

As an example, the TNO (Nederlandse Organisatie voor Toegepast Natuurwetenschappelijk Onderzoek or TNO (Netherlands Organisation for Applied Scientific Research)) twistless spinning method is shown in Figure 3.8. In the system shown there, the roving is drafted under wet conditions, which give better fibre control and an inactive starch is then applied to the drafted roving, which is also false twisted to give it temporary strength. The starch is activated by steaming the package that is then dried. The later version of TNO twistless system replaces the starch with PVA

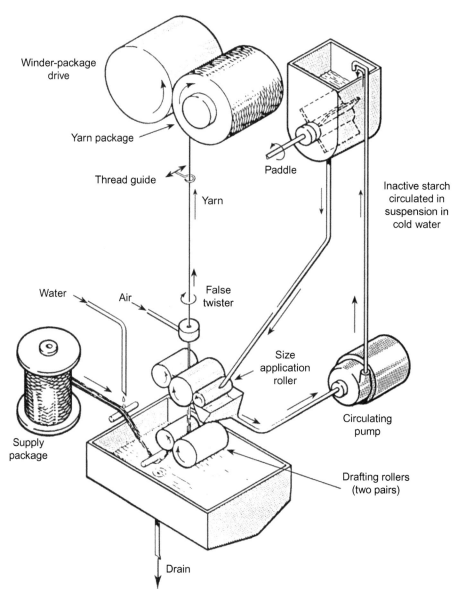

Figure 3.8 The TNO twistless yarn production method.

fibres (5–11%), which dissolves at more than 80 °C to bind the staple fibres. This is also known as the *Twilo system*.

Another twistless spinning method is the Bobtex process that can produce high-strength yarns for industrial/leisure fabrics, such as tents, work wear, and sacks. In this process, staple fibres (30–60%) are bonded to a filament core (10–50%) by a layer of molten polymer (20–50%). Production speeds of up to 1000 m/min can be achieved. The process can use all types of staple fibres including waste fibres.

3.2.7 Ply yarn

Single yarns are used in the majority of fabrics for normal textile and clothing applications, but in order to obtain special yarn features, particularly those with high strength and modulus or yarns with more stability for technical and industrial applications, ply yarns are often needed. A folded yarn is produced by twisting two or more single yarns together in one operation, and a cabled yarn is formed by twisting together two or more folded yarns or a combination of folded and single yarns.

The twisting together of several single yarns increases the tenacity of the yarn by improving the binding-in of the fibres on the outer layers of the component single yarns. This extra binding-in increases the contribution of these surface fibres to the yarn strength. Ply yarns are also more regular, smoother, and more hardwearing. Use of the appropriate single yarn and folding twists can produce a perfectly balanced ply yarn for applications that require high strength and low stretch, for example, tire cords. Additional yarn characteristics may also be introduced during the plying process.

A typical process route of a ply yarn involves the following production stages:

1. Single yarn production
2. Single yarn winding and clearing
3. Assembly winding to wind the required number of single yarns as one (doubling) on a package suitable for folding twisting
4. Twisting
5. Winding

Twisting can be done in a two-stage process or with a two-for-one twister. In the two-stage process, the ring frame is used for inserting a low folding twist in the first stage and an up-twister for inserting the final folding twist in the second stage. The ring frame uses a low twist to enable higher delivery speeds. A suitable package is formed on the ring frame for the over-end withdrawal of yarn on the up-twister. Figure 3.9 shows the principle of the up-twister. The supply package rotates with the spindle while the yarn is withdrawn over the package end from the top. The free-rotating flyer is pulled around by the yarn and inserts twist in it.

The two-for-one twister is illustrated in Figure 3.10. The supply yarn package is stationery. After withdrawal from the package, the yarn is threaded through the centre of the spindle and rotates with the spindle. Each rotation of the spindle inserts one twist in the yarn section inside the spindle centre and one twist in the yarn section outside the yarn package (the main yarn balloon). The yarn therefore gets two twists for each spindle rotation. If the supply package is rotated in the opposite direction of the spindle, the twisting rate will increase by the package rotational speed. The Saurer Tritec twister is based on this principle. Its package rotates at the same speed as the spindle but in the opposite direction, so each spindle rotation inserts three twists in the yarn. The package is magnetically driven.

The production of a ply yarn is much more expensive than of a single yarn of equivalent count. The ply yarn production not only requires the extra assembly winding and twisting processes but also the production of the finer component single yarn is much more expensive.

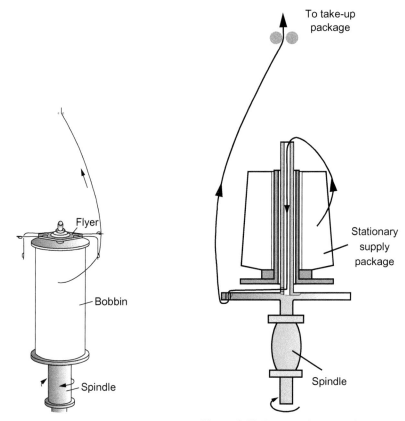

Figure 3.9 The up-twister.   **Figure 3.10** The two-for-one twister.

3.3   Filament yarns

3.3.1   Definitions

A *filament yarn* is made of one or more continuous strands called *filaments* with each component filament running the whole length of the yarn. Those yarns composed of one filament are called *monofilament yarns*, and those containing more filaments are known as *multifilament yarns*. For apparel applications, a multifilament yarn may contain as few as 2 or 3 filaments and as many as 50. In carpeting, for example, a filament yarn could consist of hundreds of filaments. Most of the synthetic fibres have been produced into the form of a filament yarn. Silk is the only major natural filament yarn.

According to the shape of the filaments, filament yarns are classified into two types, flat and bulk. The filaments in a flat yarn lie straight and neat and are parallel to the yarn axis. Thus, flat filament yarns are usually closely packed and have a smooth surface. The bulked yarns, in which the filaments are either crimped or entangled with each other, have a higher volume than the flat yarns of the same linear density.

Texturing is the main method to produce the bulked filament yarns. A textured yarn is made by introducing durable crimps, coils, and loops along the length of the filaments. As textured yarns have an increased volume, the air and vapour permeability of fabrics made from them is higher than that from flat yarns. However, for applications that require low air permeability, such as the fabrics for air bags, flat yarns may be a better choice.

3.3.2 Manufacture of filament yarns

Most synthetic fibres are extruded using either melt spinning, dry spinning, or wet spinning although reaction spinning, gel spinning, and dispersion spinning are used in particular situations. After extrusion, the molecular chains in the filaments are basically unoriented and therefore provide no practical strength. The next step is to draw the extruded filaments to orient the molecular chains. This is conventionally carried out by using two pairs of rollers, the second of which forwards the filaments at approximately four times the speed of the first. The drawn filaments are then wound with or without twist onto a package. The tow of filaments at this stage becomes the flat filament yarn. Figure 3.11 shows the melt-spinning process and the subsequent drawing process.

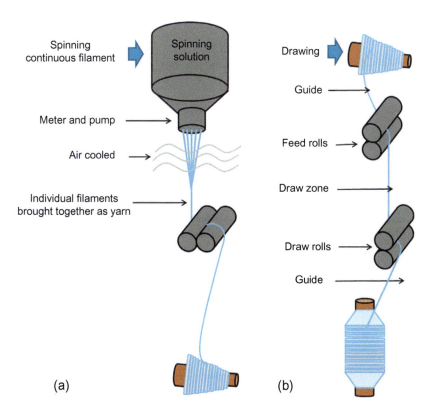

Figure 3.11 The melting-spinning process. (a) Melt-spinning and (b) fibre drawing.

For many applications, flat filament yarns are textured to gain the increased bulkiness, porosity, softness, and elasticity in some situations. Thermoplastic filament yarns are used in most texturing processes. The interfibre bonds break and re-form during the texturing process. A filament yarn is generally textured through three steps. The first step is to distort the filament in the yarn so that the interfibre bond is broken. Twisting or other means are used to distort the filaments within a yarn. The second step is to heat the yarn, which breaks bonds between polymers, allowing the filaments to stay crimped. The last step is to cool the yarn in the distorted state to enable new bonds to form between the polymers. When the yarn is untwisted or otherwise released from its distorted state, the filaments remain in a coiled or crimped condition.

The many methods for yarn texturing include false twist, air texturing, knit-de-knit, stuffer box, and gear crimp. Among these, the false twist is the most popular method. Figure 3.12 shows the principles of the main methods of yarn texturing.

3.3.3 Filament technical yarns

Many types of filament yarns have been developed for technical applications, such as reinforcing and protecting. The reinforcing technical yarns have either high modulus, high strength, or both. Yarns for protecting applications can be resistant to safety hazards such as heat and fire as well as chemical and mechanical damage. Many types of filament technical yarns are used in various applications; it is possible, therefore, to describe only a few yarns that are popularly used in the development of some technical textile products.

3.3.3.1 Aramid yarns

Aramid fibre is a synthetic fibre in which the fibre-forming substance is a long-chain synthetic polyamide that has at least 85% of the amide linkages attached directly to two aromatic rings. Nomex and Kevlar are well-known trade names of the aramid fibre owned by DuPont; Twaron and Technora are aramid fibres manufactured by Teijin. Aramid fibres have high tenacity and high resistance to stretch, to most chemicals, and to high temperature. They are well known for their relatively lightweight and resistance to fatigue and damage. Because of these properties, aramid fibres are widely used and accepted for making body armour. High-tenacity aramid fibres can be used as reinforcing material for many composite applications, including materials for boat and aircraft parts. The Nomex aramid, on the other hand, is heat resistant and is used in making firefighters' apparel and applications of a similar nature.

Aramid yarns are more flexible than many other high-performance counterparts such as glass and carbon and are thus easier to go through the subsequent fabric-making processes, be it weaving, knitting, or braiding. Care should be taken, though, because Aramid yarns are much stronger and much less extensible than the conventional textile yarns, which could make the fabric formation process more difficult. Aramid fibres are sensitive to UV light, and unprotected ones will discolour and lose strength with prolonged exposure.

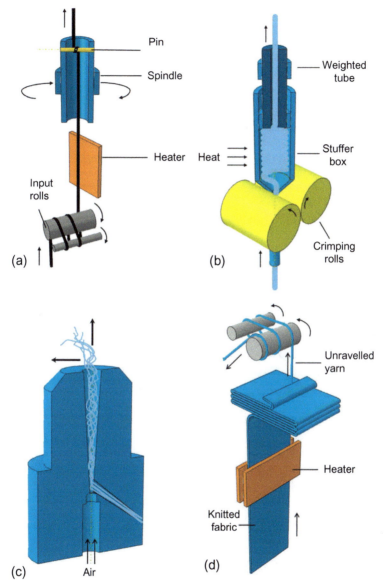

Figure 3.12 Principles of main texturing methods. (a) False twist, (b) stuffer box, (c) air-jet, and (d) knit-de-knit methods.

3.3.3.2 Glass yarns

Glass is an incombustible textile fibre that has a high tenacity. It has been used for fire-retardant applications and is commonly used in building insulation. Because of its property and low cost, glass fibre is widely used in the manufacture of reinforcement for composites. There are different types of glass fibres: E-glass, C-glass, and S-glass.

E-glass has very high resistance to moisture as well as high electrical and heat resistance. It is commonly used in glass-reinforced plastics in the form of woven fabrics. C-glass is known for its chemical resistance to both acids and alkalis. It is widely used for applications that require such resistance as in chemical filtration. The S-glass is a high-strength glass fibre used in composite manufacturing.

Glass filament yarns are brittle compared to the conventional textile yarns. It has been shown that the specific flexural rigidity of glass fibre is $0.89\,\text{mN mm}^2/\text{tex}^2$, about 4.5 times more rigid than wool. As a result, the glass yarns are easy to break in textile processing. Therefore, it is important to apply the suitable sizing to the glass yarn to minimise the interfibre friction and to hold the individual fibres together in the strand. Dextrinised starch gum, gelatin, polyvinyl alcohol, hydrogenated vegetable oils, and nonionic detergents are commonly used sizes.

When handling glass fibres, workers should wear protective clothing and a mask to prevent skin irritation and inhalation of glass fibres.

3.3.3.3   Carbon yarns

Carbon fibres are commonly made from precursor fibres such as rayon and acrylic. In the case of converting acrylic fibre to carbon, a three-stage heating process is used. The initial stage is the oxidative stabilisation, which heats the acrylic fibre at 200–300 °C under oxidising conditions. Next is the stage of carbonisation when the oxidised fibre is heated in an inert atmosphere to temperatures around 1000 °C. Consequently, hydrogen and nitrogen atoms are expelled from the oxidised fibre, leaving the carbon atoms in the form of hexagonal rings that are arranged in oriented fibrils. The final stage of the process is graphitisation, when the carbonised filaments are heated to a temperature up to 3000 °C, again in an inert atmosphere. Graphitisation increases the orderly arrangement of the carbon atoms, which are organised into a crystalline structure of layers. These layers are well oriented in the direction of the fibre axis, which is an important factor in producing high modulus fibres.

Like the glass yarns, most carbon fibres are brittle. Sizings are used so that the filaments adhere to improve the processing. In addition to protecting users against skin irritation and short fibre inhalation, consideration of protection to the processing machinery and auxiliary electric and electronic devices is necessary because carbon fibre is conductive.

3.3.3.4   HDPE yarns

HDPE refers to high-density polyethylene. Although the basic theory for making superstrong polyethylene fibres were available in the 1930s, a commercial high-performance polyethylene fibre was not manufactured until recently. Spectra, Dyneema, and Tekmilon are among the most well-known HDPE fibres. The gel-spinning process is used for producing HDPE fibres. Polyethylene with an extra high modular weight is used as the starting material. The gel-spinning process dissolves the molecules in a solvent and are spun through a spinneret. In the solution, the molecules that form clusters in the solid state become disentangled and remain in this state after the solution is cooled to give filaments. The drawing process after spinning results in a very high level of macromolecular orientation in the filaments, producing a fibre with very high tenacity and modulus. Dyneema, for example, is characterised by a parallel orientation of more than 95% and a high level of crystallinity of up to 85%. This gives unique properties to the HDPE fibres. Their most attractive properties are very high tenacity, very high-specific modulus, low elongation, and low fibre density (i.e. lighter that water).

HDPE fibres are made into different grades for different applications. Dyneema, for example, is made in SK60, SK65, and SK66 grades. Dyneema SK60 is the multipurpose grade. Its uses includes ropes, cordage, protective clothing, and reinforcement of impact-resistant composites. Dyneema SK65 has a higher tenacity and modulus than the SK60. This fibre is used when the high performance is needed and maximum weight savings are to be attained. Dyneema SK66 is designed especially for ballistic protection. This fibre provides the highest energy absorption at ultrasonic speeds.

Table 3.1 compares the properties of the filament yarns discussed here to steel.

3.3.3.5 Other technical yarns

Many other high-performance fibres, including PTFE (polytetrafluoroethylene), PBI (polybenzimidazole), and PBO (polyphenylenebenzobisoxazole), have been developed for technical applications.

PTFE fibres offer a unique blend of chemical and temperature resistance coupled with a low fraction coefficient. Because PTFE is virtually chemically inert, it can withstand exposure to extremely harsh temperature and chemical environments. The friction coefficient, claimed to be the lowest of all fibres, makes it suitable for applications such as heavy-duty bearings that involve low relative speeds.

PBI is a manufactured fibre in which the fibre-forming substance is a long-chain aromatic polymer. It has excellent thermal resistance and a soft feel, coupled with a very high

Table 3.1 Comparison of filament yarn properties

Yarns	Density (g/cm³)	Strength (GPa)	Modulus (GPa)	Elongation (%)
Aramid, regular	1.44	2.9	60	3.6
Aramid, composite	1.45	2.9	120	1.9
Aramid, ballistic	1.44	3.3	75	3.6
E-glass	2.60	3.5	72	4.8
S-glass	2.50	4.6	86	5.2
Carbon HS	1.78	3.4	240	1.4
Carbon high modulus	1.85	2.3	390	0.5
Dyneema SK60	0.97	2.7	89	3.5
Dyneema SK65	0.97	3.0	95	3.6
Dyneema SK66	0.97	3.2	99	3.7
Steel	7.86	1.77	200	1.1

Table **3.2** **High performance fibres**

Fibre	Density (g/cm³)	Tenacity (g/den)	Elongation (%)	Regain (%)
PTFE	2.1	0.9–2.0	19–140	0
PBI	1.43	2.6–3.0	25–30	15
PBO	1.54	42	2.5–3.5	0.6–2.0

moisture regain. These characteristics make the PBI fibre ideal for use in heat-resistant apparel for firefighters, fuel handlers, welders, astronauts, and race car drivers.

PBO is another new entrant to the high-performance organic fibres market. Zylon, made by Toyobo, is the only PBO fibre in production. PBO fibre has outstanding thermal properties and almost twice the strength of conventional para-aramid fibres. Its high modulus makes it an excellent material for composite reinforcement. Its low LOI (limiting oxygen index) gives PBO more than twice the flame-retardant properties of meta-aramid fibres. It can also be used for ballistic vests and helmets.

See Table 3.2 for some properties of these fibres.

References

[1]   Brearley A, Iredale J. The Worsted Industry. 2nd ed. Leeds: WIRA; 1980.
[2]   Cook JG. Handbook of Textile Fibres, Manmade Fibres. 5th ed. Sheldon, UK: Merrow; 1984.
[3]   Van Dingenen JLJ. Gel-spun High-performance Polyethylene Fibres, in High-performance fibres, Hearle JWS, editor. Cambridge: Woodhead Publishing; 2004.
[4]   Gong RH, editor. Specialist Yarn and Fabric Structures: Developments and Applications. Cambridge: Woodhead Publishing; September 2011.
[5]   Gong RH, Williamson RM. Fancy Yarns: Manufacture and Applications. Cambridge: Woodhead Publishing; October 2002.
[6]   Griffiths ET. Wrap Spun Plain and Fancy Yarns. Tomorrow's Yarns. UK: UMIST; 1984.
[7]   Kadolph SJ, Langford AL. Textiles. 8th ed. Upper Saddle River, NJ: Prentice-Hall; 1998.
[8]   Kato H. Development of MJS yarns. J Text Mach Soc Japan 1986;32(4). pp 95–101.
[9]   Klein W. The Technology of Short Staple Spinning. 2nd ed. Manchester, UK: Textile Institute; 1998.
[10] Klein W. Man-made Fibres and Their Processing. Manchester, UK: Textile Institute; 1994.
[11] Klein W. New Spinning Systems. Manchester, UK: Textile Institute; 1993.
[12] Morton WE, Wray GR. An Introduction to the Study of Spinning. 3rd ed. London: Longmans; 1962.
[13] Morton WE, Hearle JWS. Physical Properties of Textile Fibres. 3rd ed. Manchester, UK: Textile Institute; 1993.
[14] Oxtoby E. Spun Yarn Technology. Boston, MA: Butterworth; 1987.
[15] Selling HJ. Twistless Yarns. Watford: Merrow; 1971.
[16] McIntyre JE, Daniels PN. Textile Terms and Definitions. 10th ed. Manchester, UK: Textile Institute; 1995.

Technical fabric structures – 1. Woven fabrics

K.L. Gandhi[1], W.S. Sondhelm[2]
[1]Consultant, Boothstown, Worsley, Manchester, UK, [2]Consultant, Didsbury, Manchester, UK

Chapter Outline

4.1   Introduction

Technical textiles[1] are textile materials and products manufactured primarily for their technical performance and functional properties rather than their aesthetic or decorative characteristics. Most technical textiles consist of a manufactured assembly of fibres, yarns, and/or strips of material that have a substantial surface area in relation to their thickness and have sufficient cohesion to give the assembly useful mechanical strength.

Textile fabrics are most commonly woven but may also be produced by knitting, felting, lace making, net making, nonwoven processing, and tufting, or a combination of these processes. Most fabrics are two-dimensional, but an increasing number of three-dimensional woven technical textile structures are being developed and produced. Woven fabrics generally consist of two sets of yarns that are interlaced and lie at right angles to each other. The threads that run along the length of the fabric (i.e. the vertical direction is known as a *warp thread*) whereas the threads that run in the horizontal direction are *weft threads*. Frequently, in trade, they are simply referred to as *ends* and *picks*. In triaxial and three-dimensional fabrics, yarns are arranged differently.

Woven technical textiles are designed to meet the requirements of their end use, including applications in agriculture, forestry, building trade, footwear, clothing, civil engineering, geotextiles, filtration, industrial uses, medicine, transport, environment, protection, packaging, floor covering, and many more.

The strength, thickness, extensibility, porosity, and other properties of woven technical textiles, including durability, can be varied and depend on the weave used, the number of warp and weft threads per centimetre, and the raw materials, structure (filament or staple), linear density (or count), and twist factors of the warp and weft yarns. Increased strengths and stability can be obtained from woven fabrics than from any other fabric structure using interlaced yarns. Woven structures can also be varied to produce fabrics with widely different properties in the warp and weft directions.

4.2   Weave structures

The number of weave structures that can be produced is practically unlimited. In general, there are three principal classes of fabrics with different weave structures: (a) fabrics in which the ends and picks intersect one another at right angles and lie parallel in the fabric, (b) fabrics in which certain of the ends interweave alternately to the right and the left of the adjacent ends as in gauze and leno fabrics, and (c) fabrics, such as pile or plush in which a portion of the warp or weft yarns project on the foundation of the cloth and form a loop or pile on the surface of the fabric. In this section, basic structures from which all other weave structures are developed are discussed. Also briefly referred to are lenos because of their importance in selvedge constructions and triaxial fabrics because they show simple structural changes that can affect the physical properties of fabrics. Most two-dimensional woven technical fabrics are constructed from simple weaves, and of these, at least 90% use plain weave. Simple cloth constructions are discussed in greater detail by Robinson and Marks[2] and Watson[3,4] describes a large variety of simple and complicated structures in great detail.

4.2.1   Plain weave

4.2.1.1   Construction of a plain weave

Plain weave is the simplest interlacing pattern that can be produced, and diverse methods of varying the structure are employed to produce it. Used to a greater extent than any other weave, it is formed by alternatively lifting and lowering one warp thread across one weft thread, and each repeat consists of two ends and two picks. Figure 4.1 shows 16 repeats (four in the warp and four in the weft direction) of a plain weave fabric in plan view and warp way and weft way cross-sections through the same fabric. The diagrams are idealised because yarns are seldom perfectly regular and the pressure between the ends and picks tends to distort the shape of the yarn cross-sections unless the fabrics are woven from monofilament yarns or strips of film. The yarns also do not lie straight in the fabric because the warp and weft must bend around each other when they are interlaced. If in a fabric, the number of warp and weft threads per centimetre and their linear densities (thickness) are the same, the two series of threads bend equally. Such a fabric is called a *squared* or *balanced plain weave* and sometimes a *true plain cloth*. Plain weave structures give the same appearance on the face and back of the fabric. Also because of the larger number of intersections, it limits insertion of large number of picks per centimetre in the fabric. In general, plain weave constructions have a relatively low tear resistance property. The wave form assumed by the yarn is called *crimp* and is referred to in greater detail in Section 4.4.3.

4.2.1.2   Constructing a point paper diagram

Illustrating a weave either in plan view and/or in cross-section, as in Figure 4.1, takes a lot of time, especially for more complicated weaves. A type of shorthand for

Plan view

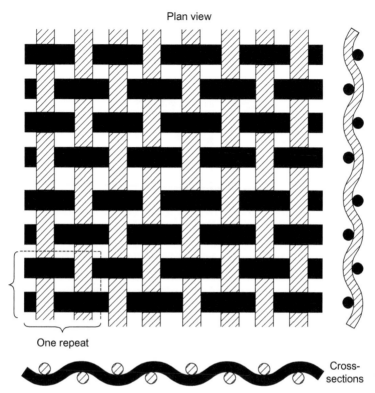

One repeat

Cross-sections

Figure 4.1 Fabric woven with plain weave – plan view (4 × 4 repeats) and warp and weft cross-sections.

depicting weave structures has therefore been evolved; the paper used for producing designs is referred to as a *squared paper*, *design paper*, or *point paper*. Generally, the spaces between two vertical lines represent one warp end and the spaces between two horizontal lines one pick. If a square is filled in, it represents an end passing over a pick whereas a blank square represents a pick passing over an end. If ends and picks must be numbered to make it easier to describe the weave, ends are counted from left to right and picks from the bottom of the point paper design to the top. The point paper design shown in Figure 4.2a is the design for a plain weave fabric. To get a better impression of how a number of repeats would look, four repeats of a design (two vertically and two horizontally) are sometimes shown. When four repeats are shown, the first repeat is drawn in the standard way, but for the remaining three repeats, crossing diagonal lines may be placed into the squares, which in the first repeats, are filled in. This method for a plain weave is shown in Figure 4.2b.

4.2.1.3 Diversity of plain weave fabrics

The characteristics of woven cloth depend on the type of fibre used for producing the yarn and whether it is a monofilament yarn; a flat, twisted, or textured (multi) filament yarn; or yarn that has been spun from natural or manufactured staple fibres.

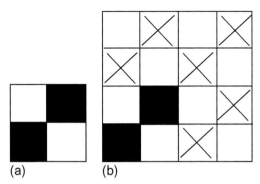

Figure 4.2 Point paper diagram of a plain weave fabric. (a) One repeat and (b) four repeats.

The stiffness of the fabric and its weaveability are also affected by the stiffness of the raw materials used and the twist factor of the yarn (i.e. the number of turns inserted in relation to its linear density). Very highly twisted yarns are sometimes used to produce special features in plain weave yarns. The resulting fabrics may have high extensibility or may be semiopaque.

The area density of the fabric can be varied by changing the linear density or count of the yarns used and altering the thread spacing, which affects the area covered by the yarns in relation to the total area. The relation between the thread spacing and the yarn linear density is called the *cover factor* and is discussed in Section 4.4. Changing the area density and/or the cover factors may affect the strength, thickness, stiffness, stability, porosity, filtering quality, and abrasion resistance of fabrics.

Square set plain fabrics, that is fabrics with roughly the same number of ends and picks per unit area and warp and weft yarns of the same linear densities, are produced in the whole range of cloth area densities and cloth cover factors. Low area-density fabrics of open construction include bandages and cheesecloth whereas light area-density high-cover factor fabrics include typewriter ribbons and medical filter fabrics; heavy open cloth includes geotextile stabilisation fabrics; and heavy closely woven fabrics include cotton awnings.

Warp-faced plain fabrics generally have a much higher warp cover factor than do weft cover factors. If warp and weft yarns of similar linear density are used, a typical is warp high and the weft crimp extremely low. The plan view and cross-sections of such fabric are shown in Figure 4.3. By using suitable cover factors and choice of yarns, most of the abrasion on such a fabric can be concentrated on the warp yarns, and the weft will be protected.

Weft-faced plains are produced by using much higher weft cover factors than warp cover factors and have higher weft than warp crimp. Because of the difference in weaving tension, the crimp difference is slightly lower than in warp-faced plain fabrics. Weft-faced plains are little used because they are more difficult and expensive to weave because of the low twist in the weft yarn and large number of picks to be inserted, thus affecting the production of the weaving machine.

Plan view

Cross-
sections

Figure 4.3 Plan view and cross-sections of warp-faced plain weave fabric with a substantially higher warp than weft cover factor.

4.2.2   Rib fabrics and matt weave fabrics

These are the simplest modifications of plain weave fabrics. They are produced by lifting two or more adjoining warp threads and/or two or more adjoining picks at the same time. This results in larger warp- and/or weft-covered surface areas than in a plain weave fabric. Because there are fewer yarn intersections, it is possible to insert more threads into a given space (i.e. to obtain a higher cover factor without jamming the weave).

4.2.2.1   Rib fabrics

Warp rib fabrics generally have more ends than picks per unit length with a high warp crimp and a low weft crimp and vice versa for weft rib fabrics. The weave structure offers a more flexible fabric than a plain weave. The simplest rib fabrics are the 2/2 warp rib and the 2/2 weft rib shown in Figure 4.4a and b, respectively. In the 2/2 warp rib, one warp end passes over two picks whereas in the 2/2 weft rib, one pick passes over two adjoining ends. The length of the floats can be extended to create 4/4, 6/6, 3/1, or any similar combination in either the warp or weft direction. See Figure 4.4c and d for 3/1 and 4/4 warp ribs.

In rib weaves with long floats, it is often difficult to prevent adjoining yarns from overlapping. Weft ribs also tend to be expensive to weave because of their relatively high picks per unit length, which reduces the production of the weaving machine unless two picks can be inserted at the same time.

The warp and weft rib structures are employed to produce a rib or a cord effect that runs from side to side of the fabric (warp ribs) or a vertical rib (i.e. the length of the fabric – weft ribs) on both faces of the fabric.

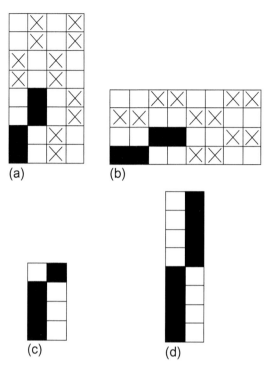

Figure 4.4 Rib fabrics. (a) 2/2 Warp rib (four repeats), (b) 2/2 weft rib (four repeats), (c) 3/1 warp rib (one repeat), and (d) 4/4 warp rib (one repeat).

4.2.2.2 Matt fabrics (or hopsack/basket)

Basically, matt fabric constructions are extensions of plain weave in which groups of two or more warp threads and picks interlace with each other to form a checkered or a dice effect.

Simple matt (or hopsack/basket) fabrics have an appearance similar to that of plain weave. The simplest of the matt weaves is a 2/2 matt shown in Figure 4.5a in which two warp ends are lifted over two picks; in other words, it is like a plain weave fabric with two ends and two picks weaving in parallel. The number of threads lifting alike can be increased to obtain 3/3 or 4/4 matt structures. Special matt weaves, such as a 4/2 matt shown in Figure 4.5b, are produced to obtain special technical effects. Larger matt structures

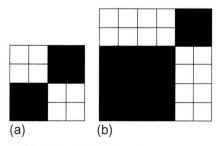

Figure 4.5 Matt fabrics. (a) 2/2 Matt and (b) 4/2 matt.

give the appearance of squares but are little used because they tend to become unstable with long floats and threads in either direction riding over each other. If large matt weaves are wanted to obtain a special effect or appearance, they can be stabilised by using fancy matt weaves containing a binding or stitching lift securing a proportion of the floats.

Matt weave fabrics can be woven with higher cover factors than is possible with ordinary plain weave and have fewer intersections. In close construction, they may have better abrasion and better filtration properties and more resistance to water penetration. In more open constructions, matt fabrics are prone to slippage and have a greater tear resistance and bursting strength. Weaving costs may also be reduced if two or more picks can be inserted at the same time.

4.2.3   Twill fabrics

A *twill* is a weave that repeats on three or more ends and picks and produces distinct diagonal lines on the face of a fabric as opposed to plain structures that give a practically uniform surface. Such lines generally run from selvedge to selvedge. The direction of the diagonal lines on the cloth's surface is generally described as a fabric that is viewed along the warp direction. When the diagonal lines are running upward to the right, they are "Z twill" or "twill right", and when they run in the opposite direction, they are "S twill" or "twill left". Changing the thread spacing and/or the linear density of the warp and weft yarns can vary their angle and definition. For any construction, twills will have longer floats, fewer intersections, and a more open construction than a plain weave fabric with the same cloth particulars.

Twill weaves are useful because they produce more ornamental weave details than the plain weave. Also, the number of threads interlacing in any twill pattern is less than in the same area of plain weave; as such, the twill weave structure allows the production of fabrics with much closer settings of ends and picks per centimetre than does the plain weave structure.

Industrial uses of twill fabrics are mainly restricted to simple twills, which are the only simple twill discussed here. Broken twills, waved twills, herringbone twills, and elongated twills are extensively used for suiting and dress fabrics. For details of such twills, see Robinson and Marks[2] or Watson.[3,4]

The smallest repeat of a twill weave consists of 3 ends × 3 picks. There is no theoretical upper limit to the size of twill weaves, but the need to produce stable fabrics with floats of reasonable length imposes practical limits.

The twill is produced by beginning the lift sequence of adjacent ends on one pick higher or one pick lower. The *lift* is the number of picks that an end passes over and under. In a 2/1 twill, an end will pass over two picks and under one, whereas in a 1/2 twill, the end passes over one pick and then under two picks. Either weave can be produced as a Z or S twill. There are, therefore, four combinations of this simplest of all twills, which are illustrated in Figure 4.6a–d. To show how pronounced the twill line is even in a 3 × 3 twill, four repeats (2 × 2) in Figure 4.6a are illustrated in Figure 4.6e with all lifted ends shown in solid point paper. The 2/1 twills are warp-faced twills (i.e. fabrics most of whose warp yarn is on the surface) whereas 1/2 twills have a weft face. Weft-faced twills impose less strain on the weaving machine than warp-faced twills because fewer ends must be lifted to allow picks to pass under them. For this

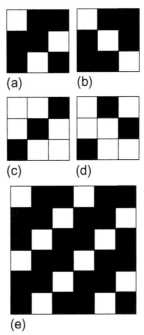

Figure 4.6 3 × 3 twill weaves. (a) 2/1 twill with Z twill line, (b) 2/1 twill with S twill line, (c) 1/2 twill with Z twill line, (d) 1/2 twill with S twill line, and (e) four repeats of (a) (2/1 twill with a Z twist line).

reason, warp-faced twills are sometimes woven upside down, that is as weft-faced twills. The disadvantage of weaving twills upside down is that it is more difficult to inspect the warp yarns during weaving.

It is also possible to create warp and weft faced twill fabric structures in which case both warp and weft threads pass over and under the same number of threads uniformly; in this case, the warp and weft must be in equal amounts on both the face and back of the fabric.

Twills repeating on 4 ends × 4 picks may be of 3/1, 2/2, or 1/3 construction and may have Z or S directions of twill. Weaves showing 4 × 4 twills with a Z twill line are shown in Figure 4.7a–c. Because the size of the twill repeat is increased, the number of possible variations also increases. In the case of a 5 × 5, the possible combinations

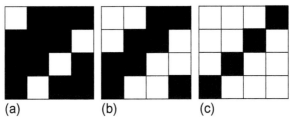

Figure 4.7 4 × 4 twill weaves. (a) 3/1 twill with Z twill line, (b) 2/2 twill with Z twill line, and (c) 1/3 twill with Z twill line.

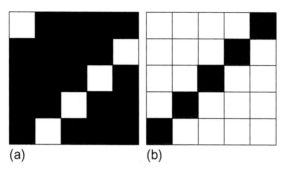

Figure 4.8 5 × 5 twill weaves. (a) 4/1 twill with Z twill line and (b) 1/4 twill with Z twill line.

are 4/1, 3/2, 2/3, and 1/4, and each of these can be woven with the twill line in either direction. Figure 4.8a and b shows 4/1 and 1/4 twills with a Z twill line.

The relative prominence of twill structures is chiefly determined by the character of the weave, the character of the yarn, and the fabric density.

4.2.4   Satins and sateens

Satin and sateen weaves are characterised by an even, smooth, and lustrous surface of either warp or weft, resulting from a perfectly regular distribution of the intersections. Each end and pick intersect only once in a repeat. It is not possible to construct a satin or sateen weave with less than five threads.

Satin is a warp-faced weave in which the binding places are arranged to produce a smooth fabric surface free from twill lines. Satins normally have many more ends than picks per centimetre. To avoid confusion, a satin is frequently described as a warp faced fabric and sateen as weft faced fabric, with binding places arranged to produce a smooth fabric free of twill lines. Sateens are generally woven with a much higher number of picks than ends. Satins tend to be more popular than sateens because weaving a cloth with a lower number of picks than ends is cheaper. Warp satins may be woven upside down (i.e. as a sateen but with a satin construction) to reduce the tension on the harness mechanism that must lift the warp ends.

To avoid twill lines, satins and sateens must be constructed in a systematic manner. To construct a regular satin or sateen weave (for irregular or special ones, see Robinson and Marks[2] or Watson[3]) without a twill effect, a number of rules must be observed. The distribution of interlacing must be as random as possible and there must be only one interlacing of each warp and weft thread per repeat per weave number. The intersections must be arranged in an orderly manner, uniformly separated from each other and never adjacent. The weaves are developed from a 1/x twill, and the twill intersections are displaced by a fixed number of steps. The steps that must be avoided are

 (i) having one or one fewer than the repeat (because this is a twill)
(ii) having the same number as the repeat or having a common factor with the repeat (because some of the yarns would fail to interlace)

The smallest weave number for either weave is 5; regular satins or sateens cannot be constructed with weave numbers of 6, 9, 11, 13, 14, or 15. The most popular weave numbers are 5 and 8; those above 16 are impractical because of the length of floats. Weave numbers of 2 or 3 are possible for five-end weaves and 3 or 5 for eight-end weaves.

Figure 4.9a and b shows five-end weft sateens with two and three steps, respectively. They have been developed from the 1/4 twill shown in Figure 4.8b. Mirror images of these two weaves can be produced. Five-end warp satins with two and three steps are shown in Figure 4.10a and b. The cloth particulars, rather than the weave pattern, generally determine which fabric is commercially described as a sateen or a satin woven upside down. Five-end satins and sateens are most frequently used because with moderate cover factors, they give firm fabrics. Figure 4.11a and b shows eight-end weft sateens with three- and five-step repeats, and Figure 4.12 shows an eight-end warp satin with a five-step repeat.

Satins and sateens are widely used in uniforms and in industrial and protective clothing. They are also used for special fabrics such as down proofs.

In North America, a satin is a smooth, generally lustrous fabric with a thick, close texture made in silk or other fibres in a satin weave for a warp-face fabric or sateen weave for a filling- (weft-) face effect. A sateen is a strong lustrous cotton fabric generally made with a five-harness satin weave in either warp or filling-face effect.

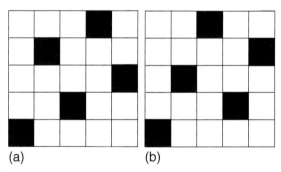

Figure 4.9 5-End weft sateen. (a) 5-End and two-step sateen and (b) 5-end three-step sateen.

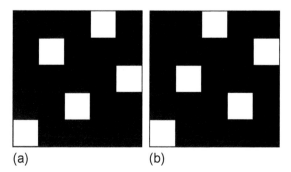

Figure 4.10 5-End warp satin. (a) 5-End two-step satin and (b) 5-end three-step satin.

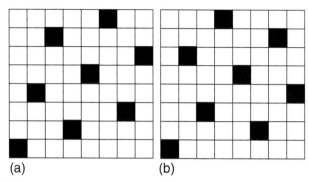

Figure 4.11 8-End weft sateen. (a) 8-End three-step sateen and (b) 8-end five-step sateen.

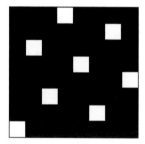

Figure 4.12 8-End five-step warp satin.

4.2.5 Lenos and gauze structures

The distinctive characteristic feature of the structures just described is the peculiar crossing of warp threads with each other, caused by pulling them out of their normal straight and parallel course, first to one side and then to the other side of the other warp thread, which cross and recross in a definite order. In gauge fabrics, the cross-weaving is applied to fabrics of an extremely light, open, and flimsy texture whereas the cross-weaving in leno fabrics is applied decoratively to a heavier texture of cotton fabrics.

Adjoining warp ends in lenos do not remain parallel when they are interlaced with the weft but are crossed over each other. In the simplest leno, one standard end and one crossing end are passed across each other during consecutive picks. Two variations of this structure are shown in plan view and cross-section in Figure 4.13a and b.[4] When the warp threads cross over each other with the weft passing between them, they lock the weft into position and prevent weft movement. Leno weaves are therefore used in very open structures, such as gauzes, to prevent thread movement and fabric distortion. When the selvedge construction of a fabric does not bind its edge threads into position, leno ends are used to prevent the warp threads at each side of a cloth length from slipping out of the fabric body. They are also used in the body of fabrics when empty dents are left in weaving because the fabrics are to be slit into narrower widths at a later stage of processing.

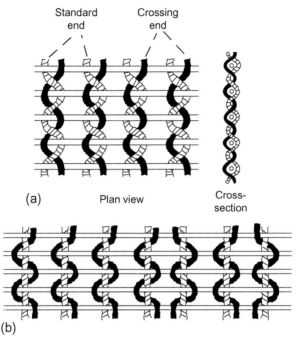

Figure 4.13 (a) Leno with standard and crossing ends of same length (woven from one beam) and (b) leno with standard and crossing ends of different length (woven from two beams).

Leno and gauze fabrics may consist of standard and crossing ends only, or pairs or multiples of such threads may be introduced according to pattern to obtain the required design. For larger effects, standard and crossing ends may also be in pairs or groups of three. Two or more weft threads may be introduced into one shed and areas of plain fabric may be woven in the weft direction between picks where warp ends are crossed over to give the leno effect. Gauze fabrics used for filtration generally use simple leno weaves.

Standard and crossing ends frequently come from separate warp beams. If both the standard and crossing ends are warped on to one beam, the same length of warp is available for both, and they must do the same amount of bending; that is, they have the same crimp. Such a leno fabric is shown in plan view and cross-section in Figure 4.13a. If the two series of ends are brought from separate beams, the standard ends and the crossing ends can have different tension, and their crimp can be adjusted separately. In such a case, it is possible for the standard ends to lie straight and the crossing ends to do all the bending. Such a fabric is shown in Figure 4.13b. This figure also shows that crossing threads can be moved either from the right to the left or from the left to the right on the same pick, and adjoining leno pairs may cross in either the same or opposite directions. The direction of crossing can affect locking, especially with smooth monofilament yarns. It is also possible with two beams to use different types or counts of yarn for standard and crossing ends for design or technical applications. The actual method of weaving the leno with doups or similar mechanisms is not discussed in this chapter.

Only one to four pairs of threads are generally used at each side when lenos are used for selvedge construction, and the leno selvedge is produced by a special leno mechanism that is independent of the shedding mechanism of the loom. The leno threads required for the selvedge then come from cones in a small separate creel rather than from the warp beam. The choice of selvedge yarns and tensions is particularly important to prevent tight or curly selvedges. The crimps selected must take into account the cloth shrinkage in finishing.

4.2.6 Triaxial weaves

Nearly all two-dimensional woven structures have been developed from plain weave fabrics and warp and weft yarns are interlaced at right angles or at nearly right angles. This also applies in principle to leno fabrics (see Section 4.2.5) and to lappet fabrics in which a proportion of the warp yarns (i.e. the yarns forming the design) is moved across a number of ground warp yarns by the lappet mechanism. The only exceptions are triaxial fabrics into which two sets of warp yarns are generally inserted at 60° to the weft, and tetra-axial fabrics in which four sets of yarns are inclined at 45° to each other. Currently, only weaving machines for triaxial fabrics are in commercial production. They were first built by the Barber-Colman Co. under license from Dow Weave and have been further developed by Howa Machinery Ltd, Japan.

Triaxial fabrics are defined as cloth in which the three sets of threads form a multitude of equilateral triangles, thus forming a more stable construction because triangles are more stable than rectangles. The weave structure forms locked intersections, giving equal strength in all directions. Two sets of warp yarn are interlaced at 60° with each other and with the weft. In the basic triaxial fabric shown in Figure 4.14a, the warp travels from selvedge to selvedge at an angle of 30° from the vertical. When a warp yarn reaches the selvedge, it is turned through an angle of 120° and then travels to the opposite selvedge, thus forming a firm selvedge. Weft yarn is inserted at right angles (90°) to the selvedge. The basic triaxial fabric is fairly open with a diamond-shaped centre. The standard weaves can be modified by having biplane, stuffed, or basic basket weaves, the latter of which is shown in Figure 4.14b. These modified weaves

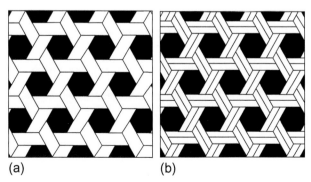

(a)   (b)

Figure 4.14 Triaxial fabrics (weaving machine: Howa *TRI-AX* Model TWM). (a) Basic weave and (b) basic basket weave.

form closer structures with different characteristics. Interlacing angles of 75° to the selvedge can be produced. Presently, thread spacing in the basic weave fabric is limited to 3.6 or 7.4 threads per centimetre.

The tear resistance and bursting resistance of triaxial fabrics are greatly superior to those of standard fabrics because a strain is always taken in two directions. The shear resistance of triaxial fabrics is also excellent because the intersections are locked. The fabrics have a wide range of technical applications including sailcloth, tire fabrics, balloon fabrics, pressure receptacles, and laminated structures.

4.3   Selvedge

The selvedges form the longitudinal edges of a fabric and are generally formed during weaving with the weft not only turning at the edges but also passing continuously across the width of the fabric from edge. The basic function of any selvedge is to lock the outside warp threads of a piece of cloth and so prevent fraying as well as to give strength to the edges of the fabric so that it will behave satisfactorily in weaving and subsequent processes.

The weave used to construct the selvedge may be the same as or may differ from the weave used in the body of the cloth. Most selvedges are fairly narrow but can be up to 20-mm wide. Descriptions may be woven into the selvedge using special selvedge jacquards, or coloured or fancy threads may be incorporated for identification purposes. For some end uses, selvedges must be discarded, but they should be constructed whenever possible so that they can be incorporated into a final product to reduce waste.

In cloth woven on weaving machines with shuttles, selvedges are formed by turning the weft at the edges after the insertion of each pick. The weft passes continuously across the width of the fabric from side to side. In cloth woven on shuttleless weaving machines, the weft is cut at the end of each pick or after every second pick. To prevent the outside ends from fraying, various selvedge motions are used to bind the warp into the body of the cloth or edges may be sealed. The essential requirements are that the selvedges should protect the edge of the fabric during weaving and subsequent processing, that they should not detract from the appearance of the cloth, and that they should not interfere with finishing or cause waviness, contraction, or creasing. Four types of woven selvedges are shown in Figure 4.15.

4.3.1   Hairpin selvedge – Shuttle weaving machine

Figure 4.15a shows a typical hairpin selvedge formed when the weft package is carried in a reciprocating media, for example a shuttle. With most weaves, it gives a good edge and requires no special mechanism. Frequently strong two-ply yarns are incorporated into the selvedge whereas single yarns are used in the body of the cloth because the edge threads are subjected to special strains during beat-up. To ensure a flat edge, a different weave may be used in the selvedge from the body of the cloth. For twills, satins, and fancy weaves, this may also be necessary to ensure that all warp threads are properly bound into the edge. If special selvedge yarns are used, it is important to

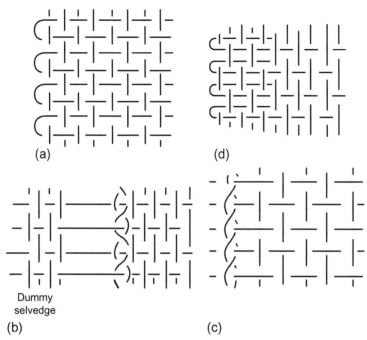

Figure 4.15 (a) Hairpin selvedge – shuttle loom, (b) leno selvedge with dummy selvedge, (c) helical selvedge, and (d) tucked-in or tuck selvedge.

ensure that they are not woven into the body of the fabric by mistake because they are likely to show after finishing or cause a reduction in the tear and/or bursting resistance of industrial fabrics.

When two or more different weft yarns are used in a fabric, only one weft yarn is being inserted at a time and the other yarn(s) are inactive until the weft is changed. In shuttle looms, the weft being inserted at any one time will form a normal selvedge; the weft yarn(s) not in use will float along the selvedge. If long floats are formed in weaving because one weft is not required for a considerable period, the floats may have to be trimmed off after weaving to prevent them from causing problems during subsequent processing. When a pirn (a weaver's bobbin, spool, or reel) is changed or a broken pick is repaired, a short length of yarn will also protrude from the selvedge and must be trimmed off after weaving.

Industrial fabrics with coarse weft yarn are sometimes woven with loop selvedges to ensure that the thread spacing and cloth thickness remain the same right to the edge of the cloth. To produce a loop selvedge, a wire or coarse monofilament yarn is placed 3 or 4mm outside the edge warp end, and the pick reverses around the wire to form a loop. Because the wire is considerably stiffer than a yarn, it will prevent the weft from pulling the end threads together during beat-up. The wire extends to the fell of the cloth and is automatically withdrawn during weaving. Great care is required to ensure that no broken-off ends of wire or monofilament remain in the fabric because they can cause serious damage to equipment and to the cloth during subsequent processing.

4.3.2   Leno and helical selvedges

In most shuttleless weaving machines, a length of weft must be cut for every pick. For looms not fitted with tucking motions, the warp threads at the edges of the fabric must be locked into position to prevent fraying. With some weft insertion systems, the weft is cut only after every second pick, and it is possible to have a hairpin selvedge on one side of the fabric and a locked selvedge on the other side. Leno and helical selvedges are widely used to lock warp yarns into position; see Figure 4.15b and c, respectively. When shuttleless weaving machines were first introduced, there was considerable customer resistance to fringe selvedges, but they have been found to meet most requirements.

With leno selvedges (see also Section 4.2.5), a set of threads at the edge of the fabric is interlaced with a gauze weave that locks around the weft thread and prevents the warp from ravelling. Because the precut length of the picks always varies slightly, a dummy selvedge is sometimes used at the edge of the cloth as shown in Figure 4.15b. This makes it possible to cut the weft close to the body of the cloth, resulting in a narrow fringe that has a better appearance than the selvedge from which weft threads of varying lengths protrude. Leno selvedge has the advantage in finishing and coating because there are no long lengths of loose weft that can untwist and shed fly. Because of the tails of weft and of the warp in dummy selvedges, more waste is generally made in narrow shuttleless woven fabrics than in fabrics woven on shuttle looms. For wide cloth, the reverse frequently applies because there is no shuttle waste. Leno selvedges, sometimes referred to as *centre selvedges*, may be introduced into the body of the cloth if the intention is to slit the width of cloth produced on the loom into two or more widths either on the loom or after finishing.

Helical selvedges consist of a set of threads that make a half or complete revolution around one another between picks. They can be used instead of leno selvedges and tend to have a neater appearance.

4.3.3   Tucked-in selvedges

A tuck or tucked-in selvedge is shown in Figure 4.15d. First developed by Sulzer Brothers[5] (now Sulzer-Textil) in Winterthur, Switzerland, for use on the company's weaving machines, this is a very neat and strong selvedge. Its appearance is close to that of a hairpin selvedge, and it is particularly useful when cloth with fringe selvedges would have to be hemmed. High-speed tucking motions are now available, and it is possible to produce tucked-in selvedges even on the fastest weaving machines.

Mechanical or pneumatic tucked-in devices are available on the market. The mechanically tucked-in selvedge device produces a special hooked needle driven by a cam; after being cut, the protruding weft yarn end is inserted into the subsequent shed, thus forming a stronger edge. This system is generally used for light- to medium-weight fabrics when the weave and the fabric density permit. The principle of pneumatic tucking-in is the use of the air to hold the weft end and then forcing the weft end to be tucked in the next shed by the air.

Tucked-in selvedges, even when produced with a reduced number of warp threads, are generally slightly thicker than the body of the cloth. When large batches of cloth with such selvedges are produced, it may be necessary to travel against the cloth on the cloth roller to build a level roll. The extra thickness also must be allowed for when fabrics are coated. When tucked-in selvedges can be incorporated into the finished product, there is no yarn waste because no dummy selvedges are produced, but the reduced cost of waste may be counterbalanced by the relatively high cost of the tucking units.

4.3.4 Sealed selvedges

When a fabric is produced from yarns with thermoplastic properties, its edge may be cut and sealed by heat by pressing a hot mechanical element on the fabric edge. The edge ends of fringe selvedges are frequently cut off in the loom and the edge threads with the fringe are drawn away into a waste container. Heat cutting may also be used to slit a cloth, in or off the loom, into a number of narrower fabrics or tapes.

For special purposes, ultrasonic sealing devices are available. These devices are fairly expensive and can cut more cloth than can be woven on one loom. They can be mounted on the loom, but because of their cost, they are more frequently mounted on a separate cutting or inspection table.

4.4   Fabric specifications and fabric geometry

Fabric specifications[6] provide information regarding a cloth but frequently need an experienced user to interpret them correctly. The most important elements are cloth width, threads per centimetre in the warp and weft directions, linear density (count), and type of warp and weft yarns (raw material, filament or staple, construction, direction of twist, and twist factors), weave structure (see Section 4.2), and finish. If these weaving machine particulars and finishing instructions are known, the cloth area's density can be calculated.

It has been assumed that other cloth particulars, such as warp and weft cover factors, crimp, cloth thickness, porosity, and drape, must be either estimated or measured by various test methods. Peirce[7] showed that standard physical and engineering principles can be applied to textiles and cloth specifications to forecast whether they allow for the effect of interlacing the warp and weft threads and the distortion of the yarns.

4.4.1   Fabric width

The width of fabrics is generally expressed in centimetre and is measured under standard conditions to allow for variations caused by moisture and tension. It is necessary to know whether the cloth width required is measured from edge to edge or excludes the selvedges. Before deciding on the weaving specifications of a fabric, allowance must be made not only for shrinkage from reed width to grey width during weaving but also for contraction (if any) during finishing. If the cloth width changes, other parameters such as cloth area density and threads per centimetre will be affected.

4.4.2   Fabric area density

Fabric area density is generally expressed in grams per square metre although sometimes it is reported as grams per linear metre. It is essential to specify whether the area density required is loomstate or finished. The loomstate area density depends on the weaving specification, that is, the yarns, thread spacing and weave, and any additives, such as size, that are used to improve the weaving process. Fabric area density also affects the geometry and volume of the voids in the fabric structure. During finishing, the cloth area density is frequently altered by tension and chemical treatments or compressive shrinkage that affects cloth width and length by removing additives needed in weaving and by substances added during finishing.

4.4.3   Crimp

The waviness or crimp[1] of a yarn in a fabric, shown in the cross-sections of Figure 4.1, is caused during weaving and may be modified in finishing. It results from the yarns being forced to bend around each other during beat-up. The crimp depends on the characteristics of the yarns and the weaving and finishing tensions applied to the yarn during weaving, the yarn's linear density, the weave structure, and the warp and weft cover factors, which are described in Section 4.4.4.

The crimp is measured by the relation between the length of the fabric sample and the corresponding length of yarn when it has been straightened after being removed from the cloth (Figure 4.16), which, as drawn, is idealised and applies only to monofilament yarns. The pressure exerted by the warp on the weft or vice versa depending on yarn particulars and cloth construction deform multifilament and staple yarn to some extent at intersections. The most convenient way to express crimp is as a percentage, which is 100 divided by the fabric length and multiplied by the difference between the yarn length and the fabric length.

Warp crimp is used to decide the length of yarn that must be placed on a warper's beam to weave a specified length of fabric. Allowance must be made for the yarn's stretching during weaving, which is generally low for heavily sized yarns but can be of importance when unsized or lightly sized warps are used. For fabrics woven on shuttleless weaving machines, the estimated weft crimp must be adjusted for the length of yarn used to produce a selvedge and, sometimes, a dummy selvedge.

Fabric thickness is affected by the crimp ratio of the two yarn systems. When the warp and weft crimps are equal, the fabric thickness is at a minimum.

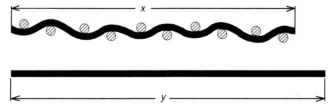

Figure 4.16 Crimp – relation of length of yarn in cloth to length of cloth. x is the width of sample, y is the length of yarn extracted from sample and percentage crimp = $(y - x)/x \times 100$.

4.4.4   Cover factors

The *cover factor*[1] indicates the extent to which the area of a fabric is covered by one set of threads. All fabrics have two cover factors: the warp and the weft. The cloth cover factor is obtained by adding the weft cover factor to the warp cover factor. The cover factors can be adjusted to allow for yarns of different relative densities because of either the yarn structure or the raw material used. The cloth cover factor is important with respect to water, water vapour, or air permeability. For example, if the fabric is to be coated, the degree of openness influences the penetration of the coating into or through the fabric and in part affects adhesion of the coating substrate to the base fabric.

The cover factors in SI units[8] are calculated by multiplying the threads per centimetre by the square root of the linear density of the yarn (in tex) and dividing by 10. The resulting cover factor differs by less than 5% from the cotton cover factor pioneered by Peirce[7] that is expressed as the number of threads per inch divided by the square root of the cotton yarn count.

For any given thread spacing, plain weave has the largest number of intersections per unit area. All other weaves have fewer. The likely weave ability of all fabrics woven with the same weave and from similar yarns can be forecast from their cover factors. Plain weave fabrics with warp and weft cover factors of 12 in each direction are easy to weave. Thereafter, weaving becomes more difficult and for cover factors of 14 + 14, fairly strong weaving machines are required. At a cover factor of 16 + 16, the plain structure jams, and a very strong loom with heavy beat-up is needed to deform the yarns sufficiently to obtain a satisfactory beat-up of the weft. Some duck and canvas fabrics woven on special looms achieve much greater cover factors. Three clothes with cover factors of 12 + 12 are shown in Table 4.1, which shows how thread spacing and linear density must be adjusted to maintain the required cover factor and how cloth area density and thickness are affected.

When widely varying cover factors are used for warp and weft, a high cover factor in one direction can generally be compensated for by a low cover factor in the other direction. In Figure 4.3, which shows a poplin-type weave, the warp does all the bending and the weft lies straight. With this construction, warp yarns can touch because

Table 4.1   Comparison of fabrics with identical warp and weft cover factors woven with yarns of different linear densities (in SI units)

Cloth	Threads per cm		Linear density		Cover factor		Cloth weight[a] (g/m²)	Thickness[b] (mm)
	n_1	n_2	N_1	N_2	K_1	K_2		
A	24	24	25	25	12.0	12.0	130	0.28
B	12	12	100	100	12.0	12.0	260	0.56
C	6	6	400	400	12.0	12.0	520	1.12

n_1 are warp threads, n_2 are weft threads, N_1 is the linear density of warp, N_2 is the linear density of weft, K_1 is the warp cover factor, K_2 is the weft cover factor.
[a]Allowing for 9% of crimp.
[b]Allowing 25% for flattening and displacement of yarns.

Table 4.2 **Cover factor adjustment factors for weave structure**

Weave	Adjustment factor[a]
Plain weave	1.0
2/2 weft rib	0.92
1/2 and 2/1 twills	0.87
2/2 matt	0.82
1/3, 3/1, and 2/2 twills	0.77
5-end satin and 5-end sateen	0.69

Source: Rüti Ref. [10].
[a]Multiply actual cover factor by adjustment factor to obtain equivalent plain weave cover factor.

they bend around the weft and cloth cover factors of well above 32 can be woven fairly easily. Peirce calculated the original adjustment factors for a number of weave structures. Rüti[9] published an adjustment factor that has been found to be useful to establish whether fabrics with various weave structures can be woven in their machines (see Table 4.2). Sulzer prepared graphs showing how easy or difficult it is to weave fabrics with different weave structures, thread spacing, and linear densities in various types of weaving machines.

4.4.5 Thickness

Yarn properties are as important as cloth, particularly when forecasting cloth thickness. It is difficult to calculate the cloth thickness because it is greatly influenced by yarn distortion during weaving and by pressures exerted on the cloth during finishing. It is also difficult to measure thickness because the results are influenced by the size of the presser feet used in the test instrument, the pressure applied, and the time that has elapsed before the reading is taken. It is therefore essential to specify the method of measurement of thickness carefully for many industrial fabrics.

4.4.6 Parameters of woven fabric

The main parameters of a woven fabric are warp and weft setts, warp and weft linear densities (count), warp and weft crimps, fabric area density, warp and weft cover factors, and the order of threads interlacing (weave).

- Sett: This is normally expressed as ends or picks per centimetre (or per inch).
- Linear density: This provides a means of indicating the yarn thickness or diameter. The popular yarn count systems are cotton system, worsted system, denier, tex, d.tex, and the metric system.
- Crimp: This refers to the amount of bending that is done by a thread because it interlaces with the threads that are lying in the opposite direction in the fabric. Crimp is important because it affects the appearance and properties and is used to calculate the length of warp and weft required to produce a certain area of fabric. It is normally expressed as a percentage:

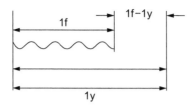

$$\text{Crimp}\left(\%\right)=\left(\frac{\text{ly}-\text{lf}}{\text{lf}}\right)\times100$$

- Fabric area density:

$$\text{Mass of fabric}\left(\text{g m}^{-2}\right)=\text{mass of warp}\left(\text{g m}^{-2}\right)+\text{mass of weft}\left(\text{g m}^{-2}\right).$$

$$\text{Mass of fabric}\left(\text{g m}^{-2}\right)=\frac{\text{ends cm}^{-1}\times100\times\text{warp tex}\times\left(100+C_{1}\right)}{1000\times100}$$
$$+\frac{\text{picks cm}^{-1}\times100\times\text{weft tex}\times\left(100+C_{2}\right)}{1000\times100}$$

where C_{1} and C_{2} are warp and weft percentage crimps, respectively.
- Cover factor: It is a measure of the relative closeness of the yarns in the warp or weft of a woven fabric. This ratio also expresses the fraction of the cloth area covered by the warp or weft yarn.

$$K=nN^{1/2}\times10^{-1}\quad\text{where }n=\text{sett, }N=\text{tex}$$

$$K_{1}=n_{1}N_{1}^{1/2}\times10^{-1};\quad\text{and}\quad K_{2}=n_{2}N_{2}^{1/2}\times10^{-1}$$

where n_{1}, n_{2}, N_{1}, and N_{2} are warp sett, weft sett, warp tex, and weft tex, respectively.

$$K_{c}=K_{1}+K_{2}$$

where K_{1}, K_{2}, and K_{c} are warp, weft, and fabric cover factors, respectively.
Most woven fabrics have K_{c} between 0 and 26.7
- Weaving machine productivity: The production of any weaving machine can be calculated as follows:

$$\frac{\text{picks per minute}\times60\times\text{machine efficiency}\left(\%\right)\times\text{fabric width}\left(\text{cm}\right)}{\text{picks per cm}\times100\times100\times100}=\text{m}^{2}\text{h}^{-1}$$

4.5 Warp preparation

The success of the weaving operation depends on the quality of the weaver's beam presented to the weaving machine because each fault in the warp will either stop the machine and require correction or cause a fault in the cloth that is woven from it.

Before most fabrics can be woven, a weaver's beam (or beams) holding the warp yarn must be prepared. For very coarse warp yarns or very long lengths of filament woven without modification of the warping particulars, a separate cone creel placed behind each weaving machine can be used economically. It improves the weaving efficiency by avoiding frequent beam changes but requires a large amount of space. For most yarns, especially sized yarns, preparing and using a weaver's beam in the weaving machine is more economical.

The object of most warp preparation systems is to assemble all ends needed in the weaving machine on one beam and to present the warp with all ends continuously present and the integrity and elasticity of the yarns as wound fully preserved.

In order to produce a weaver's beam with a few hundred to a few thousand warp yarns and their length on the beam, which is positioned at the back of the loom, the yarn produced during the spinning process undergoes the preparation processes of (a) winding, (b) warping, and (c) sizing.

During the winding process, faults or imperfections from yarn such as thick/thin places, neps, and so on, which for many reasons occur during the spinning process, are cleared, and the fault-free yarn is wound onto cones or cheeses of suitable weight or length.

In warping, the second stage of yarn preparation, cones of the desired number are positioned on a creel situated at the back of a warping machine. Yarn from each cone is brought forward in a sheet form wound onto a beam called a *warper's beam*. The desired number of such beams, which depends on the total number of warp yarns required in the total width of the fabric, is then placed onto a sizing machine for preparing the weaver's beam. During the sizing process, the yarns from the warper's beams in sheet form are coated with a cooked sizing paste containing suitable starch (adhesive) and other ingredients; after drying and separating all the ends, these are wound onto the weaver's beam.

The main purpose of warp sizing is to improve the abrasion resistance of the warp yarn and to enable it to withstand the complex stresses to which it is subjected during the weaving process. Sizing is an optional process; some coarse ply yarns and strong filament yarns can be woven without being sized (i.e. directly after the warping process).

Details of the various warp preparation processes are described by Ormerod and Sondhelm[14] and Gandhi.[20]

4.5.1   Weaving – Machines (looms) and operations

Although the principle of weft interlacing in weaving has not changed for thousands of years, the methods used and the way that weaving machines are activated and controlled has been significantly modified. During the last few decades of the twentieth century, developments in weaving machine design occurred continuously; weft insertion rates (WIRs), which control machine productivity, increased 10-fold from around 1950 to 2000. Weaving, which in the past was a labour-intensive industry, is now capital intensive because it uses the most modern technology.

The essential (primary) operations in the weaving of a cloth are

1. Shedding (the separation of the warp threads into two (or more) sheets according to a pattern to allow for weft insertion
2. Weft insertion (picking)
3. Beating-up (forcing the pick, which has been inserted into the shed, up to the fell of the cloth) (line where the cloth terminates after the previous pick has been inserted).

A typical cross-section through a shuttle weaving machine[10] showing the main motions, together with the arrangement of some ancillary motions is shown schematically in Figure 4.17. Most single-phase machines, regardless of their weft insertion system, use similar motions and a nearly horizontal warp sheet between the back-rest (2) and the front rest (9). Although this is the most common layout, other successful layouts have been developed. One of these, for an air jet,[11] is shown in Figure 4.18. An interesting double-sided air jet with two vertical warp lines back to back is being developed by Somet.[12]

Weaving machines can be subdivided into (a) single-phase machines and (b) multiphase machines. On single-phase machines, one weft thread is laid across the full width of the warp sheet followed by the beat-up and the formation of the next shed in preparation for the insertion of the next pick. Multiphase machines perform several phases of the working cycle at any instant so that several picks are being inserted simultaneously. Single-phase machines are further subdivided according to their weft insertion system, and multiphase machines are classified according to their method of shed formation. The classification of weaving machines is summarised in Table 4.3.

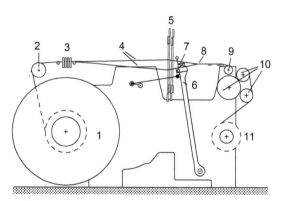

Figure 4.17 Schematic section through Rüti shuttle weaving machine (machine motions and parts shown in *italics*). 1, *Weaver's beam* holding warp yarn (controlled by *let-off*); 2, *back rest* guiding yarn and controlling warp tension; 3, *drop wires* of *warp stop motion*; 4, shed formed by top and bottom warp sheet; 5, *healds* controlled by *shedding motion (crank, cam, or dobby)* (Jacquard shedding motions do not use healds); 6, *sley* carrying *reed* and *shuttle*. It reciprocates for beat-up. (In many shuttleless machines the weft insertion system and the reed have been separated.) 7, *Shuttle* carrying the *pirn*; 8, fell where cloth is formed; 9, *front rest* for guiding cloth; 10, *take-up motion* controlling the pick spacing and cloth wind-up; 11, cloth being wound on to *cloth roller*.

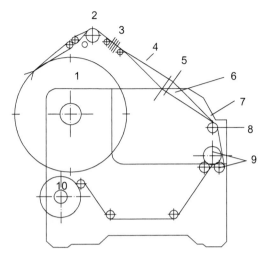

Figure 4.18 Schematic section through Elitex air jet weaving machine (machine motions and parts shown in *italics*). 1, *Weaver's beam* holding warp yarn (controlled by *let-off*); 2, *back rest* guiding warp yarn; 3, *drop wires* of *warp stop motion*; 4, shed formed by top and bottom warp sheet; 5, *healds* controlled by *cam shedding motion*; 6, *weft insertion area* (the reed for beating-up located in this area is not shown); 7, fell where cloth is formed; 8, *front rest* for guiding cloth; 9, *take-up motion* controlling the pick spacing and cloth wind-up; 10, cloth being wound on to *cloth roller*.

In this chapter, the details of the different types of weaving machines and their functions can only be outlined and the most important points highlighted. Weaving machines and their operation are described and discussed by Ormerod and Sondhelm[13] and Gandhi.[20]

4.5.2   Shedding

Regardless of whether cloth is woven on an ancient handloom or produced on the most modern high-speed multiphase weaving machine, a shed must be formed before a pick can be inserted prior to beat-up and cloth formation. The shed must be clean, that is slack warp threads or hairy or taped ends must not obstruct the passage of the weft or of the weft carrier. If the weft cannot pass without obstruction either the machine will stop for rectification, a warp end may be cut or damaged or a faulty weave pattern may be produced.

4.5.2.1   Shedding in single-phase machines

In most single-phase weaving machines before pick insertion commences, an upper and a lower warp sheet is formed. The lifting pattern is not changed until the weft thread has been inserted across the full width of the warp. The shedding mechanism is used to move individual ends up or down in a prearranged order governed by the weave pattern. To maintain a good separation of the ends during weaving and to avoid adjoining ends from interfering with each other, the ends in each warp sheet can be

Table 4.3 **Classification of weaving machines**

Single-phase weaving machines
Machines with shuttles (looms):
  Hand operated (hand looms)
  Nonautomatic power looms (weft supply in shuttle changed by hand)
  Automatic weaving machines
  • Shuttle changing
    Rotary batteries, stack batteries, box loaders, or pirn winder mounted on machine (Unifil)
  • Pirn changing
Shuttleless weaving machines
  Projectile
  Rapier
  • Rigid rapier(s)
    Single rapier, single rapier working bilaterally, or two rapiers operating from opposite
    sides of machine
  • Telescopic
  • Flexible
  Jet machines
  • Air (with or without relay nozzles)
  • Liquid (generally water)
Multiphase weaving machines
  Wave shed machines:
  Weft carriers move in straight path
  Circular weaving machines (weft carriers travel in circular path)
  Parallel shed machines (rapier or air jet)

staggered, but in the area of weft insertion, an unobstructed gap through which the weft can pass must be maintained. The shedding mechanism chosen for a given weaving machine depends on the patterns that will be woven on it. Most shedding mechanisms are expensive; the more versatile the mechanism is, the higher is its cost. Some weaving machines also have technical limitations governing the shedding mechanisms that can be fitted.

When crank, cam (or tappet), or dobby shedding is used, ends are threaded through eyes in healds that are placed into and lifted and lowered by heald frames. All healds in one heald frame are lifted together so that all ends controlled by the frame lift alike. The weave pattern therefore controls the number of healds required. To prevent overcrowding of healds in a heald frame or to even the number of ends on it, more than one frame may lift to the same pattern. To weave a plain fabric, for example, two, four, six, or eight heald frames may be used with an equal number of frames lifting and lowering the warp on each pick. Crank motions are generally limited to 8 shafts, cams to 10 or 12, and dobbies to 18 or 24. When the necessary lifting pattern cannot be obtained by the use of 24 shafts, a jacquard shedding mechanism in which individual ends can be controlled separately must be used.

The crank shedding mechanism is the simplest and most advantageous available. It can be used only for a plain weave. It is cheap and simple to maintain, and in many high-speed machines, increases WIRs by up to 10%. This mechanism has not been used as much as would be expected because of its lack of versatility. It is, however, particularly useful for many technical fabrics that, like the majority of all fabrics, are woven with plain weave.

Cam or tappet shedding motions on modern high-speed machines use grooved or conjugate cams because they give a positive control of the heald shafts. Negative profile cams are still widely used especially for the weaving of fairly open fabrics of light- and medium-area density. The cam profile is designed to give the necessary lifting pattern to healds in the sequence required by the lifting plan that is constructed from the weave structure.

The main advantage of the third method of controlling heald shafts by dobby is that there is practically no limit to the size of pattern repeat that can be woven whereas it is difficult and expensive with cam motions to construct a repeat of more than 8 or 10 picks. It is also easier to build dobbies for a large number of heald shafts. Until the end of the twentieth century, dobbies were controlled by pattern chains containing rollers or pegs that accentuated the lifting mechanism of the heald shafts. Punched rolls of paper or plastic were used instead of the heavier wooden pegs or metal chains for long repeats. During the 1990s, electronic dobbies replaced mechanical ones, enabling weaving machines to operate at much higher speeds and reducing the cost of preparing a pattern and the time required for changing to a new design. Following the development of electronic dobbies, cam machines are becoming less popular because those for high-speed machines are expensive. If large numbers of them are required because of the weave structure or of frequent pattern changes, it may be more economical to buy machines fitted with dobbies.

Shedding mechanisms are still undergoing intense development, and electronic control of individual shafts may soon reduce the difference in cost between crank, cam, and dobby machines. Developments are likely to simplify shedding units, reduce the cost of shedding mechanisms and their maintenance, and make weaving machines more versatile.

When the patterning capacity of dobbies is insufficient to weave the required designs, jacquards must be used. Modern electronic jacquards can operate at very high speeds and impose practically no limitation on the design. Every end across the full width of the weaving machine can be controlled individually, and the weft repeat can be of any desired length. Jacquards are expensive and, if a very large number of ends must be controlled individually rather than in groups, the jacquard may cost as much as the basic weaving machine above which it is mounted.

4.5.2.2 Shedding in multiphase weaving machines

In all multiphase weaving machines,[1] several phases of the working cycle take place at any instant so that several picks are being inserted simultaneously. In wave shed machines, different parts of the warp sheet are in different parts of the weaving cycle at any one moment. This makes it possible for a series of shuttles or weft carriers to move along in successive sheds in the same plane. In parallel shed machines, several sheds

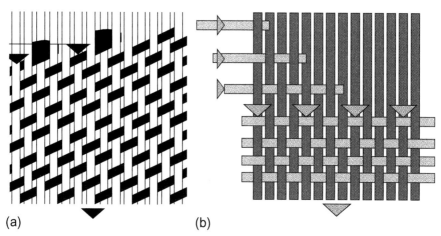

(a)   (b)

Figure 4.19 Weft insertion on multiphase weaving machines (plain weave). (a) Wave shed principle and (b) multiphase parallel shed principle. Reproduced by kind permission of Sulzer Textil.

are formed simultaneously with each shed extending across the full width of the warp and with the shed moving in the warp direction. The difference between the methods of shed formation is shown schematically in Figure 4.19.[15]

Weaving machines in which shuttles travel in a circular path through the wave shed are generally referred to as *circular weaving machines*. They have been widely used for producing circular woven polypropylene fabrics for large bags for heavy loads. The layout of a Starlinger circular weaving machine[16] is shown in Figure 4.20. The creels containing cheeses of polypropylene tape are located on either side of the weaving unit that is in the centre. The warp yarns are fed into the weaving elements from the bottom, interlaced with weft carried in either four or six shuttles through the wave shed in the weaving area, and the cloth is drawn off from the middle top and wound on to a large cloth roll by the batching motion located on the right.

No wave shed machine using weft carriers moving in a straight path has proved commercially successful because the shedding elements have no upper constraint, making it impossible to guarantee a correct cloth structure and because maintaining

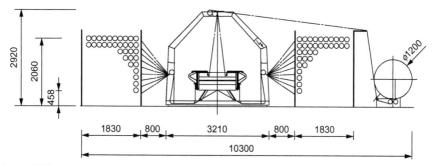

Figure 4.20 Schematic drawing of Starlinger circular loom SL4. Reproduced by kind permission of Starlinger & Co. GmbH (all dimensions in mm).

Combined shed-forming reed/insertion channel

Beat-up reed

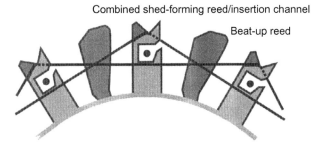

Figure 4.21 Shed formation of Sulzer Textil multiphase parallel shed M8300. Shed-forming reed with weft insertion channel and beat-up reed mounted on drum. Reproduced by kind permission of Sulzer Textil.

a clean shed has proved very difficult. Their attraction was a WIR of up to 2200 m/ min, but they became obsolescent when simpler single-phase air-jet machines began to exceed this speed.

The first parallel shed machine to win limited commercial acceptance was the Orbit 112,[13] which on a two-head machine, weaving used rigid rapiers for weft insertion and reached WIRs of up to 3600 m/min when weaving bandage fabrics. A few years ago Sulzer-Textil released a new multiphase linear air-jet machine that is capable of operating at WIRs of up to 5000 m/min with a reed width of 190 cm.

This machine, which needs no healds, has a rotating drum with 12 channels running at 212 rpm, on which the shed-forming reed and beat-up reed are mounted (Figure 4.21)[17]. A number of sheds that are opened simultaneously are arranged in parallel, one behind the other in the direction of the warp, and four weft yarns are inserted at the same time. Shedding on this machine is gentle, and individual picks are transported across the warp sheet at relatively low speeds. Weft yarns do not have to be accelerated and decelerated continuously, thus reducing the stresses imposed on the yarns. This reduces demands on yarn and, for the first time, we have a machine with a high insertion rate that should be able to weave relatively weak yarns at a low stop rate and with high efficiency. The machine also offers the lowest power consumption per square metre because of small rotating masses, weft insertion by low pressure air, and yarn draw-off speed of 20–25 m/s.

4.5.3   Weft insertion and beat-up (single-phase machines)

All single-phase weaving machines are classified in accordance with their weft insertion system. The different types have been summarised in Table 4.3. The main methods of single-phase weft insertion are by shuttle (Figure 4.22), projectile (Figure 4.23), rapier, and air or water jet (Figure 4.24).

4.5.3.1   Weft insertion by shuttle

Looms using shuttles for carrying the weft through the warp sheet dominated cloth production until the 1980s even in high-wage countries like the United States. They are now obsolete except for use in weaving a few highly specialised fabrics. In spite of this,

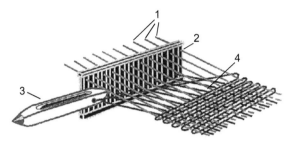

Figure 4.22 Weft insertion by shuttle (schematic). 1, Warp yarns; 2, reed; 3, shuttle carrying pirn (entering shed); 4, fell of cloth. Picking motions and race board not shown. Reproduced by kind permission of Sulzer Textil.

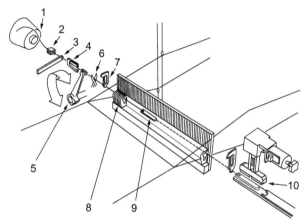

Figure 4.23 Weft insertion by projectile (Sulzer system) (schematic). 1, Weft (on cone); 2, yarn brake (adjustable); 3, weft tensioner; 4, weft presenter; 5, torsion rod; 6, weft cutter (scissors); 7, gripper (to hold cut end); 8, guide teeth; 9, projectile; 10, projectile brake (receiving side). Reproduced with kind permission of Sulzer Textil.

large numbers of automatic bobbin-changing looms are still in use but are being rapidly replaced by shuttleless weaving machines. These machines produce more regular fabrics with fewer faults and need less labour for weaving and maintenance. Millions of hand looms that are protected by legislation are still in operation in Southeast Asia.

Figure 4.22 shows schematically the production of cloth on a shuttle loom. The shuttle carrying the pirn, on which the weft is wound, is reciprocated through the warp by a picking motion (not shown) on each side of the machine. For each pick, the shuttle must be accelerated very rapidly and propelled along the race board. While the shuttle is crossing through the shed, one pick of weft is released and when the shuttle reaches the second shuttle box, it must be stopped very rapidly. After each pick has been inserted, it must be beaten-up, that is moved to the fell of the cloth. The reed and the race board are mounted on the sley and during the weaving cycle are reciprocated backward and forward. While the shuttle passes through the shed, the sley is close to the healds to enable the shuttle to pass without damaging

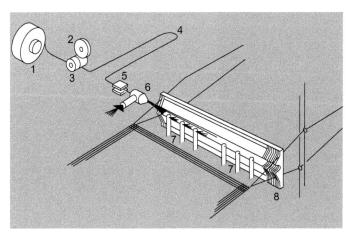

Figure 4.24 Weft insertion by airjet (Sulzer Rüti L5000). 1, Supply package; 2, measuring disc; 3, rollers; 4, storage tube; 5, clamp; 6, main nozzle; 7, relay nozzles; 8, reed with tunnel. Reproduced by kind permission of Sulzer Textil.

the warp yarns. The sley is then moved forward for beat-up. There needs to be an open shed for weft insertion during a considerable part of the picking cycle, and the weight of the sley carrying the race board and the reed imposes restrictions on the picking speed (i.e. the number of revolutions at which the loom can operate).

The basic weakness of fly shuttle weaving machines is the unsatisfactory ratio between the large projected mass of the shuttle and the weft bobbin in relation to the small variable mass of the weft yarn carried in the shuttle.[13] Only about 3% of the energy imparted to the shuttle is used for the actual weft insertion. Further limitations on machine speed are imposed by the need to reciprocate the heavy sley. Although it is theoretically possible to attain WIRs of up to 450 m/min for wide machines, few shuttle machines in commercial use exceed 250 m/min.

Each time a pirn is nearly empty in a nonautomatic power loom, the weaver must stop the loom and replace it. Pirns should be replaced when there is still a little weft left on them to prevent the weft running out in the middle of the shed and creating a broken pick that must be repaired. In industrialised countries, most power looms have been replaced by automatic pirn-changing weaving machines that, in turn, are now being replaced by shuttleless weaving machines. The pirns in automatic weaving machines are changed without weaver attention and stopping the loom. The replacement pirns are periodically placed into a magazine by an operator so that the machine can activate the replacements whenever necessary. The magazine fillers can be replaced by a "box loader" attachment when the pirns are brought to the loom in special boxes from which they are transferred automatically to the change mechanism. Shuttle change looms in which the shuttle rather than the pirn is changed whenever the pirn empties are available for very weak yarns. All of these methods require pirns to be wound prior to being supplied to the loom. Alternatively, a Unifil attachment from Leesona Corporation, USA can be fitted to wind the pirns on the weaving machine and feed them into the change mechanism.

There are practically no restrictions on the widths or area density of fabrics that can be woven on shuttle machines. Automatic looms can be fitted with extra shuttle boxes and special magazines so that more than one weft yarn can be inserted in accordance with a prearranged pattern. Compared with similar equipment for shuttleless machines, this equipment is clumsy and labour intensive.

4.5.3.2 Projectile machines

Projectile machines can use either a single projectile that is fired alternately from each side of the machine and requires a bilateral weft supply or a unilateral weft supply and a number of projectiles that are always fired from the same side and are returned to the picking position on a conveyor belt. Since Sulzer began series production of its unilateral picking multiple projectile machines in the 1950s, it has dominated the market and sold more shuttleless machines than any other manufacturer. Sulzer-Textil has continuously developed its machines, improved WIRs and machine efficiency, and extended the range of fabrics that can be woven on them. The machines are now used to weave not only a vast range of standard fabrics but also heavy industrial fabrics of up to 8 m wide, such as sailcloth, conveyor belts, tire cord fabrics, awnings, geotextiles, airbags, and a wide range of filter fabrics of varying area density and porosity.

One of the major advantages of all shuttleless machines is that the weft on the cone does not have to be rewound before it is used. This eliminates one process and reduces the danger from mixed yarn and ensures that weft is used in the order in which it is spun. Weft on shuttle looms is split into relatively short lengths, each of which is reversed during weaving, which can show long periodic faults in a yarn.

The standard projectile is 90 mm long and weighs only 40 g, a fraction of the mass of a shuttle. For pick insertion on a Sulzer machine (see Figure 4.23), weft is withdrawn from the package through a weft brake and a weft tensioner to the shuttle feeder, which places it into the gripper of the projectile, which is then thrown from the picking side to the receiver side across the machine. The energy used to throw the gripper holding the weft yarn is derived from the shear energy stored in a torsion bar that is twisted to a predetermined amount and triggered off to provide the means of propulsion. The torsion bar can be adjusted to deliver the energy required to propel the projectile through the shed in guides and draws the weft yarn into the shed. It is thought that the gripper projection energy is about one-third to one-half of that normally needed in conventional shuttle picking. Other advantages offered by the projectile method of weft insertion include flexibility to weave more than one width of fabric at a time; low power consumption; reduced waste of weft yarn because of its unique, clean tucked-in selvedge; the ability to weave multicolour weft for any sequence of up to six different weft yarns, and low number of spare parts consumed. Sulzer redesigned the reed and the beat-up mechanism to obtain a stronger and more rapid beat-up, thus making a higher proportion of the picking cycle available for weft insertion.

Narrow machines can operate at WIRs of up to 1000 m/min whereas 3600 mm wide machines can insert weft at up to 1300 m/min. Models are available for weaving heavy fabrics, for weaving coarse and fancy yarns, and for up to six weft colours.

The machines can be fitted with a variety of shedding motions and are equipped with microprocessors to monitor and adjust machine performance. Because of the increase in WIRs with increases in reed width and because of the decrease in capital cost per unit width for wider projectile machines, it is often attractive to weave a number of widths of fabric side-by-side in one projectile machine.

For even wider and heavier fabrics, Jürgens[18] built a machine using the Sulzer Rüti system of pick insertion. This machine can propel a heavier projectile carrying a weft yarn of up to 0.7 mm diameter across a reed width of up to 12 m. These machines take warp from one, two, or three sets of warp beams and maintain weaving tensions of up to 30,000 N m^{-1}, accommodate up to 24 shafts controlled by an extra heavy dobby, and deliver the cloth in a large batching motion.

Jäger has developed a projectile-weaving machine for fabrics of medium area density and up to 12 m wide using a hydraulically propelled projectile.

4.5.3.3 Rapier machines

Rapier looms, which produce a huge variety of fabric styles and designs, including many varieties of technical textiles, are considered to be the most flexible weaving machines.

At ITMA Paris 1999, of 26 machinery manufacturers showing weaving machines, 17 offered rapier machines, and some offered machines of more than one type. Rapier machines were the first shuttleless machines to become available but at first were not commercially successful because of their slow speed. With the introduction of precision engineering and microprocessor controls, the separation of the weft insertion from the beat-up and the improved rapier drives and heads, the WIRs of these machines have increased rapidly. Those with up to 2500 mm reed space equal projectile machines with which they are now in direct competition.

Machines may operate with single or double rapiers. Single rapier machines generally use rigid rapiers and resemble refined versions of ancient stick looms. These machines have proved attractive for weaving fairly narrow cloth from coarse yarns. Wide single rapier machines are too slow for most applications. The rapier of a circular cross-section in these machines, equal to the width of the reed, enters the shed from one side, traverses the full width of the shed, picks up the tip of the weft yarn on the other side, and draws it through the shed on its return. Consequently, the rapier carries the yarn in only one direction, and half of the rapier movement is wasted, thus limiting the loom speed significantly. Therefore this type of weft insertion is not very popular. A variation of the single rigid rapier is the single rapier working bilaterally, sometimes referred to as a *two-phase rapier.* Because it has not been used to any extent for industrial fabrics, it is not considered here but is discussed in the book by Ormerod and Sondhelm.[13]

Most rapier machines use double rapiers, a "giver" and a "taker". The giver takes the yarn from the supply package on the left of the loom and takes it to the middle of the shed where it meets the taker, which retracts and transfers the weft yarn to the other side of the warp. The Gabler system weft makes inserts alternately from both sides of the machine and cuts yarn every second pick with hairpin selvedges being formed alternately on both selvedges. The Gabler system has now been largely superseded by the Dewas system that inserts weft from one side only and is cut after every pick.

Double rapier machines use either rigid or flexible rapiers. The rigid ones need more space than those fitted with any other weft insertion system. Rigid rapier machines, of which the Dornier HTV and P[19] series are prime examples, are capable of weaving most types of industrial fabrics with weft linear densities of up to 3000 tex, in widths of up to 4000 mm, and at WIRs of up to 1000 m/min. Typical fabrics being produced on these machines range from open-coated geotextile mesh to heavy conveyor belting. A variation of the rigid rapier is the telescopic rapier.

By far, the largest number of rapier machines uses double flexible rapiers that are available in widths up to 4600 mm with even wider machines being custom built for industrial applications. Standard machines have a relatively low capital cost and can be used to weave a wide range of low- and medium-area density fabrics. They are ideal for weaving short runs and for fabrics woven with more than one weft because their weft change mechanism for up to eight colours is simple and cheap. These machines are widely used for furnishing and fashion fabrics, often with jacquards. Standard machines are also used for some industrial cloth.

4.5.3.4   Jet looms: water jet, and air jet[20]

On the jet looms, the weft is inserted by means of a nozzle and is carried across the shed by a jet of working substance (i.e. fluid, air, or water). The relative velocity difference between the jet and the weft thread produces a force on the weft that results in its insertion in the shed. The jet looms do not need a weft carrier or a rapier for weft insertion and therefore have fewer moving parts and less mass to move. Because the tractive force applied to the weft is not very high, the weft thread cannot be unwound directly from a cross-wound package but must be prepared for picking by a metering device.

Water-jet looms are mechanically simple, require much lower energy than air-jet looms, and are ideal for weaving fabrics of light and medium weight in standard constructions. Each water-jet loom is equipped with an individual miniature pump to feed the water under pressure to a single nozzle at the picking side (Figure 4.25)[21] to propel the yarn through the full width of the shed, limiting their width. A water jet of only 0.1 cm is sufficient to carry a yarn across a 121.9 cm wide shed. The amount of water required for each weft yarn is less than 2.0 cm.[22]

For nearly two decades in the 1970s and 1980s, water-jet looms were regarded as being superior because of their impressive WIRs of up to 2700 m/min. However, they are confined to handling only 100% hydrophobic yarns and are not readily adoptable to deal with multicoloured or multitypes of weft yarns. With these limitations and from the 1980s onwards, dramatic improvements in the productivity of air-jet and rapier weaving machines started to overshadow their popularity so much so that during the last few years, the demand of water-jet looms has been dramatically reduced.

Air-jet looms

Air-jet weaving is considered to be the most efficient and productive way of inserting weft across the shed, for the production of light to medium weight fabrics, such as terry towels, furnishing fabrics, denim, and some technical textiles. Compared to the

Figure 4.25 Water-jet outlet with weft from the nozzle.

rapier and projectile methods of weft insertion, the mass of the insertion medium in air-jet weaving is very small. Over the years, there have been considerable technological advances in air-jet weaving, as a result of which WIRs of 3000 m/min and fabric widths of 340 cm have been achieved.

Compressed air is expensive to produce and its flow is difficult to control. The airflow in the shed therefore must be restricted either by special air guides or "confusers" or by passing the weft through a channel in a special "profile" reed. The former method was pioneered by Elitex and is used in most of its "P" type weaving machines (Figure 4.18) of which a large number are in use for weaving light- and medium-weight fabrics. Sulzer Rüti developed the "te strake" profile reeds with relay nozzles (see the schematic in Figure 4.24). There is one main nozzle per colour and one set of relay nozzles, which are integral to the sley and spaced at regular intervals across the reed width. The yarn from the constant-speed weft feeder units is presented to the main jet nozzles to which the air is delivered by the pressure-regulating valve. The main nozzles provide the initial acceleration, while the relay nozzles provide high air velocity to carry the weft across the width of the warp. The weft is measured to length in the weft feeder and then carried by the air stream of the main nozzle into the weft duct, accelerated and transported further by air discharged from the relay nozzles. After insertion, a stretch nozzle at the receiving side holds the pick under tension until it is bound into the cloth.

Since air jets came into large-scale commercial use in the 1970s, they have been developed rapidly. They can now weave the majority of fabrics and are dominant for the mass production of fairly simple cloth. The machines have reached WIRs of up to 3000 m/min, twice that of any other single-phase weft insertion system, and are still under intense development. Their capital cost per metre of weft inserted is highly competitive. Their operating costs depend largely on the local cost of electricity and whether low-grade waste heat from the compressors can be used for other operations in a plant.

Air-jet machines fitted with an automatic weft fault repair system can correct the majority of weft faults, including part picks, which occur between the main nozzle and the selvedge on the receiving side. The fault repair system removes the broken thread from the shed without disturbing the warp ends and then restarts the machine. It signals for attention only if it cannot locate and repair a fault. Because weft stops represent the majority of stops on an air jet, the system greatly reduces the weaver's workload, frequently by more than 50%. It also reduces machine interference and improves the quality of many fabrics.

Weft insertion rates (WIR) on shuttleless looms

Based on the WIR, weaving machine productivity measured in metres per minute is directly proportional to the loom speed (picks/min) if no loom stoppages occur. WIR indicates the amount of weft yarn that is converted into a fabric or inserted in 1 min during weaving. The WIR is a product of the loom speed and the width of warp in the reed. Table 4.4 presents the WIR and fabric widths for the looms exhibited at ITMA 2011 in Barcelona, Spain.

4.5.4   Other motions and accessories for single-phase weaving machines

4.5.4.1   Warp let-off and fabric take-up motions

Warp yarns are generally supplied to the weaving machine on one or more weaver's beams. In special cases, as previously mentioned, cone creels can be used. The warp yarns assembled on the weaver's beam should be evenly spaced and under standard tension to ensure that all are of exactly the same length when they are unwound for weaving. The larger the diameter of the weaver's beam is, the longer is the length of warp that can be wound on it, and the fewer warp changes are required, but the tension variations are higher and must be compensated for. Different weaving machines

Table 4.4   **Weft insertion rates and fabric widths that can be achieved on different looms**

Loom	Weft insertion rate (WIR) (m/min)	Fabric width (cm)
Rapier loom	1500	520
Projectile	1400	530
Air-jet	3030	330
Water-jet	1400	230
Multiphase[17]	4780	190
Multiphase linear	3000	190

accommodate beams of different maximum diameter, but most modern machines can take beams of up to 1000 mm in diameter. If even larger diameter beams are needed for very coarse warp yarns, for example, industrial fabrics or denims, the beams can be placed into a separate beam creel outside the loom frame. Such beam creel units cater for beams of up to 1600 mm diameter.

The width of yarn on the weaver's beam must be at least as wide as the yarn in the reed. When the warp width required exceeds 2800 mm, more than one beam is frequently used to simplify sizing and warp transport. If yarns from more than one beam are used in one cloth, preparing both beams is essential to prevent variations that can cause cloth faults after finishing. The let-off tension of different beams must be carefully controlled, which with the introduction of electronic sensors has become simpler. When the cloth design requires the use of warp yarns of widely differing linear densities or results in different warp yarns weaving with widely differing crimps, two or more warp beams may have to be used in parallel. They can be placed either above each other or behind each other in the weaving machine. Comparison of Figures 4.17 and 4.18 shows how the layout of machines can be modified without impairing their efficiency.

Let-off is the delivery of the warp yarn from the weaver's beam at a constant rate and constant suitable tension by unwinding it from the beam as weaving continues. Let-off and take-up motions play a very important part in ensuring a high degree of fabric quality throughout the weaving process. Any malfunctioning of these motions creates corresponding unacceptable defects in the fabric, such as start and stop marks and variation in the pick density. This results in almost all modern shuttle less looms being equipped with the state-of-the-art electronically controlled let-off and take-up motions. Electronic let-off motion maintains consistent warp tension from full to empty beam.

During weaving, the let-off motion releases the required amount of yarn for each pick cycle. It must also hold the warp yarns under even tension so that they separate easily into two or more sheets during shedding and prior to weft insertion to maintain the required tension during beat-up when the reed moves the newly inserted weft to the fell of the cloth. Let-off motions were previously mechanically controlled with tension measured by the deflection of the backrest, but electronic sensors are used now to take tension measurements, and the let-off is frequently controlled by separate servo motors. As the weaver's beam diameter decreases, the rpm of the motor is controlled with the help of a load cell.

Take-up motions are required to withdraw the cloth at a uniform rate from the fell and wind it in most of the weaving machines onto the cloth roller as weaving continues. However, if large diameter fabric rolls must be made, as is common for heavy fabrics, a separate batching motion is placed outside the loom frame. Batching motions can be on a different level from the weaving machine, either below or above, to reduce the area required for the weaving shed. To prevent weaving long lengths of fabric that will be rejected, batching motions can incorporate cloth inspection facilities, sometimes with an intermediate cloth storage unit.

Take-up motion controls the speed of withdrawal of the fabric (i.e. picks per centimetre in the fabric), and it plays a significant role in maintaining the fabric quality. All shuttleless looms manufacturers now equip their machines with electronic take-up motions that can generate a pick density resolution of 0.1 picks/cm.[23]

4.5.4.2   Automatic stop motions

The supervision of warp threads during the weaving process is absolutely an essential factor for maintaining the fabric quality; to do this, all modern weaving machines are equipped with warp protector, warp stop, and weft stop motions. These motions generally employ mechanical, electrical, or electronic detection systems.

The first group, warp protector motion, is necessary only on machines that use a free-flying shuttle or projectile. They are designed to prevent the forward movement of the reed if the shuttle fails to reach the receiving side. This prevents damage to the machine and the breakage of large numbers of ends if the shuttle is trapped.

Warp stop motions stop the machine if an end breaks or any warp yarn has a tension level below normal level for some reason and it becomes slack. These motions are activated when a drop wire through which an end has been threaded drops because a broken end no longer supports the drop wire. Drop wires can be connected to mechanical or electrical stop motions. Yarns must be properly sized to prevent their being damaged by the drop wires. Electronic warp stop motions, which do not require physical contact with the warp, are now being introduced, especially for fine filament yarns. Electronic warp stop motions also help the weaver to locate the broken end as well as provide an automatic analysis of the warp stops through a data detection system.[24]

The primary function of the weft stop motion is to stop the loom in case of the weft breakage or if it is too short to reach the other side of the shed (short pick).

Weft stop motions are also used to activate weft changes in automatic shuttle looms and to stop weaving machines if the weft breaks during weft insertion. Electronic motions are available that will stop the machine even if a broken end catches again before it reaches the receiving side. In air-jet machines fitted with automatic repair facilities, the weft stop motion also starts the weft repair cycle.

In the case of rapier and projectile weaving machines, a weft sensor is used for weft detection. The most popularly used electronic device is provided with piezoelectric crystals.[24] During the weft insertion, the yarn is drawn from the thread guide of the sensor. This yarn movement is converted into an electrical signal by piezo element inside the sensor. The electrical signal is amplified and evaluated with the flag position to detect the broken weft and stop the machine. The weft sensor is available in four, six, and eight selectors with an "anti" feature to prevent insertion of more than one yarn.[25]

In the case of fluid-jet machines, it is preferable not to hinder the weft fly; therefore, optical sensors that do not touch the yarn are used. These sensors are primarily infrared photocells suited to detect the presence of the yarn or the quantity of yarn accumulated in a prefixed zone. Air-jet machines have a device that automatically restores the broken weft and restarts the machine.[24]

4.5.4.3   Quick-style change and quick warp change

Almost all major shuttleless weaving machine manufacturers offer their own version of quick-style change (QSC) and quick warp change (QWC). QSC equipment, first shown by Picanol in 1991 and now available from most manufacturers, greatly reduces the time a weaving machine must be stopped for a warp change. The warp beam, backrest, warp stop motion, heald frames, and reed are located on a module that separates from the weaving machine's main frame and is transferred by a special transport unit to and from the entering and knotting department where the module becomes the preparation frame for a replacement warp.[14] Thus, 90% of the workload, which is normally done on the stopped weaving machine, is eliminated from the warp replenishment cycle and the weaving machine, improving efficiency. The QSC also results in cleaner reeds and healds and in better machine performance and improved cloth quality. The entire operation can be performed by one person in about 30 min and is most suited for fabric manufacturers that produce small quantities per style (high-quality clothing). The operation significantly improves the weaving efficiency by optimising the style change.

The QWC consists of (a) replacement of only an empty beam with a ready full beam (of the same quality), (b) knotting the new warp ends with those of run-out warp with an automatic tying-in machine, and (c) forward movement of the warp beyond the reed to eliminate the knotted zone.[26]

Reduction of the number of warp and style changes

In addition to QSC and QWC, another possibility for reducing setup times is to reduce the number of warp changes. This can be done by using large diameter warp/weaver's beams and take-up devices for large batch rolls installed outside the weaving machine. This arrangement is most suited for cost-effective production of standard and particular fabric categories meant for some technical textile applications.

4.5.5 Shuttleless looms offered by different weaving machinery manufacturers for the production of fabrics for technical textiles

Shuttleless looms suited for the production of a wide variety of technical textiles are available from well-known weaving machinery manufacturers. A brief review of some of these machines and other related innovative improvements is given in the following sections.

4.5.5.1 Picanol – Belgium

Picanol offers its two major innovative machines, OMNIplus 800 air-jet and Optimax flexible rapier weaving machines[27] for mid- and high-end market segments. These machines produce technical textiles for tire cord, conveyor belts, canvas, industrial glass, one-piece woven air bags, awnings, spinnaker sailcloth, medical textiles, coating fabrics, high-speed leno processes, and car seats. These machines are highly modular when, for example, jacquards are needed instead of dobbies. This and similar retrofits can be made without much effort, and the rapier machine that weaves leno can be converted to standard harness movements in a matter of hours. The Optimax[28] is equipped with a new guided positive gripper available in width up to 5.4 m and offers opportunities for effortlessly dealing with technical fabrics, such as coating fabrics, primary and secondary carpet backing, and geogrids.

The primary aspects of Picanol's air-jet machine OMNIplus Summum include a new insertion system and the upgraded blue box electronic platform, which increases the calculation speeds by a factor of 10. This results in far greater accuracy of many controls and functions. These machines also allow for increasing fabric quality requirements for production of some vital technical textiles lower energy consumption.

4.5.5.2 Sultex weaving machines

Sultex Ltd.'s P7300HP projectile weaving machine[29–31] offers a very flexible solution for weaving technical fabrics. It is designed to weave very dense, heavy, and extra wide technical fabrics. It is available in a wide variety of customised versions and meets virtually every conceivable market demand. The machine is capable of weaving a multitude of top-quality demanding fabrics economically and reliably by using linear yarn densities up to 9000 dtex. This is of utmost importance with membrane fabrics for architecture, fire retardance, and land erosion prevention.

Because the technical fabrics are very diverse, ranging from light to heavy or dense, Sultex produces different machines to meet requirements such as beat-up forces of up to 15,000 N m. Other features are low energy consumption, ability to weave fabrics up to a width of 655 cm, and unparalleled ability to adapt and thrive. The machine is best suited for a wide variety of technical fabrics, including leno, propylene tape, and novelty denim.

In the case of rapier weaving machines,[29,32,33] Sultex offers model G6500 with a unique shed geometry, adaptable gripper head, and the rotary Rotocut weft cutter. The model is ideally suited for the production of aramid fabrics used in the production for bullet-proof fabrics, ultrafine filter fabrics, labels, and especially the production of plain and one-piece-woven (OPW) airbags, agro textiles, geotextiles, sailcloth, cinema and theatre screens, tarpaulins, and wire fabrics. The machine is equipped with unguided (i.e. "free flight") rapier tapes. Its shed has no guides, hooks, or any other element to interfere with the warp yarn and, thus, can produce no damaged filaments. The rapier heads are smaller and are near the reed; the shed opening and warp tension are kept at a minimum. The extra small shed angle of 19° results in the machine's high performance. These machines have active rapier technology capable of inserting an unlimited range of weft yarns into the same shed. Sultex's versatile L5500 high-performance air-jet machine with compact dimensions can offer the appropriate weaving system for almost every application, including the production of a wide variety of some technical textiles.

4.5.5.3 Dornier weaving machines for technical textiles

The rapier and air-jet Dornier weaving machines offer all necessary elements for the production of technical textiles. In the technical textile sector, an orthogonal, anisotropic textile fabric is one that has characteristics that are not homogenous in all directions.[34] The company's PS rapier weaving machine[35] offers a gentle filling process with very low tension peaks, which enables the processing of extremely fine or weak yarns. During the open shed insertion, the yarn is tightly held by the positively controlled rapier head and is gently inserted into the shed without any guiding elements. The machine is designed to cope with a very high or low warp tension, which allows for weaving very dense fabrics, such as conveyor belts or aramid fabrics as well as open mesh structures of glass or carbon. Fabrics from dobby machines and multicoloured patterned jacquard fabrics with a wide spectrum of weft yarns, nep yarn, chenille, or Lurex can also be easily woven on the PS rapier weaving machine.

Specifically for technical textiles, Dornier offers the PTS 4/S2 C model rapier machine[28] with a reed width of 2.2 m that is capable of weaving heavy-duty filter fabric. The machine's reed is able to withstand high impact forces and to handle very stiff and thick monofilaments.

Lindauer Dornier GmbH in Germany markets a 540-cm wide air-jet machine[36] for cost-effective production of screen fabrics, spinnaker sailcloth, and coating fabrics used for airbags, geotextiles, and tire cord. The versatile and robust Dornier rapier weaving machine offers open shed weaving technology that processes coarse and heavy yarns that cannot be used by weft insertion systems on any other conventional weaving machine. The unique positive-controlled rapier head can process almost any material that can be wound on a bobbin. The machine can produce fabrics without faults reliably and cost-effectively from the finest wire (0.012 mm) or monofilament (11 dtex) up to 50 K carbon or BCF yarn (0.5 Nm) with a single or fivefold multiple weft insertion in a single layer or up to eight layer fabrics.

Dornier offers an air-jet stitch-weaving machine[37] in which the reed is open at the top, and laterally displaceable yarn carriers are aligned and mounted onto a pole so that the warps are displaced in weft direction and worked into the fabric. These characteristics are especially useful for fabrics with stitch design and for the diagonal alignment of warps for technical textiles in a technically modified type as an open-reed-weaving (ORW) technology.

The development of this technology created the foundation for completely integrating new functions into the weaving process. It integrates additional threads into the weaving process by using special thread guides arranged between the reed and heald shafts. The linear drives shift the thread guides horizontally on a profile depending on the pattern. The additional pattern threads create a weft effect on the fabric surface that can be controlled as desired.

ORW technology is also a process for creating multiaxial fabrics for technical textiles. In addition to the option of integrating additional thread systems in the plain weave across the complete fabric width, ORW technology is especially suitable for this partially to integrate reinforcements required for specific fabrics. Modular integration of the ORW unit retains the full productive and unrestricted application spectrum of the weaving process.

4.5.5.4   Smit Textile SpA – Italy

Smit Textile SpA, Schio, Italy, offers a range of technologies and solutions for the production of technical fabrics. The GS900 rapier[38] and the JS900 air-jet weaving machine ensure high fabric quality and productivity. The modular concept developed for these machines and their weft transport technology permits working widths of 140–360 cm. The machines allow the production of heavy industrial, airbag, and bolting fabrics for various sectors of technical textile fabrics.

4.5.5.5 Tsudakoma – Jet weaving machines

The Japanese company Tsudakoma Corp. has a ZAX 9100 air-jet weaving machine[39] in wide-ranging widths that weaves various technical fabrics and incorporates a wide range of factors, such as competitive cost, QSC, and energy savings. It can handle six-colour weft selections, and the PSC programmable speed control supports the weaving of value-added fabrics of various kinds and thicknesses of the weft.

4.5.5.6 Van de Wiele & Bonas – Weaving machines for technical fabrics

Michel Van de Wiele NV in Kortrijk, Belgium, manufactures a robust rapier weaving machine,[40] the VTR42, that has dobbytronic, integrated gear boxes, and rigid rapiers for the production of carpet, velvet, and other typical technical fabrics such as three-dimensional or spacer fabrics, multilayer weaves, and even woven artificial grass. Distance fabrics have many applications including vehicle sandwich fabrics for noise or heat barriers (construction), packaging textiles such as cement bags, and flexible fabrics for inflatable air cushions for heavy loads and tumbling mats. Distance fabrics have a high potential in many fields. Woven grass is a revolutionary development whose upright pile position, highest tuft lock, and recycling possibilities have led to opportunities.

Bonas Textile Machinery NV, a member of the Van de Wiele group, offers a machine that can weave airbags as a bag fabric that avoids additional assembly afterwards by using the company's jacquard machines. Airbags are no longer used only for the driver and front passenger seats but have been extended to passenger side, side impact, curtain, and knee protection (Figure 4.26).[40]

Airbags

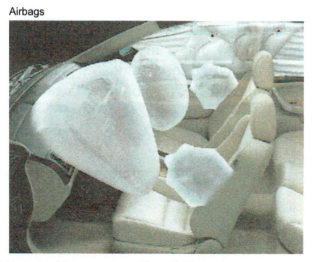

Figure 4.26 Airbags for driver, passenger-side, and side-impact.
Permission for inclusion of this figure has been received from the Trade Journal Technical Textiles.

Michel Van de Wiele

Woven artificial turf is becoming increasingly popular for sport fields and landscaping. The SR/02 weaving machines[41] from Michel Van de Wiele produce woven artificial grass up to 14 cm/pile height after cutting the pile in the middle.

4.5.5.7   Bonas machines for technical textiles

Bonas Textile Machinery NV in Kortrijk, Belgium, offers a versatile jacquard weaving machine[42] for producing technical fabric used, for example, for vehicle air bags. The increased use of airbags has led to jacquards that break convention and push performance criteria well beyond previous limits. A typical bag in most modern cars requires a jacquard of approximately 12,000 hooks to weave at speeds in excess of 600 ppm.

4.5.5.8   Jurgens – Germany

Jurgens GmbH & Co. produces a projectile weaving machine[37] with a working width of 10 m and a machine speed of 140–160 rpm. It is best suited for producing heavy technical fabrics such as filter cloth for the paper industry and conveyer belt fabrics. The machines are marketed in working widths of up to 35 m.

4.5.5.9   Staubli Textile machinery for technical textiles

The textile division of the Staubli Group in Pfaffikon, Switzerland, is a leading global supplier of shedding systems for weaving machines and weaving preparation systems. The group also includes Schonherr Textilmaschinenbau, GmbH, in Chemnitz, Germany, which is well known for its Alpha 400 and 500 weaving systems.[43] Its Unival 100 single-end jacquard machine offers a wide range of possibilities in shed opening, enabling engineers to test new fabric constructions with glass, carbon, and Kevlar® fibres. Advanced Unival technology is proven for automotive and aeronautic textiles and is expected to be used for technical textiles in the sports, industrial, and medical sectors.[43] Other weaving machines that are best suited for producing technical textiles fabrics include Alpha 400, Alpha 500 Tech, and D_Tech.

4.5.5.10   TITV Greiz – Germany

Textile Research Institute, Thuringia Vogland (TITV) Greiz, Germany has developed narrow fabrics[44] by using metal wires as warp, weft, and auxiliary yarns produced on high-performance weaving needle looms. These narrow fabrics possess the same properties in the entire fabric including the selvedges in contrast to ribbons cut out from broadly woven fabrics. The developed narrow fabrics can be produced in widths of less than 10 mm for use as conveyor, drive belts, and filters for technical end uses. The developed metal narrow fabrics can be applied in different industries and are suitable for the exposure to high temperature, weathering, chemicals, liquid solutions, and so on. Because the fabrics are easy to clean and meet strict hygiene requirements, they can even be used in the medical and food industries.

4.5.6   Machine width

The reed width of a machine must be equal to or greater than the width in reed of the fabric to be woven. Width in reed must allow for the width of selvedge and dummy ends. If the machine's reed width is narrower than the cloth width, the cloth cannot be woven in the machine. Generally it is not possible to increase the available reed width of a machine.

Although machine width cannot be exceeded, it is generally possible to weave fabrics that are narrower. Sulzer projectile machines are capable of weaving up to 50% narrower fabrics if desired. Different manufacturers and models have various arrangements for weaving narrower fabrics, and some allow for only 200 mm, which is often insufficient because of likely changes in materials and styles. It is most economic to use a high proportion of the reed width because weaving a narrower fabric is likely to reduce the WIR. Wider machines also tend to have high capital and running costs. In some instances, weaving a number of fabrics side-by-side in one machine is economical. A wide Sulzer projectile machine can produce five, six, or seven roller towels, each with its own tucked selvedges.

4.5.7   The future

In response to rapid technological advances, modern weaving machines have become highly automated with most functions electronically controlled. Machine settings can be adjusted and transferred, and many repairs can be made automatically. The frequency of machine stops and their duration has been reduced. The cost of labour has decreased, and the cost of production has been reduced while product quality has improved. The cost of modern machines makes using the appropriate equipment for the job and operating machines at high efficiency more important than ever.

At present, projectile weaving machines are the most versatile conventional machines. By bolting the appropriate equipment to a machine, it can weave virtually any fabric well. Most simple fabrics can be woven at much lower capital cost on air jets fitted with simple shedding systems. With the availability of modern appropriate equipment their economic range is being extended continuously. In-between fabrics can be woven most cheaply on rapiers. Multiphase machines, such as Sulzer Textil's linear air-jet M8300, are likely to become cheap producers of simple fabrics in the not too distant a future.

References

[1]   The Textile Institute Terms and Definitions Committee. Textile Terms and Definitions. 10th edition. Manchester: The Textile Institute; 1995.
[2]   Robinson AT, Marks R. Woven Cloth Construction. Manchester: The Textile Institute; 1973.
[3]   Watson W. Textile Design and Colour–Elementary Weaves and Figured Fabrics. London: Longmans; 1912 [later editions revised by Grosicki and published by Butterworth].
[4]   Watson W. Advanced Textile Design. London: Longmans; 1912 (later editions revised by Grosicki and published by Butterworth).
[5]   Sulzer Brothers (now Sulzer-Textil), Winterthur, Switzerland.

 [6] Marks R. An Introduction to Textiles—Workbook: Fabric Production—Weaving. In: Sondhelm WS, editor. Manchester: Courtaulds Textiles plc; 1989.
 [7] Peirce FT. The geometry of cloth structure Manchester. J. Text. I. 1937;28:T45–96.
 [8] The Textile Institute Terms and Definitions Committee. Textile Terms and Definitions. 10th edition. Manchester: Table of SI Units and Conversion Factors, The Textile Institute; 1995.
 [9] Rüti Machinery Works Ltd, Verkauf, Rüti, Switzerland; 1977.
[10] Rüti Machinery Works Ltd, Switzerland, 'C' Model Shuttle Weaving Machine.
[11] Investa, Czechoslovakia, Elitex Air Jet, 'P' Type.
[12] Somet, Colzate (BG), Italy, 'Mach 3' Air Jet.
[13] Ormerod A, Sondhelm WS. Weaving—Technology and Operations, Chapter 5: 'Weaving'. Manchester: The Textile Institute; 1995 [reprinted 1999].
[14] Ormerod A, Sondhelm WS. Weaving—Technology and Operations, Chapter 2: 'Warping' and Chapter 3: 'Sizing'. Manchester: The Textile Institute; 1995 [reprinted 1999].
[15] Sulzer Rüti, Wilmslow, Fabric, Structure, 3085/06.10.95/Legler/sei, 1995.
[16] Maschinenfabrik Starlinger, Vienna, Austria, Circular Loom SL4, Drwg. 0304A3, 1997.
[17] Sulzer Rüti, Wilmslow, Shed Formation, 3138/10.10.95/Legler/kaa, 1995.
[18] Jürgens GmbH, Emsdetten, Germany, Projectile weaving machine JP-2000, 1996.
[19] Lindauer Dornier GmbH, Germany, 1999.
[20] Gandhi KL. Woven Textiles, Principles, Developments and Applications, Chapters, 2, 3, 4 Yarn Preparation, Chapter 5 Weaving Technology. Woodhead Publishing Ltd, Sawston, Cambridge, UK, 2012.
[21] Talavasck O. Vladimir Svaty—Shuttleless Weaving Machines. Oxford: Elsevier Science; 1981.
[22] www.textilelearner.blogspot.co.uk/2012/05/water-jet-weaving-machine-water-jet.html.
[23] www.textilesindepth.com/advances-in-weaving-technolgy-and-looms/chnical, maintenance,personnel,etc.
[24] www.textileschool.com/articles/169/weft-and-warp-control.
[25] www.yarnbreakdetector.com/weft_sensor.html.
[26] www.textileschool.com/articles/167/weaving-equipments.
[27] Weaving solutions for technical fabrics, Technical Textiles, 2009, p. E102; May, 2011, p. E130.
[28] Daniel Palet, Textiles, 2011, No. 3/4, pp. 24–27.
[29] Weaving Machines for Technical Fabrics, Technical Textiles, 2007, p. E115.
[30] Projectile Weaving Machine for Technical Textiles, Technical Textiles, 2008, p. E122.
[31] ITEMA Weaving, Technical Textiles, 2011, p. E187.
[32] ITEMA Weaving Machines for Technical Fabrics, Technical Textiles, 2009, p. E110.
[33] Weaving of Aramid fibres, Technical Textiles, 2009, p. E53.
[34] Adnan Wahhoud, Dornier GmbH, Germany, August, Technical Textiles, 2011, p. E165.
[35] Dornier Weaving Machines for Technical Textiles, Technical Texitles, 2010, p. E88.
[36] Dornier Weaving Machines for Technical Textiles, Technical Textiles, 2009, p. E112.
[37] Hausding J, et al. ITMA 2011 Machines for Technical Textiles, Technical Textiles, 2012, p. E28, www.lindauerdornier.com/en/weaving-machine/open-reed-weave-orw-technology.
[38] Smit Textile, Technical Textiles, 2007, p. E112.
[39] Tsudakoma—Jet Weaving Machines, Technical Texitles, 2010, p. E92.
[40] Van de Wiele & Bonas—Weaving Machines for Technical Textiles, Technical Textiles, 2011, p. E133.
[41] Van de Wiele, Woven Artificial Turf, Technical Textiles, 2012, p. E91.
[42] Bonas Jacquard Machines for Technical Textiles, Technical Textiles, 2009, p. E106.
[43] Staubli-Textile Machinery for Technical Fabrics, Technical Textiles, 2013, p. E150.
[44] Simone Schwabe, Heike Oschatz, Uwe Möhring, Textile Research Institute, Thuringia Vogtland, Greiz/Germany, Technical Textiles, 2012, p. E 56.

Technical fabric structures – 2. Knitted fabrics

S.C. Anand

Institute of Materials Research and Innovation, The University of Bolton, Bolton, UK

Chapter Outline

5.1   Terms and definitions

- *Warp knitting* is a method of making a fabric by making loops from each warp formed substantially along the length of the fabric. It is characterised by the fact that each warp thread is fed more or less in line with the direction in which the fabric is produced. Each needle within the knitting width must be fed with at least one separate and individual thread at each course. This method of converting yarn into fabric is faster than weaving and weft knitting (Figure 5.1).
- *Weft knitting* is a method of making a fabric by making loops from each weft thread that are formed substantially across the width of the fabric. It is characterised by the fact that each weft thread is fed more or less at right angles to the direction in which the fabric is produced. It is possible to knit with one thread only, but up to 144 threads can be used on one machine. This method is the more versatile of the two in terms of the range of products produced and the type of yarns used (Figure 5.2).
- *Single-jersey fabric* is a weft knitted fabric made on one set of needles.
- *Double-jersey fabric* is a weft knitted fabric made on two sets of needles, usually based on rib or interlock gaiting, in a manner that reduces the natural extensibility of the knitted structure. These fabrics can be non-jacquard or jacquard.
- *Course* is a row of loops across the width of the fabric. Courses determine the length of the fabric and are measured as courses per centimetre.
- *Wale* is a column of loops along the length of the fabric. Wales determine the width of the fabric and are measured as wales per centimetre.
- *Stitch density* is the number of stitches per unit area of a knitted fabric (loops cm^{-2}). It determines the area of the fabric.

Handbook of Technical Textiles. http://dx.doi.org/10.1016/B978-1-78242-458-1.00005-4

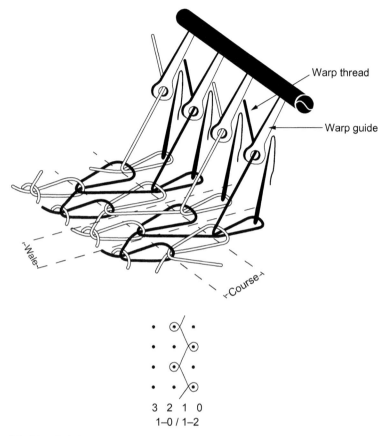

Figure 5.1 Warp knitting.

- *Stitch length* is the length of yarn in a knitted loop. It is the dominating factor for all knitted structures. In weft knitting, it is usually determined as the average length of yarn per needle; in warp knitting, it is normally determined as the average length of yarn per course.
- *Yarn linear density* indicates the thickness of the yarn and is normally determined in tex, which is defined as the mass in grams of 1 km of the material. The higher the tex number, the thicker is the yarn and vice versa.
- *Overlap* is the lateral movement of the guide bars (GB) on the beard or hook side of the needle. This movement is normally restricted to one needle space. In the fabric, a loop or stitch is also termed the overlap.
- *Underlap* is the lateral movement of the guide bars on the side of the needle remote from the hook or beard. This movement is limited only by mechanical considerations. It is the connection between stitches in consecutive courses in a warp knitted fabric.
- *Tightness factor K* is a number that indicates the extent to which the area of a knitted fabric is covered by the yarn. It is also an indication of the relative looseness or tightness of the knitting ($K = \text{tex}^{1/2} l^{-1}$), where l is the stitch length.
- *Area density* is a measure of the mass per unit area of the fabric (gm^{-2}).

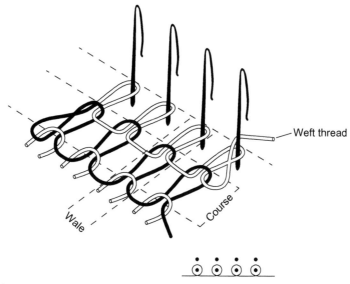

Figure 5.2 Weft knitting.

5.2 Weft knitting machines

Figure 5.3 shows a simplified classification of weft knitting equipment. Notice in Figure 5.3 that the latch needle is the most widely used needle in weft knitting because it is self-acting or loop controlled. It is also regarded as more versatile in terms of the range of materials that can be processed on machines with this needle. *Bearded needles* are less expensive to manufacture, can be produced in finer gauges, and supposedly knit tighter and more uniform stitches compared with latch needles but have limitations with regard to the types of material that can be processed and the range of structures that can be knitted on them. Bearded needle machines are faster than the equivalent latch needle machines. A *compound needle* has a short, smooth, and simple action, and because it requires a very small displacement to form a stitch in both warp and weft knitting, its production rate is the highest of the three main types of needle. Compound needles are now the most widely used needles in warp knitting; a number of manufacturers offer circular machines equipped with compound needles. The operation speeds of these machines are up to twice those of the equivalent latch needle machines.

The main parts of the bearded, latch, compound needle (fly needle frame), and compound needle (Kokett) are shown in Figures 5.4–5.7, respectively. Variations of latch needles include rib loop transfer needles and double-ended purl needles, which can slide through the old loop in order to knit from an opposing bed and thus draw a loop from the opposite direction.

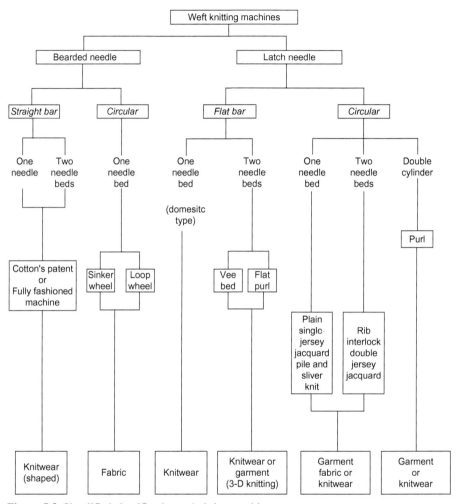

Figure 5.3 Simplified classification or knitting machinery.

5.2.1 Loop formation with latch needles

Figure 5.8a illustrates the needle at tuck height, that is, high enough to receive a new yarn but not high enough to clear the old loop below the latch. The needle is kept at this position because the loop formed at the previous course (A) lies on the open latch and stops the latch from closing. Note that once a latch is closed, it can be opened only by hand after stopping the machine.

In Figure 5.8b, the needle has been lifted to the clearing position and the new yarn (B) is presented. The old loops (A) are below the latch and the new yarn (B) is fed into the needle hook. A latch guard normally prevents the latch from closing at this point.

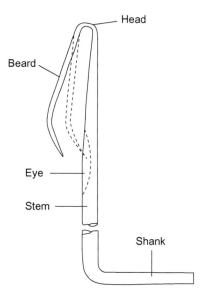

Figure 5.4 Bearded needle.

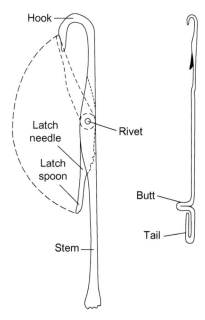

Figure 5.5 Latch needle.

In Figure 5.8c, the needle has moved to its lowest position or knitting point and has drawn the new loop (B) through the old loop (A), which is now cast off or knocked over. The needle now rises, and the sequence of movements is repeated for the next course.

The loop formation on a latch needle machine is also illustrated in Figure 5.9a, and a typical cam system on such a machine is shown in Figure 5.9b.

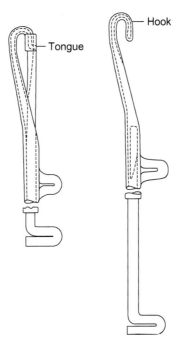

Figure 5.6 Compound needle (fly needle frame).

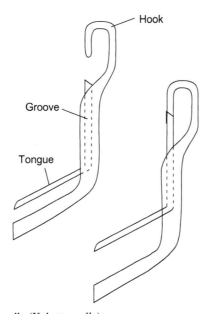

Figure 5.7 Compound needle (Kokett needle).

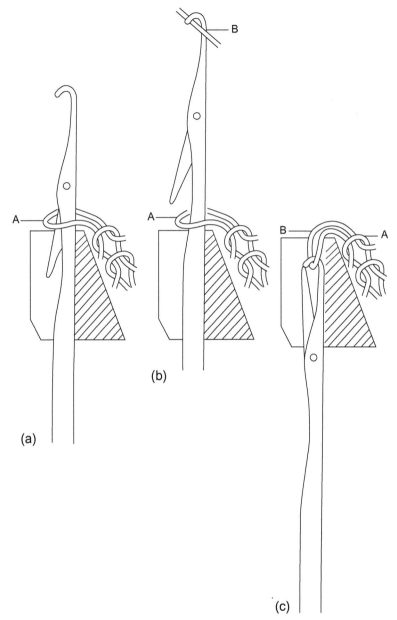

Figure 5.8 Loop formation with latch needles.

5.2.2 Single-jersey latch needle machines

This type of machinery is employed throughout the world, either as a basic machine or with certain refinements and modifications, to produce fabric ranging from stockings to single-jersey fabric for dresses and outerwear and a wide range of fabrics and products for technical applications. The machine sizes vary from one feed 1-in. diameter

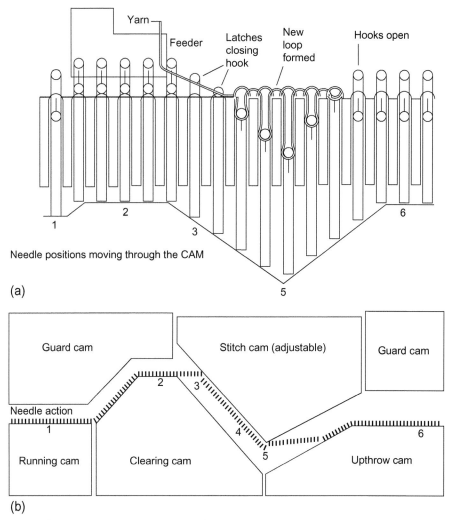

Figure 5.9 (a) Loop formation and (b) typical cam system on a latch needle machine.

to 144 feeds 30-in. diameter. Most single-jersey machines are the rotating cylinder type, although a few rotating cam box machines are still used for specialised fabrics. These machines are referred to as *sinker top machines* and use web holding sinkers. Comprehensive published reviews of single- and double-jersey knitting machinery and accessories exhibited at ITMA'95 and ITMA'99 illustrate the versatility and scope of modern weft knitting equipment.[1,7]

5.2.2.1   Knitting action of a sinker top machine

Figure 5.10a–d shows the knitting action of a sinker top machine during the production of a course of plain fabric.

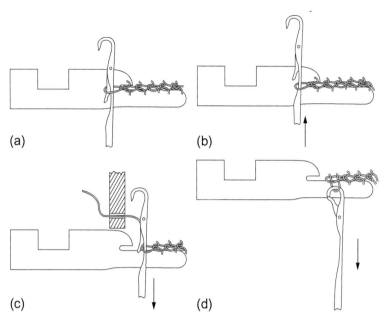

Figure 5.10 Knitting action of a sinker top machine.

Figure 5.10a illustrates the rest position. It shows the relative position of the knitting elements between the feeders with the needle at tuck height and the fabric loop held on the needle latch by the forward movement of the sinker towards the centre of the machine.

Figure 5.10b illustrates the clearing position. The needle has been raised to its highest position by the clearing cam acting on the needle butt; the old loop slides down from the open latch on to the needle stem.

Figure 5.10c illustrates the yarn feeding position. The sinker is partially withdrawn, allowing the feeder to present its yarn to the descending needle hook while freeing the old loop so that it can slide up the needle stem and under the open latch spoon.

Figure 5.10d illustrates the knock-over position. The needle has now reached its lowest position and has drawn the new loop through the old loop that is now knocked over the sinker belly.

Stitch length may be controlled in a number of ways. Machines without positive feed mechanism control it mainly by the distance the needle descends below the sinker belly. Other factors such as input tension (T_i), yarn to metal coefficient of friction (μ), and takedown tension also influence the final stitch length in the fabric. When a positive feed device is used, the length of yarn fed to the needles at a particular feed determines the stitch length. Other factors such as input tension T_i, μ, stitch cam setting, and take-down tension influence the yarn or fabric tension during knitting and, hence, determine the quality of the fabric. Stitch length is fixed by the positive feed device setting.

The sinker has two main functions:

- To hold the fabric loop in a given position whenever the needles rise.
- To provide a surface over which the needles draw the loops.

Other advantages of using sinkers include:

- The control exerted by the sinker allows minimum tension on the fabric, thus producing a good quality fabric with even loops.
- Fine adjustments in quality and those required in knitting certain difficult yarns and structures are possible.
- The sinker facilitates the setup of the machine after a partial or full press-off (after the latches have been opened manually).

5.2.3 Double-jersey machines

Figure 5.11 shows the needle layout of rib and interlock machines. Both types are available as circular or flat machines whereas straight bar or fully fashioned machines are available in rib type only.

Rib and interlock double-jersey machines are used either as garment length machines or for producing rolls of fabric. They can be either plain or equipped with a

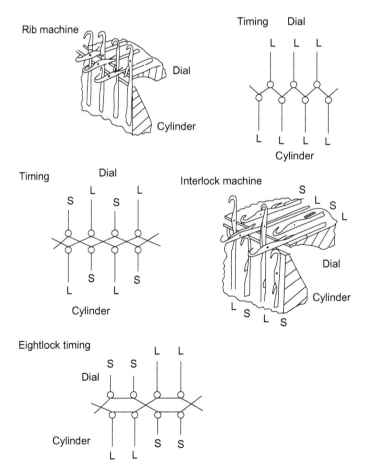

Figure 5.11 Double-jersey machine needle layouts.

wide range of mechanical patterning mechanisms at each feed in the cylinder. Both types can also be equipped with electronic needle mechanisms to produce large area jacquard patterns at high speeds.

5.2.3.1   Rib machines

Rib machines use two sets of needles and can be flat, circular, or fully fashioned. The needles in the two beds are staggered or have spaces between them (rib gaiting). Most machines have revolving needle cylinders, but in some cases, the cams rotate past stationary needles. Patterning is obtained by altering cam and needle setout or by using various needle selection mechanisms including individual needle electronic selection with computer-aided design systems. Machine diameters range from 7½ in. to 20 in. for garment length and from 30 to 33 in. for fabric machines. An example of a modern double-jersey machine is the Monarch V-7E20, a 30-in. diameter, E20, with 72-feeder eight-lock machine with RDS on dial and ACT II motorised automatic friction takedown system. The machine has 2×2 cam tracks and can be converted from rib to interlock or eight-lock timing in minutes. All basic non-jacquard double-jersey structures can be knitted at a speed factor of 900 (machine diameter (in.) × machine rpm).

5.2.3.2   Interlock machines

These are latch needle circular machines of the rib type with a cylinder and a dial. Unlike rib machines where the tricks of the dial alternate with the tricks of the cylinder (rib gaiting), the needle tricks of the cylinder are arranged exactly opposite those of the dial (interlock gaiting). Long- and short-stemmed needles that are arranged alternately, one long, one short in both cylinder and dial are used as shown in Figure 5.11. An example of a modern high-speed interlock machine is Sulzer Morat Type 1 L 144, which has a 30-in. diameter and 144 feeds, 28- or 32-gauge needles per inch (NPI), and 28 rpm, producing at 100% efficiency $86.4 \, \text{m h}^{-1}$ ($15.55 \, \text{kg h}^{-1}$) of 76 dtex polyester with 14 cpc (courses per centimetre) and with an area density of $180 \, \text{gm}^{-1}$ (60-in. wide) finished fabric.

To accommodate the long- and short-stemmed needles, the cam system has a double cam track. The long dial needles knit with the long cylinder needles at feeder 1, and the short cylinder needles knit with the short dial needles at feeder 2. Thus, two feeders are required to make one complete course of loop.

5.2.3.3   Needle timing

Two different timings can be employed on 1×1 rib and 1×1 interlock machines.

- *Synchronised timing* is the timing of a machine that has two sets of needles; the point of knock-over of one set is aligned with the point of knock-over of the other set.
- *Delayed timing* sets the point of knock-over of one set of needles on a two-bed knitting machine out of alignment with that of the other set to permit the formation of a tighter stitch. Broad ribs (i.e. 2×2, 3×3) and rib jacquard fabrics cannot be produced using delayed timing because there are not always cylinder needles knitting either side of the dial needles from which to draw yarn. Up to a nine-needle delay is possible, but a four- to five-needle delay is normal.

5.2.3.4   Knitting action of V-bed flat machine

Figure 5.12 illustrates the different stages of loop formation on a V-bed flat knitting machine and Figure 5.13 shows the cam system used on a simple single-system flat machine. Power V-bed flat machines are used mainly for the production of knitwear for children, women, and men. They range from simple through mechanical jacquard machines to fully electronic and computerised flat machines even equipped with a presser foot. The developments in the automation of fabric designing, pattern preparation,

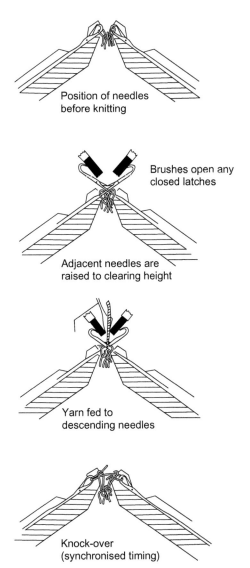

Figure 5.12 Loop formation on a V-bed flat knitting machine.

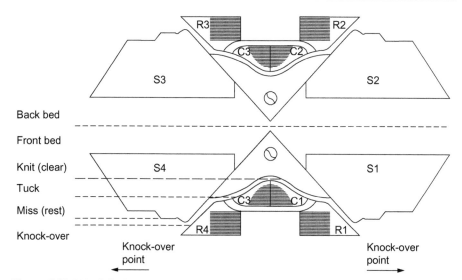

Figure 5.13 V-bed single-system cams. S are stitch cams, R are raising cams, and C are clearing cams.

and electronic needle selection and in the range of structures and effects that can be produced have been tremendous; flat machines and their products are now regarded as extremely sophisticated. High-quality garments can now be produced at competitive prices as the result of revolutionary garment production systems that are feasible with a presser foot. Two- and three-dimensional structures and complete garments without any seams or joins can be produced on the latest electronic flat knitting machines and associated design systems.

5.3   Weft knitted structures

The basic weft knitted structures and stitches are illustrated in Figure 5.14a–g, and the appearance, properties, and end-use applications of plain, 1×1 rib, 1×1 purl, and interlock structures are summarised in Table 5.1. These basic stitches are often combined in one fabric to produce an enormous range of single- and double-jersey fabrics or garments. Weft knitted fabrics are produced commercially for apparel and household and technical products and are used for an extremely large array of products ranging from stockings and tights to imitation furs and rugs.

The importance and diversity of warp and weft knitted fabrics used for various technical applications has been discussed by Anand,[2] who highlighted the fact that knitted fabrics are being increasingly designed and developed for technical products ranging from scouring pads (metallic) to fully fashioned nose cones for supersonic aircrafts. Warp and weft knitted products are becoming popular for a wide spectrum of medical and surgical products.[3]

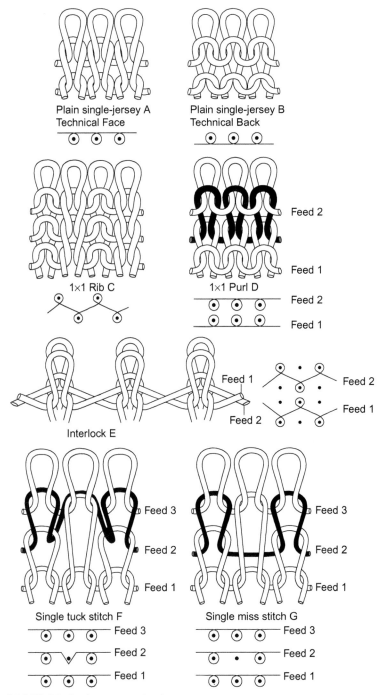

Figure 5.14 Weft knitted structures (a–g).

Table 5.1 Comparison of appearance and properties

Property	Plain	1×1 Rib	1×1 Purl	Interlock
Appearance	Different on face and back; V-shapes on face, arcs, on back	Same on both sides, like face of plain	Same on both sides, like back plain	Same on both sides, like face of plain
Extensibility				
Lengthwise	Moderate (10–20%)	Moderate	Very high	Moderate
widthwise	High (30–50%)	Very high (50–100%)	High	Moderate
Area	Moderate–high	High	Very high	Moderate
Thickness and warmth	Thicker and warmer than plain woven made from same yarn	Much thicker and warmer than plain woven	Very much thicker and warmer than plain woven	Very much thicker and warmer than plain woven
Unroving	Either end	Only from end knitted last	Either end	Only from end knitted last
Curling	Tendency to curl	No tendency to curl	No tendency to curl	No tendency to curl
End-uses	Ladies' stocking Fine cardigans Men's and ladies' shirts Dresses Base fabric for coating	Socks Cuffs Waist bands Collars Men's outerwear Knitwear Underwear	Children's clothing Knitwear Thick and heavy Outerwear	Underwear Shirts Suits Trouser suits Sportswear Dresses

5.4   Process control in weft knitting

5.4.1   Main factors affecting the dimensional properties of knitted fabrics or garments

- Fabric structure: Different structures relax differently.
- Fibre(s) type: Fabrics or garments made from different fibre(s) relax differently.
- Stitch length: The length of yarn in a knitted loop is the dominating factor for all structures.
- Relaxation/finishing route: The fabric dimensions vary according to relaxation/finishing sequence.
- Yarn linear density: This affects the dimensions (slightly), fabric tightness, area density (gm^{-2}), and other physical properties.

5.4.2   Laboratory stages of relaxation

- On machine – strained state: This is predominantly length strain.
- Off machine – dry relaxed state: The fabric moves to this state with time. The dry relaxed state is restricted by fabric structure and fibre type. Only wool can attain this state.
- Static soak in water and dry flat – wet relaxed state: Tight structures do not always reach a "true" relaxed state. Only wool and silk can attain this state.
- Soak in water with agitation, agitation in steam, or static soak at selected temperatures (>90 °C) plus; dry flat – Finished relaxed state: The agitation and/or temperature induces a further degree of relaxation, producing a denser fabric. This includes wool, silk, textured yarn fabrics, and acrylics.
- Soak in water and tumble dry at 70 °C for 1 h – Fully relaxed state: This involves three-dimensional agitation during drying for all fibres and structures.

5.4.3   Fabric geometry of plain single-jersey structures

1. Courses per cm (cpc) $\alpha 1 / l = \dfrac{k_c}{l}$

2. Wales per cm (wpc) $\alpha 1 / l = \dfrac{k_w}{l}$

3. $s = (\text{cpc} \times \text{wpc})\ \alpha 1 / l^2 = \dfrac{k_s}{l^2}$

4. $\dfrac{\text{cpc}}{\text{wpc}} \alpha c = \dfrac{k_c}{k_w} (\text{shape factor})$

   k_c, k_w, and k_s are dimensionless constants, l is the stitch length, and s is the stitch density.

5.4.4   Practical implications of fabric geometry studies

- Relationship between yarn tex and machine gauge is given by Equation (5.1):

$$\text{Optimum tex} = \frac{\text{constant}}{(\text{gauge})^2} \qquad\qquad (5.1)$$

For single-jersey machines, the optimum tex $= 1650/G^2$, and for double-jersey machines, the optimum tex $= 1400/G^2$, where G is measured in needles per centimetre (npc).
- Tightness factor is given by Equation (5.2):

$$K = \sqrt{\frac{\text{tex}}{l}} \tag{5.2}$$

where l is the stitch length measured in millimetres. For single-jersey fabrics: $1.29 \leq K \leq 1.64$. Mean $K = 1.47$. For most weft knitted structures (including single- and double-jersey structures and a wide range of yarns): $1 \leq K \leq 2$. Mean $K = 1.5$. The tightness factor is very useful in setting up knitting machines. At mean tightness factor, the strain on yarn, machine, and fabric is constant for a wide range of conditions.
- Fabric area density is given by Equation (5.3):

$$\text{Area density} = \frac{s \times l \times T}{100} \text{gm}^{-2} \tag{5.3}$$

where s is the stitch density/cm^2, l is the stitch length (mm), and T is the yarn tex or Equation (5.4):

$$\frac{k_s}{l} \times \frac{T}{100} \text{gm}^{-2} \tag{5.4}$$

where k_s is a constant and its value depends on the state of relaxation, that is, dry, wet, finished, or fully relaxed (Table 5.2). The area density can also be given by Equations (5.5) and (5.6):

$$\text{Area density} = \frac{n \times l \times \text{cpc} \times T}{10,000} \text{gm}^{-1} \tag{5.5}$$

where n is the total number of needles, l is the stitch length (mm), and T is the yarn tex or

$$\frac{n \times k_c \times T}{10,000} \text{gm}^{-1} \tag{5.6}$$

where k_c is a constant and its value depends on the state of relaxation, that is dry, wet, finished, or fully relaxed (Table 5.2).
- Fabric width is given by Equation (5.7):

Table 5.2 k-Constant values for wool plain single jersey[a]

	k_c	k_w	k_s	k_c/k_w
Dry relaxed	50	38	1900	1.31
Wet relaxed	53	41	2160	1.29
Finished relaxed	56	42.2	2360	1.32
Fully relaxed	55 ± 2	42 ± 1	2310 ± 10	1.3 ± 0.05

For a relaxed fabric cpc/wpc = 1.3. cpc/wpc > 1.3 indicates widthwise stretching. cpc/wpc < 1.3 indicates lengthwise stretching, k_s > 2500 indicates felting or washing shrinkage. Relaxation shrinkage is the change in loop shape. Felting/washing shrinkage is the change in loop length.
[a]Courses and wales are measured per centimetre and l is measured in millimetres.

$$\text{Fabric width} = \frac{n \times l}{k_w} \text{cm} \tag{5.7}$$

where k_w is a constant and its value depends on the state of relaxation, that is, dry, wet, finished, or fully relaxed (Table 5.2). It can also be given by Equation (5.8):

$$n \times l = L(\text{course length}) \quad \therefore \text{Fabric width} = \frac{L}{k_w} \text{cm} \tag{5.8}$$

Fabric width depends on course length, not on the number of needles knitting.
- Fabric thickness (t) in the dry and wet relaxed states depends on fabric tightness, but in the fully relaxed state, it is more or less independent of the fabric tightness factor. In the fully relaxed state, $t \approx 4d$, where d is the yarn diameter.

5.4.5   Quality control in weft knitting

The dimensions of a weft knitted fabric are determined by the number of stitches and their size, which in turn is determined by stitch length. Most knitting quality control therefore reduces to the control of stitch length; differences in mean stitch length give pieces of different size; variation of stitch length within the piece gives appearance defects, by far the most common of which is the occurrence of widthwise bars or streaks resulting from variation in stitch length between adjacent courses.

5.4.5.1   Measurement of stitch length, l

- Off machine (in the fabric):
  - HATRA course length tester
  - Shirley crimp tester.
- On machine (during knitting):
  - Yarn speed meter (revolving cylinder only)
  - Yarn length counter (both revolving cylinder and cambox machines)

5.4.5.2   Control of stitch length, l

- Positive feed devices:
  - Capstan feed: cylindrical or tapered
  - Nip feed: garment length machines
  - Tape feed (Rosen feed): circular machines producing plain structures
  - Ultrapositive feed: IPF or MPF
- Constant tension device: Storage feed device: flat, half-hose, hose, and circular machines producing either plain or jacquard structures
- Specialised positive feed devices:
  - Positive jacquard feeder MPF 20 KIF
  - Striper feeder ITF
  - IROSOX unit (half-hose machines)
  - Elastane feed MER2
  - Air-controlled feeds for flat and fully fashioned machines

See a tape feed in Figure 5.15. A modern ultrapositive feed and a yarn storage feed device are illustrated in Figures 5.16 and 5.17, respectively.

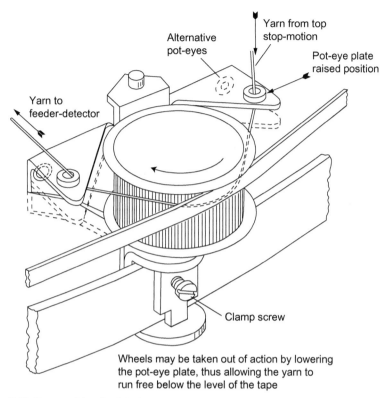

Figure 5.15 Tape positive feed device.

5.5 End-use applications of weft knitted fabrics

Weft knitted fabrics are used for apparel and household and technical products. The main outlets for the different types of weft knitted fabrics are shown below. The knitting equipment used to produce these fabrics is also given.

5.5.1 Flat bar machines

- Machine gauge: normally needles per inch, 3–18 NPI
- Machine width: up to 78.7 in.
- Needle type: latch (compound needle machines are being developed)
- Needle bed type: single (hand machines), but mainly rib type
- Products: jumpers, pullovers, cardigans, dresses, suits, trouser suits, trimmings, hats, scarves, accessories, and ribs for straight-bar (fully fashioned) machines. Cleaning clothes, three-dimensional and fashioned products for technical applications, and multiaxial machines are under development

5.5.2 Circular machines

- Machine gauge: normally needles per inch, 5–40 NPI
- Machine diameter: up to 30 in., now available in machines with up to 60-in. diameter

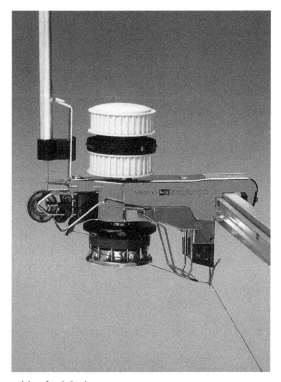

Figure 5.16 Ultrapositive feed device.

- Needle type: latch (bearded on sinker wheel and loop wheel, some compound needle machines)
- Needle bed type: single, rib, interlock, double cylinder
- Products
  - Hose machines: seamless hose, tights, industrial-use dye bags, knit-de-knit yarns, industrial fabrics
  - Half-hose machines: men's and boys' half-hose, ladies' stockings, children's tights, sport socks
  - Garment blank machines: underwear, T-shirts, jumpers, pullovers, cardigans, dresses, suits, trouser suits, vests, briefs, thermal wear, cleaning cloths, technical fabrics
  - Fabric machines: rolls of fabric with the following end uses: jackets, ladies' tops, sport and T-shirts, casual wear, suits, dresses, swimwear, bathrobes, dressing gowns, track suits, jogging suits, furnishing, upholstery, automotive and technical fabrics, household fabrics

5.5.3 Straight bar machines (fully fashioned machines)

- Machine gauge: normally needles per 1½-in., 9–33 (up to 60-gauge machines have been produced)
- Machine width: from 2 to 16 section machines; each section up to 36 in. wide (up to 40-section machines have been produced)
- Needle type: bearded or bearded and latch
- Needle bed type: single and rib
- Products: jumpers, pullovers, cardigans, dresses, suits, trouser suits, fully fashioned hose, sport shirts, underwear, thermal wear

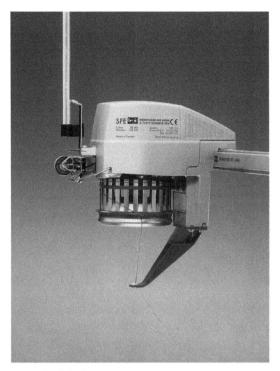

Figure 5.17 Yarn storage feed device.

5.6 Warp knitting machines

5.6.1 Introduction

The first weft knitting machine was built by William Lee in 1589. In 1775, just under 200 years later, the first warp knitting machine was invented by Crane, an Englishman. It was a single-guide bar machine to make blue and white zig-zag striped silk hosiery. The fabrics were named after Van Dyck the painter. With the advent of acetate continuous-filament yarns after World Ward I, the first bulk production of tricot fabrics began with British Celanese on German Saupe 2-guide bar, 28-gauge machines. Locknits replaced the single-guide bar atlas fabrics for lingerie, which are difficult to finish and laddered easily.

From 1950 to 1970, the growth of the warp knitting industries in the United Kingdom and other western countries was phenomenal. The main reasons for this colossal expansion are summarised below (although developments in the various fields mentioned here were taking place concurrently). The state-of-the-art and current developments in tricot and raschel machinery are summarised below.

Anand published a review of warp knitting equipment exhibited by Karl Mayer at ITMA in 1995[4] and in 1999.[8]

5.6.1.1   Yarn developments

- The discovery of thermoplastic yarns and their suitability, even in very low linear densities (deniers) and in flat or low-twist form, to be knitted with very low yarn breakage rates on modern high speed tricot and raschel machines
- The extra design scope offered by differential dye yarns
- Improved cover and comfort attained through textured and producer-bulked yarns
- Elastomeric yarns, which have given a tremendous fillip to the raschel powernet industry

5.6.1.2   Machinery developments

- Higher machine speeds (up to 3500 cpm)
- Finer gauges (up to 40 needles per inch)
- Wider machines (up to 260 in.)
- Increased number of guide bars (up to 78)
- Special attachments such as cut presser, fall plate, and swanwarp
- Some specialty raschel machines such as Co-we-nit and jacquard machines and, more recently, redesigned full-width weft insertion raschel and tricot machines
- High speed direct-warping machines and electronic yarn inspection equipment during warping
- Electronic stop motions for the knitting machines
- Larger knitting beams and cloth batches
- Modern heat-setting and beam-dyeing machinery
- Electronic warp let-off, electronic patterning, electronic jacquard, and electronic fabric takeup mechanisms
- Loop-raised fabrics
- Stable constructions, such as sharkskins and queens cord
- Various net constructions using synthetic yarns
- Mono-, bi-, tri-, and multiaxial structures for technical applications
- Three-dimensional and shaped (fashioned) structures for medical and other high-technology products

It is well known that the warp knitting sector, particularly tricot knitting, has grown in step with the expansion of manufactured fibres. In 1956, 17.8 million pounds (lbs) of regenerated cellulosic and synthetic fibre yarns were warp knitted; the figure reached a staggering 70.6 million lbs in 1968.

In the mid-1970s, the tricot industry suffered a major setback, mainly because of a significant drop in the sale of nylon shirts and sheets, which had been the major products of this sector. It is also true that the boom period of textured polyester double-jersey was also a contributing factor in the sudden and major decline in the sale of tricot products. A change in the fashion and the growth in the demand for polyester/cotton woven fabrics for shirting and sheeting was another cause of this decline. The two major manufacturers of warp knitting equipment, Karl Mayer and Liba, both in West Germany, have been actively engaged in redesigning their machinery to recapture some of the lost trade. The compound needle is the major needle used on both tricot and raschel machines, and many specialised versions of warp knitting machines are now available for producing household and technical products. One of the major developments in warp knitting has been the commercial feasibility of using staple-fibre yarns for a wide range of products. It is also significant to note that market base of the warp knitting sector has broadened and expanded into household and technical fabric markets, such as lace, geotextile, automotive and sportswear, and a wide spectrum of surgical and healthcare products. The current and future potential of warp knitted structures in engineering composite materials has been discussed by Anand.[5]

5.6.2   Tricot and raschel machines

The principal differences between tricot and raschel machines are:

1. Latch needles are generally used in raschel machines; bearded or compound needle machines are referred to as *tricot machines*. Compound needle raschel machines are also now fairly common. The compound needle is the most commonly used needle on warp knitting equipment.
2. Raschel machines are normally provided with a trick plate whereas tricot machines use a sinker bar.
3. Raschel machines take up the fabric parallel to the needle stems; the tricot machines, however, take it up at approximately right angles to the needles.
4. Raschel machines are normally in a coarser gauge; they are also slower because more guide bars are frequently used, and they require a longer and slower needle movement.
5. Raschel machines are much more versatile in terms of their ability to knit most types of yarns such as staple yarns and split films. Only continuous-filament yarns can be successfully knitted on most tricot machines.
6. Generally, warp beams are on the top of the machine on raschel machines; on tricot machines, they are generally at the back of the machine.

A simplified classification of warp knitting equipment is given in Figure 5.18; note that apparel and household and technical fabrics are produced on modern warp knitting machinery. It is in fact in the technical applications that the full potential of warp knitting is being exploited. It is virtually possible to produce any product on warp knitting equipment but is not always most economical.

The simplest warp knitted structures are illustrated in Figure 5.19. It can be seen that both closed- and open-loop structures can be produced, and there is normally very little difference in the appearance and properties between the two types of loops.

5.6.3   Knitting action of compound needle warp knitting machine

Figure 5.20a shows that the sinkers move forward holding the fabric down at the correct level in their throats. The needles and tongues rise with the needle rising faster until the hook of the needle is at its highest position and is open. Figure 5.20b illustrates that the guides then swing through to the back of the machine and Figure 5.20c shows the guides shog for the overlap and swing back to the front of the machine.

Figure 5.20d shows the needles and the tongues starting to descend more slowly, thus closing the hooks. The sinkers start to withdraw as the needles descend so that the old loop lands on the closed hook, and the new loops are secured inside the closed hook.

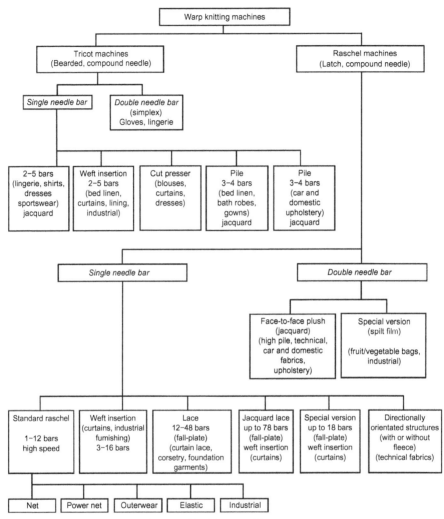

Figure 5.18 Simplified classification of knitting machinery.

Figure 5.20e shows the needle descending below the sinker belly, and the old loop is knocked over. At this point, the underlap occurs and as in Figure 5.20f, the sinkers move forward to hold down the fabric before the needles commence their upward rise to form a fresh course.

5.6.4 Knitting action of standard raschel machine

In Figure 5.21a, the guide bars are at the front of the machine completing their underlap shog. The web holders move forward to hold the fabric down at the correct level while the needle bar starts to rise from knock-over to form a fresh course.

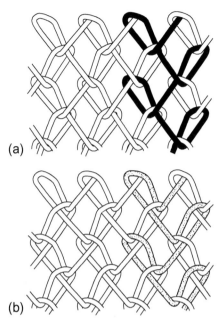

(a)

(b)

Figure 5.19 Single-guide bar warp knitted fabrics. (a) Closed lap fabric and (b) open lap fabric.

Figure 5.21b shows that the needle bar has risen to its full height and the old loops slip down from the latches onto the stems after opening the latches. The latches are prevented from closing by the latch guard. The web holders then start to withdraw to allow the guide bars to form the overlap movement.

In Figure 5.21c, the guide bars swing to the back of the machine and then shog for the overlap, and in Figure 5.21d, the guide bars swing back to the front, and the warp threads are laid into the needle hooks. Note that only the front guide bar threads have formed the overlap movement, the middle and back guide bars thread return through the same pair of needles as when they swung towards the back of the machine. This is called *laying in motion.*

In Figure 5.21e, the needle bar descends so that the old loops contact and close the latches, trapping the new loops inside. The web holders start to move forward.

Figure 5.21f shows the needle bar continuing to descend, its head passing below the surface of the trick plate, drawing the new loops through the old loops, which are cast off, and as the web holders advance over the trick-plate, the underlap shog of the guide bar commences.

The knitting action of bearded needle warp knitting machines is not described here because in general, the machines likely to be used for technical textile products use either latch or compound needles. Also, the proportion of new bearded needle machines sold has decreased steadily over the years. This is mainly due to the lack of versatility of these machines in terms of the variety of yarns that can be processed and the range of structures that can be normally knitted on them. The displacement curves for the three main types of needle are shown in Figure 5.22.

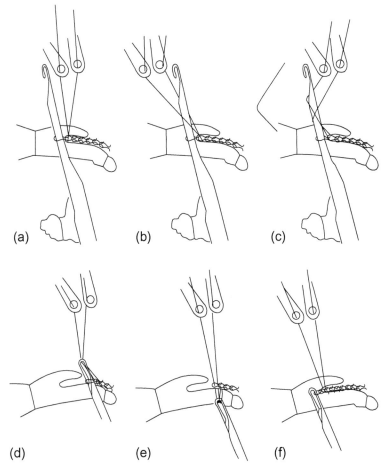

Figure 5.20 Knitting action of compound needle warp knitting machine.

It is obvious that compound needle machines would operate at faster rates provided that all other factors are similar.

5.7   Warp knitted structures

5.7.1   Stitch notation

Some of the more popular stitches used in the production of warp knitted fabrics are given in Figure 5.23. These stitches combined with the number of guide bars used, a comprehensive range of types and linear densities of yarns available, fancy threading, controlling individual run-ins and run-in ratios, and various finishing techniques are modified to construct an endless variety of fabrics. The lapping movements of the individual guide bars throughout one repeat of the pattern are normally indicated on special paper called *point paper*. Each horizontal row of equally spaced dots represents the

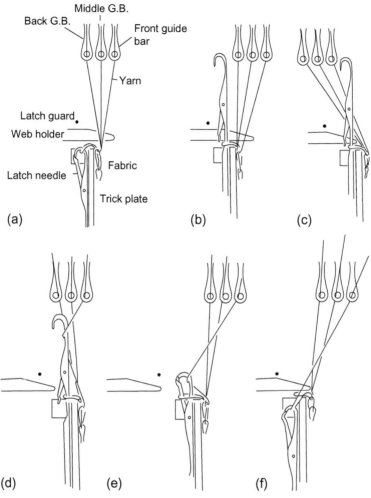

Figure 5.21 Knitting action of a standard raschel machine. (a) Start of new course, (b) start of overlap, (c) guide bar swinging motion, (d) return swing after overlap, (e) old loop closing latch, and (f) knock-over and underlap movements.

same needle at successive courses. The spaces between the dots, or needles, are numbered 0, 1, 2, 3, 4, and so on and show the number of needles crossed by each guide bar. Although three links per course are normally employed, only two are actually required; the third (last link) is used only to effect a smoother movement of the guide bar during the underlap. The first link determines the position of the guide bars at the start of the new course. The second link determines the direction in which the overlap is made. The links, therefore, are grouped together in pairs, and the lapping movements at each course are separated by a comma. For instance, the lapping movements shown in Figure 5.23c are interpreted as follows:

- (1–0) is the overlap at the first course.
- (0, 1) is the underlap at the same course but made in the opposite direction to the overlap.

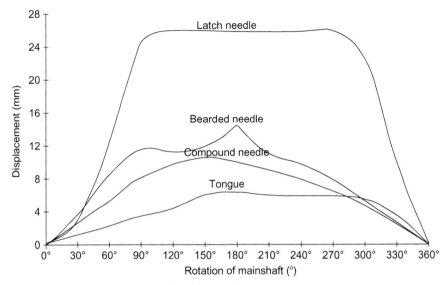

Figure 5.22 Displacement curves of various needles.

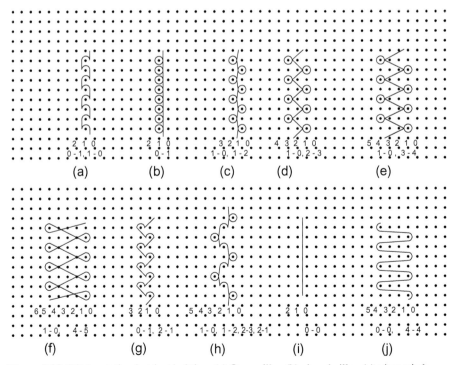

Figure 5.23 Stitch notation in tricot knitting. (a) Open pillar, (b) closed pillar, (c) tricot stitch, (d) 2 × 1 closed lap, (e) 3 × 1 closed lap, (f) 4 × 1 closed lap, (g) open tricot stitch, (h) two-course atlas, (i) misslapping, and (j) laying-in.

- (1–2) is the overlap at the second course.
- (2,1) is the underlap at the second course but made in the opposite direction to the previous underlap.

It can also be observed in Figure 5.23 that when the underlap is made in the opposite direction to the immediately preceding overlap, a closed loop is formed. However, when the underlap is made in the same direction as the immediately preceding overlap or no underlap is made, then an open loop results.

It is vital to ensure when placing a pattern chain around the drum that the correct link is placed in contact with the guide bar connecting rod; otherwise, the underlap occurs on the wrong side of the needles, or open loops may be formed instead of the intended closed loops.

5.7.2 Single-guide bar structures

Although it is possible to knit fabrics using a single fully threaded guide bar, such fabrics are now almost extinct because of their poor strength, low cover, lack of stability, and pronounced loop inclination on the face of the fabric. Three examples of single-guide bar structures are shown in Figure 5.24.

5.7.3 Two-guide bar full-set structures

The use of two guide bars gives a much wider pattern scope than is possible when using only one, and a large proportion of the fabrics produced in industry are made

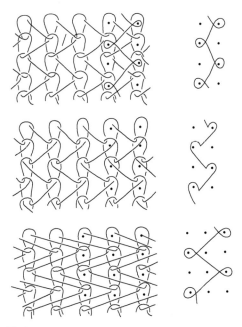

Figure 5.24 Single-guide bar structures.

with two-guide bars. The first group of fabrics to consider are those made with fully threaded guide bars because many different effects may be obtained by altering the lapping movements; these effects may be increased further by mixing different yarns or using colour, linear densities, or different yarn types, such as yarns with different dyeing characteristics, textured yarns, and so on.

5.7.3.1   Loop plating

With two fully threaded guide bars, each loop in the fabric contains two threads, one supplied by each bar. The underlaps made by the front guide bar are plated on the back of the fabric and the loops from this bar are plated on the face of the fabric, whereas the loops and the underlaps formed by the back guide bar are sandwiched between those from the front guide bar (see Figure 5.25). It can be observed in Figure 5.20c that when the guide bars swing through the needles to form the overlap, the ends are crossed on the needle hook (normally the two bars form overlaps in opposite directions). As the guide bars return to the front of the machine, the threads of the front guide bar are the first to strike the needles and are wrapped around the needle hook first whereas the back guide bar threads are placed later and above those from the front guide bar. If the tensions of the two warp sheets are similar and the heights of the guide bars are correctly adjusted, the front bar loops are always plated on the face of the fabric. Any coloured thread in the front guide bar thus appears prominent on both fabric surfaces, an important factor to be remembered in warp knitted fabric designing (see Figure 5.25 for loop plating).

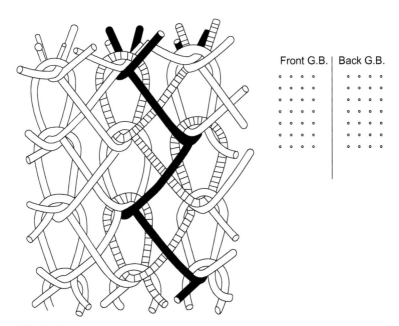

Figure 5.25 Full tricot.

5.7.3.2   Different structures

The two guide bars may make their underlaps in the same or opposite directions. If they are made in the same direction, the fabric shows distortion similar to that of the single-bar fabric – see Figure 5.29a – because the loops are inclined. If, however, the underlaps are made in opposite directions, an equal tension is imposed in both directions, and loops are upright.

The structure of the simplest fabric made with two guide bars is shown in Figure 5.25; it is known as *full tricot*. The appearance of full tricot may be varied by threading the guide bars with different coloured threads to give vertical stripes of colour.

The most common fabric of all is locknit (see Figure 5.26 for its structure and the lapping movements). When correctly knitted, the fabric shows even rows of upright loops on its face, and the two needle underlaps on the back of the fabric give a smooth sheen. It has a soft handle and is very suitable for lingerie. If the lapping movements for the bars are reversed to give reverse locknit, the fabric properties are completely changed; see Figure 5.29e. The short underlaps then appear on the back of the fabric and trap the longer ones to give a more stable and stiff structure with far less width shrinkage from the needles than ordinary locknit. The underlaps of the back guide bar may be increased to give even more stability and opacity with practically no width shrinkage from the needles. An example of this is sharkskin, whose structure and lapping movements are shown in Figure 5.27. Another stable structure known as *queenscord* is shown in Figure 5.28. The long back guide bar underlaps are locked firmly in the body of the fabric by the chain stitches of the front guide bar. Both sharkskin and queenscord structures can be made more stable, heavier, and stronger by increasing the back guide underlaps to four or five needle spaces. The vertical chains of loops from the front guide bar may be used to give single-wale vertical stripes of colour, such as pin stripes in men's suiting.

If the guide bars making a sharkskin are reversed, that is, if the front bar makes the longer underlaps, the resultant fabric is known as *satin*, which is a lustrous soft

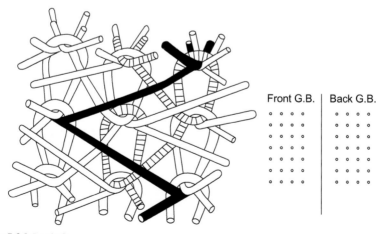

Figure 5.26 Locknit.

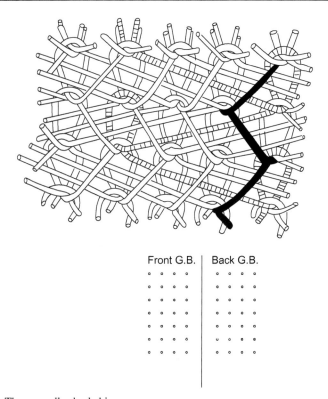

Front G.B.	Back G.B.
o o o o	o o o o
o o o o	o o o o
o o o o	o o o o
o o o o	o o o o
o o o o	o o o o
o o o o	o o o o
o o o o	o o o o

Figure 5.27 Three-needle sharkskin.

fabric similar to the woven satin. Because of the long floats on the back of the fabric, satin laps are used to make loop-raised fabrics. The raising machine is set so that the underlaps are raised into loops without actually breaking any filaments. In order to achieve the maximum raising effect, the two-guide bars in a loop-raised fabric are normally made to cross in the same direction, and open loops may also be used. The lapping movements of three-needle satin are shown in Figure 5.29b and those for a three-needle loop-raised fabric are shown in Figure 5.29a. The density and height of pile can be increased by increasing the front guide bar underlaps to four, five, or six needle spaces.

Yarns may be introduced into the fabric without actually knitting. Figure 5.30 shows the structure lapping movements and pattern chains of a laid-in fabric. The laid-in thread is trapped between the loop and the subsequent underlap of the guide bar, which must be situated in front of the laying-in bar. To lay in a yarn, therefore, that yarn must be threaded in a guide bar to the rear of the guide bar (knitting bar), and it must make no overlaps. Laying-in is a useful device because a laid-in thread never goes round the needle, introducing very thick or fancy yarns into the fabric, such as heavy worsted yarn or metallic threads. Figure 5.31 shows the laid-in thread being trapped in the fabric by the front guide bar threads knitting an open tricot stitch (0–1, 2–1).

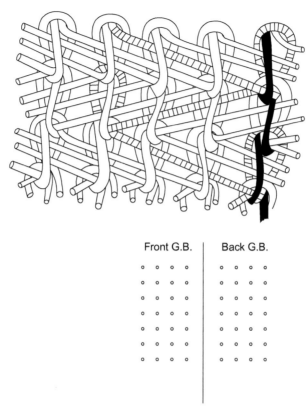

Front G.B.	Back G.B.
o o o o	o o o o
o o o o	o o o o
o o o o	o o o o
o o o o	o o o o
o o o o	o o o o
o o o o	o o o o
o o o o	o o o o

Figure 5.28 Three-needle queenscord.

5.7.4   Grey specification of a warp knitted fabric

A complete grey specification of a warp knitted fabric should include the following details:

1. Gauge of machine in needles per inch
2. Number of guide bars in use
3. Number of ends in each warp
4. Types and linear densities of yarns used
5. Run-in per rack for each warp
6. Knitted quality of the fabric in courses per cm
7. Order of threading in each guide bar
8. Lapping movements of each guide bar during one repeat of the pattern or details of the pattern wheels or pattern chains
9. Relative lateral positions of the guide bars at a given point in the lapping movements
10. Any special knitting instructions

5.7.5   Fabric quality

The main parameter controlling the quality and properties of a given structure is the run-in per rack, or the amount of yarn fed into the loop. *Run-in per rack* is defined as the length of warp fed into the fabric over 480 courses (1 rack=480 courses). In two-guide

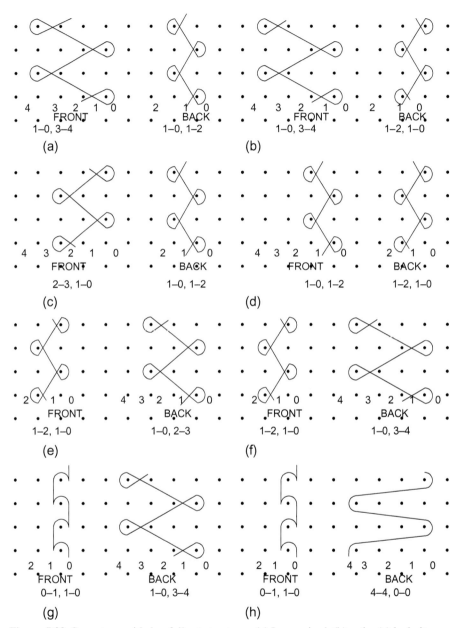

Figure 5.29 Some two-guide bar full set structures. (a) Loop raised, (b) satin, (c) locknit, (d) full tricot, (e) reverse locknit, (f) sharkskin, (g) queenscord, and (h) laid-in fabric.

bar fabrics, the run-in per rack for each guide bar may be the same or different, depending on the fabric structure. For example, in full tricot structures (front: 1–2, 1–0; back: 1–0, 1–2), using the same run-in per rack from both beams or 1:1 is normal, whereas in three-needle sharkskin fabrics (front: 1–2, 1–0; back: 1–0, 3–4), the run-in per rack required from the back beam would be more than the front beam, for example, 1:1.66.

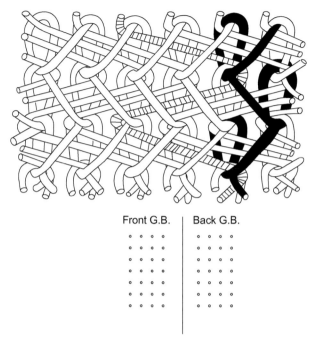

Front G.B. | Back G.B.

Figure 5.30 Laid-in structure.

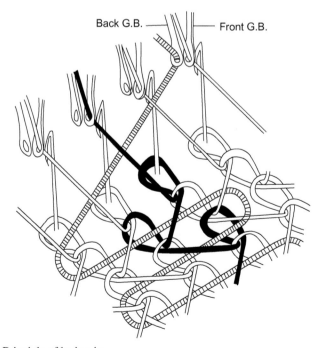

Figure 5.31 Principle of laying-in.

The run-in may be altered in two different ways, first by altering the total run-in of the bars and second by altering the ratio or difference between the bars. Altering the total run-in affects the finished number of courses per cm and hence the area density of the fabric, the stability, and the cover but not the general shape of the loop. Altering the difference between the guide bars changes the balance of the fabric and affects the inclination of the loops and because it puts more or less strain on the individual yarns, changes the strength.

Fabric takeup on the machine is adjusted to attain trouble-free knitting and to effect ease of finishing.

5.7.6 Tightness factor

The tightness factor K of a knitted fabric is defined as the ratio of the fabric area covered by the yarn to the total fabric area. It is regarded as a measure of looseness or tightness of the structure and influences dimensions such as the length, width, thickness, and many other fabric characteristics such as area density, opacity, abrasion resistance, modulus, strength, and shrinkage.

If the tightness factor of a single-guide bar fabric is defined as in Equation (5.2):

$$K = \sqrt{\frac{\text{tex}}{l}} \tag{5.9}$$

where l is the stitch length, measured in millimetres, and tex is the yarn linear density, then the tightness factor of a two-guide bar, full-set fabric is given by Equation (5.10):

$$K = \frac{\sqrt{\text{tex}_f}}{l_f} + \frac{\sqrt{\text{tex}_b}}{l_b} \tag{5.10}$$

where suffixes f and b refer to front and back guide bars, and l is the stitch length equal to (run-in/rack)/480 and is measured in millimetres. If the same tex is employed in both bars, then

$$K = \sqrt{\text{tex}} \left(\frac{1}{l_f} + \frac{1}{l_b} \right) \tag{5.11}$$

For most commercial two-guide bar full-set fabrics, $1 \leq K \leq 2$ with a mean tightness factor value of 1.5.[6]

5.7.7 Area density

The area density of a single-guide bar fabric can be determined from Equation (5.12):

$$\begin{aligned} \text{Mass of the fabric} &= \text{cpc} \times \text{wpc} \times l \times T \times 10^{-2} \, \text{gm}^{-2} \\ &= s \times l \times T \times 10^{-2} \, \text{gm}^{-2} \end{aligned} \tag{5.12}$$

where s is the stitch density (cm^{-2}) or (cpc $\times$ wpc), l is the stitch length (mm), and T is the yarn tex.

Similarly, the area density of a two-guide bar full-set fabric would be Equation (5.13):

$$\text{Mass of the fabric} = s\left[\left(l_f \times T_f\right) + \left(l_b \times T_b\right)\right] \times 10^{-2}\,\text{gm}^{-2} \tag{5.13}$$

where suffixes f and b refer to the front and back guide bars. If the same tex is used in both guide bars, then Equation (5.13) can be written as Equation (5.14):

$$\text{Mass of the fabric} = s \times T \times 10^{-2}\left(l_f + l_b\right)\text{gm}^{-2}$$

or

$$= s \times T \times 10^{-2}\left(\frac{\text{Total run-in}}{480}\right)\text{gm}^{-2} \tag{5.14}$$

If the stitch density, that is, the number of loops cm^{-2}, stitch length in millimetres of the individual guide bars and tex of yarns used in individual beams are known, the fabric area density can be readily obtained using Equation (5.13) in any fabric state, that is, on the machine, dry relaxed or fully relaxed.

The geometry and dimensional properties of warp knitted structures have been studied by a number of researchers including Anand and Burnip.[6]

5.7.8   End-use applications of warp knitted fabrics

Specification for tricot machines is:

- Type of needle: compound or bearded
- Machine gauge: from 18 to 40 needles per inch (E18–E40)
- Machine width: from 213 to 533 cm (84–210 in.)
- Machine speed: from 2000 to 3500 courses per minute (HKS 2 tricot machine operates at 3500 cpm)
- Number of needle bars: one or two
- Number of guide bars: two to eight
- Products: lingerie, shirts, ladies' and men's outerwear, leisurewear, sportswear, swimwear, car seat covers, upholstery, technical fabrics, bed linen, towelling, lining, nets, footwear fabrics, medical textiles.

Specification for raschel machines is:

- Type of needle: latch or compound
- Machine gauge: from 12 to 32 needles per inch (E12–E32).
- Machine width: from 191 cm to 533 cm (75–210 in.)
- Machine speed: from 500 to 2000 courses per minute
- Number of needle bars: one or two
- Number of guide bars: 2–78
- Products: marquisettes; curtains; foundation garments; nets (including fishing, sports, and power); technical fabrics; curtain lace; tablecloths; bed covers, elastic bandages; cleaning cloths; upholstery, drape, and velvet fabrics; carpets; ladies' underwear; fruit and vegetable bags; geotextiles and medical textiles.

5.8   Recent advances in warp and weft knitting

The latest advances and developments in both warp and weft knitting equipment and fabric structures were exhibited at ITMA 2011 in Barcelona, Spain. The author has published a paper covering these developments, which have been described in detail in the following sections.[9]

5.8.1   Warp Knitting Equipment

5.8.1.1   Karl Mayer, Germany

Tricot machines
Karl Mayer exhibited two tricot machines, models HKS 2-3E and HKS 3-1.

HKS 2-3E
This is a two-guide bar tricot machine for producing elastic fabrics for lingerie and lightweight stretch fabrics for apparel applications. It is an E50, 130-in. wide machine, running at 3200 rpm and is capable of producing 63 kg per hour at 35 courses per cm in the fabric.

This two-guide bar, high-speed tricot machine has been available in a maximum gauge of E44, and this latest upgrade is now setting the standard in product design, especially when microfibre yarns are combined with extremely fine Elastane yarns.

HKS 3-1
This is an E28, 210-in. wide machine running at 3000 rpm and producing 65.4 kg per hour at 22 courses per cm in the fabric. The machine is ideally suited for producing elastic and rigid fabrics for sportswear, outerwear, household and automotive textiles, coating substrates among others.

Such extremely high speeds on a three-guide bar, 210-in. wide tricot machine are possible by developing and using carbon-fibre reinforced plastic (CRP) guide bars, which are much lighter and stronger as well as much more stable to fluctuations in temperature and humidity levels in the knitting room, as illustrated in Figure 5.32.

The maximum number of needles underlap possible on guide bars are 1, 2, and 3, are 8, 6, and 3, respectively.

Raschel machines
Karl Mayer exhibited raschel machine model RSJ 4/1, which has high production output and high flexibility, can produce a wide range of patterns, and has the capability of easy and fast pattern change.

The machine has a split jacquard front guide bar (JB1), three ground guide bars (GB2 to 4), knock-over comb, and stitch comb bars. In combination with the piezo jacquard system, it is possible to manufacture patterned corsetry, elastic and rigid lingerie, tulle structures, sportswear, and decoration fabrics. Electronic single-speed yarn let-off drives to individual guide bars, five-pattern disc drives,

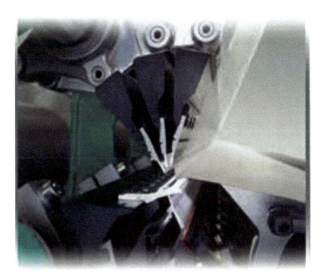

Figure 5.32 Carbon-fibre reinforced plastic (CRP) bars.

electronic fabric takeup and batching devices are features that make work more efficient and accurate.

The machine uses KAMCOS (Karl Mayer Command System), which consists of these features: motion control/single-speed for the control of basic functions; pattern control for the control of jacquard bar; 3.5-in. floppy disc drive for the reading and writing patterns on floppy discs; operator interface; 12.1-in. colour touch screen for production display; network interface (Ethernet) for data collection; teleservice and integrated yarn inspection laserstop. The model shown at ITMA was E28, 130-in. wide and operated at 1500 rpm. The maximum production of the machine is 13.6 running metres per hour with 66 courses per cm in the finished-state fabric.

JL40/1F jacquardtronic raschel machine

Karl Mayer showcased their latest development for the sector of functional lingerie with typical lace elements. The interplay of the stitch forming jacquard bar, 36 pattern guide bars and two ground bars (GB) for Elastane opens up entirely new perspectives for the designer, at the same time offering new applications for the textile market.

The machine exhibited at ITMA was an E28, 134 in. wide, which was operating at 800 rpm. The machine produces around 10.0 running metres of finished-state fabric per hour. The machine consists of one ground guide bar (GB1), one jacquard bar in split execution (JB2 and 3) stitch forming, 36 pattern guide bars (string bars 4–39), and two ground bars for Elastane (GB 40 and 41).

Electronic let-off for the jacquard and ground guide bars, electronic guide bar control (EL4 and EL3 types), electronic fabric takeup, and batching devices are the standard fitments on this extremely versatile and high-production machine.

TL 59/1/24: Multibar-jacquard raschel machine for the production of high-quality elastic and rigid lace and allover lace fabrics

Karl Mayer also exhibited textronic lace machine TL59/1/24. This machine was E28, 132 in. wide, running at 550 rpm. The machine is capable of producing 6.4 linear metres per hour of fabric containing 51.6 courses per cm in the finished fabric. This machine offers extensive patterning possibilities and increased productivity. This machine is designed to bridge the gap between the TL43/1/24 and the TL71/1/36. In conjunction with a shog distance of 180 mm, the machine offers additional patterning means as compared with the existing lace products, thus, contributing to the enhancement of these articles in the "Textronic" lace sector.

The user can select the yarn feeding either from the creel or the pattern beams in combination with the newly developed positive pattern beam drive (PPD). Pattern guide bars are driven by servomotors with high operational reliability and low maintenance and are setting the trend for the drive concept employed. There are 24 pattern guide bars (string bars 1–24) in front of the fall plate, 2 pattern guide bars for picot edge (string bars 27 and 28), one ground guide bar (GB 32), one jacquard bar in split execution (JB33/34), 30 pattern guide bars (string bars 35–64), and one ground guide bar (GB 66 for Elastane).

The machine is equipped with all standard electronic and most up-to-date control devices starting from yarn let off, pattern control, KAMCOS, electronic fabric takeup, and batching devices.

DJ6/2EL: Double needle-bar raschel machine for the production of jacquard seamless fabrics

A DJ6/2EL machine with integrated PPD was exhibited at the ITMA show. The machine produced three-dimensional shapewear panties in a gauge of E28 and incorporated additional Elastane yarn into specific zones.

The machine can produce cylindrical goods up to almost complete products with less need of a sewing process. Considerably improved pattern effect provides the production of a variety of products ranging from innerwear such as seamless foundation wear and pantyhose to outerwear. A variety of mesh structures and pattern effects by a piezo-jacquard system offer a comfortable feeling to the wearer and a fascinating and exciting appearance. The machine consists of four ground guide bars, two piezo-jacquard bars in split execution, two individual latch needle bars, with a trick plate distance of 0.65 mm, two knock-over comb bars, and two stitch comb bars. Figure 5.33 shows some examples of garments produced on this machine.

Malitronic multiaxial stitch bonding machine

Karl Mayer exhibited the latest innovation from its multiaxial range of machines for the technical textiles market. These machines are used for rotor blades in wind power, sporting goods, boat and ship building, and automotive industries. The wind power capacity in China has increased by 62% last year alone as compared to 2009. In Germany, the increase has averaged 18% every year during the last decade.

Figure 5.33 Some examples of garments produced on the DJ6/2EL machine.

Karl Mayer exhibited its Malitronic Multiaxial C & L (Cut and Lay), a highly efficient machine for producing carbon-based multiaxial non-crimp fabrics (NCF). The
key feature of this new system is an external fibre spreading line (FSL), which actively
pulls the carbon tapes from the bobbins on the creel, homogenises the several rovings
with respect to width and grammage, merges them as defined together and winds the
originated tape onto a double-flanged bobbin. This yarn package, which is similar to a
sectional warp beam, is then fed to the Malitronic Multiaxial C & L system.

Among numerous innovative features on these machines is the soft stitch mode.
During the vertical needle stroke, this special feature makes it possible for the compound needle to move horizontally in the direction of fabric production. This fact
results in a reduced load and thus less wear of the needle elements. In addition, this
feature can also have a favourable influence on the fabric quality during knitting.

5.8.1.2   Liba, Germany

Only two machines were showcased at ITMA by Liba, Germany. These were a two-in-one high performance tricot machine and a tricot machine with multiaxial weft insertion.

Copcentra 2 M twin

In this concept of two machines in one, one main drive installed in one machine bed drives two individual knitting systems and two sets of let-off and takeup systems. One Copcentra Twin simultaneously produces two separate fabrics. An extremely high productivity combined with low energy costs and space requirement is achieved. These machines are used to produce plain, rigid tulle; raised and elastic fabrics for underwear, outerwear, sportswear, and swimwear; and a wide range of technical textiles fabrics. It is claimed that one of these machines is capable of saving 45% floor space, 49% energy consumption, and 20% initial investment costs as compared to two conventional Copcentra 2 M machines. The machine also uses the CNC central machine control panel, which incorporates a 10.4-in. touch screen with the computer control of all the major functions of the machine.

Copcentra Max 4 CNC

This tricot machine with multiaxial weft insertion can be used for many different applications, especially when high strength and stiffness combined with low weight are required. Among many important applications, the following industries use this application: aircraft, automotive, ship and boat building, sports equipment, and rotor blades for wind power generation. Figure 5.34 illustrates some products incorporating multiaxial stitch bonded fabrics.

Figure 5.34 Multiaxial stitch bonded fabrics used in various applications.

The machine is offered in gauges ranging from E5 to E14 and in widths of 50, 101 130, and 152 in. The knitting elements consist of two guide bars, one needle bar, one tongue bar, one sinker bar, and one pillar thread bar. There are three weft insertion systems with digital servo drives, which can be independently programmed between −20° to +20° layer angles. Subsequent extension of up to seven weft insertion systems is possible.

Similar to the Karl Mayer multiaxial machine, Liba has also introduced a walking needle concept, in which, in addition to the standard vertical knitting motion, the compound needle is also operated in the horizontal direction – in direction of the fabric takeup motion. This "walking" needle motion ensures reduced wear on the needle system, improved fabric quality, and higher production speed.

5.8.1.3 Texma, Spain

Texma, Spain, exhibited the double-needle bar raschel machine, model BRDF 6/30, which can be used to produce spacer fabrics, and cut pile blankets. For the latter product, the company supplies a warping machine, double-needle bar raschel machine, and a cutting machine to produce two cut pile fabrics. These machines are offered in widths from 75 to 150 in. in gauges E14, E16, and E18. The machine has six guide bars and is capable of producing from 6 to 30 mm thick spacer fabrics at a maximum speed of 800 rpm.

5.8.1.4 Jakob Muller, Switzerland

At this ITMA, the Jakob Muller Group exhibited the following two warp knitting/ crochet machines, in addition to a wide range of narrow weaving machines:

Raschelina RD3.8
This machine is used for the production of women's and men's underwear, bandages, and many other technical textiles articles. Numerous technological advantages of these high-precision and high-tech machines include: (a) patented needles that offer high running performance and can be employed for fine yarns, exclusively for closed loops; (b) staple-fibre yarns and twistless filaments that can be employed in the weft without any problem; and (c) a chain drum that controls up to eight weft bars mechanically. The profile-milled, steel chain links ensure smooth running and precise laying even at high speeds. The model at ITMA was an E15, 25-in. wide machine with a maximum of eight weft bars. The machine was producing women's underwear tape in polymide filament, polyester filament, Lycra, and rubber covered polyester filament yarns at a speed of 1200 rpm and producing tapes at 18×27.7 m per hour.

MDC 3/830 E with electronic controls
The MDC 3/830 E fully electronic machine is capable of producing elastic and non-elastic medical bandages, abdominal bandages, kidney belts, and narrow fabrics for clothing and household textiles. The machine at the show was producing a heating mat in copper wire and polyamide and polyester filament yarns. Figure 5.35 shows a wide range of products produced on this machine.

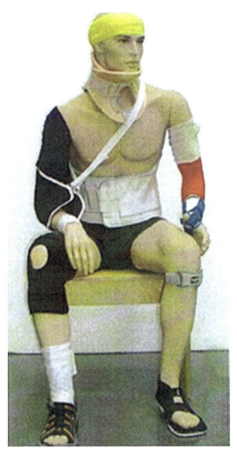

Figure 5.35 Various products manufactured on the MDC 3/830 E.

The important features of this machine are that the weft guide bars are electronically controlled, which makes a major contribution to pattern design diversity. The weft guide bars are driven by three differing systems:

1. Two long wefts with 170, 340, or 450 mm variations by using a servomotor.
2. One rubber guide bar with a mechanical cam drive.
3. One weft section guide bar with a maximum stroke of 25 mm by using a linear motor.

The entry of pattern design and other production data are effected by use of the MUDATA C200 control module, which is fitted with a touch screen.

5.8.1.5 Harry Lucas, Germany

Among a number of highly specialised circular weft knitting machines, Harry Lucas exhibited its circular warp knitting machine, model VEPA-E02. This machine can be used for the production of tubular netting for medical net bandages and packaging.

It can process most types of yarns, including cotton, polyamide, polyester, and, of course, elastomeric yarns. The machine is offered in cylinder diameters ranging from 2 to 7 in. and in a wide range of gauges. The machine shown at ITMA was equipped with four full electronic yarn-laying devices, and its needle strokes per minute ranged from 400 to 1200, depending on the material used on the machine.

5.8.1.6 Comez, Italy

Comez exhibited a number of electronic single- and double-needle bar warp knitting machines in addition to crochet equipment.

SNB/EL-800

This machine has been developed for the production of outerwear (scarves and stoles), medical products (dressings, bandages, and surgical products), and technical fabrics (netting for sports equipment and automotive industry). It is a single needle bar machine E10, 31.5 in. wide with latch needles and seven electronically controlled pattern guide bars. The electronic drive of the thread feeders and the fabric takedown motion allows changing the course density of a single product and the let-off rates of the weft, warp, and elastomeric yarns independently. A new data control controller manages all the necessary machine functions, monitors production data, and produces lengthy pattern repeats.

DNB-32

This Comez machine is an electronic double-needle bar warp knitting machine for the production of flat, spacer, and tubular fabrics for an extremely wide spectrum of products and applications. It is an E18, 32 in. wide machine in which the distance between the two needle bars can be adjusted from 1 to 18 mm. The machine incorporates all features described in the previous machine and uses a laser stop device for the detection of broken threads and photoelectric barriers for prompt machine stoppage in case the operator accesses the machine working area.

Comez also offer two other models of electronic double-needle bar warp knitting machines, the DNB/EL-800-8B, and the DNB/EL-1270. The former model is offered in gauges from E5 to E20, it is equipped with four, six, or eight guide bars with electronic control, and the distance between the needle bars can be altered up to 9 mm maximum. The latter model is offered in E5 to E20 and has 12 pattern guide bars with electronic control. The distance between the needle bars can be altered up to a maximum of 16 mm.

5.8.1.7 Rius, Spain

Rius also demonstrated their high efficiency electronic double-needle bar warp knitting machine, model MINI-TRONIC 1200, for the production of flat knitted and tubular fabrics, technical meshes, knot-free nets, and three-dimensional spacer structures. The application areas of these fabrics range from medical, food processing, automotive, aeronautical, and building and construction industries. The machine is E12, 50 in. wide with eight electronically controlled pattern bars.

The machine was knitting a three-dimensional spacer structure in 555 dtex polyester yarn on both faces and a monofilament polyester yarn of 0.2 mm diameter spacer yarn.

5.8.1.8   Chinese warp knitting machinery companies

It was interesting to observe that at least two Chinese companies exhibited their products during ITMA 2011.

They were Changzhou Runyuan Warp Knitting Machinery Co. Ltd and Changzhou Longlongsheng Warp Knitting Machinery Co. Ltd, both based in Changzhou City, Jiangsu Province, China. Both companies manufacture the full range of raschel machines, and the former company also manufactures stitch bonding equipment, weft insertion machines, multiaxial machines, lace machines, double-needle bar machines, and the full range of jacquard raschel machines.

The former company also exhibited its RS4 raschel machine during ITMA. The machine was E28, 170 in. wide with four guide bars and operated at 1950 rpm. The machine uses four EBA electronic let-off motions and new hollow-material knitting elements, together with precise compound needles and electronic draw-off and take-down mechanisms.

The other company, Changzhou Longlongsheng supplied only a DVD of its equipment at its stand.

5.8.2   Weft knitting equipment

The overall impression of the weft knitting machinery was that manufacturers at ITMA 2011 concentrated on three major aspects: high quality, high technology, and high efficiency.

Circular knitting machines
5.8.2.1   Mayer & Cie, Germany

Relanit 4.0 U H S (ultrahigh speed)
Mayer & Cie exhibited its famous single-jersey ultrahigh-speed Relanit 4.0 machine. The relative movement technology with its lower number of knitting elements in the knitting zone takes optimum care of the yarn. The E24 machine has a speed factor (rpm × diameter) or SF of 2100, which is unprecedented in practical application. The machine is 30 in. diameter with 120 feeders and operates at 70 rpm. It produced plain single-jersey fabric in 20 tex OE cotton yarn at 57.5 kg per hour with 18 courses per cm at 85% efficiency.

S4–3.2 R 11
This single-jersey machine is another example of high technology; it has six colour stripers with 3.2 feeders per inch, which also sets new standards in striped structures. The machine is E28, diameter of 30 in. with 96 feeders and operating at 18 rpm. It can knit 14.6 tex, 50% cotton, and 5% model yarn plated with 22/1 dtex Lycra at a production of 12.7 kg per hour with 18 courses per cm at 85% efficiency. The arrangement of the six-striper unit at every feed is illustrated in Figure 5.36.

Figure 5.36 Six-colour striper units on the S4-3.2 R11.

MJ 3.2 E DNS

This model is the first ultrafine gauge (E60) fully electronic jacquard machine in the world. It uses the standard three-way selection technique in which the needle selection is effected by a Mayer mono magnet and is equipped with automatic temperature control. The diameter of the machine on display was 30 in. with 96 feeders, E60, and the machine was running at 18 rpm. It produced a full-jacquard single-jersey fabric in polyamide 6, polyamide 66, and Lycra yarns at a production rate of 1.5 kg per hour with 32 courses per cm in the fabric at 85% efficiency.

OVJA 1.6 EE

The OVJA 1.6 EE machine on show had a 42-in. diameter, 68 feeders, E18, and a running speed of 14 rpm. The machine produced a full-jacquard mattress fabric in viscose and three different types of polyester yarns. The electronic single needle selection in cylinder and dial cams gives unlimited creatively in designing and finishing fabric properties. This machine knits rib jacquard in all possible constructions and patterns because the dial needles also knit the pattern. There is a single-needle track in the cylinder and dial with two-way technique for miss-knit, miss-tuck, and tuck-knit selection in both cylinder and dial needles. A wide range of optional attachments and equipment is available, including the computer control and transmission of patterns, laying-in feeders, open-width fabric batching, and so on.

The spinit systems machine

This highly innovative machine, the first of its kind in the world, was kept undercover and was shown to only a limited number of major knitting companies that were already principal customers of Mayer & Cie company. The roving packages used on the creel and the double-roving feed are shown in Figure 5.37.

This system uses roving packages on the creel of the knitting machine, and the yarn spinning and knitting take place in a single process. There is online quality control of the mass variation and other defects in the roving because it is unwound from the packages, has permanent roving count control, and above all, automatically eliminates the roving and yarn faults without knitting machine stoppage. A three-over-three roller drafting unit with double end feeding suitable for short staple spinning is used at each feeder. The spinning nozzle twists the drafted fibre assembly, and the level of false twist inserted in the yarn is adjustable depending on the input raw material. During the process between the roving and yarn, all fibre waste, foreign matter, dust, and so on are automatically removed from the knitting area (Figure 5.38).

The major innovation of this technology is the process integration, which has enormous saving potential regarding requirements for (a) floor space; (b) energy; (c) personnel; (d) quality control; (e) raw material; and (f) investment in machinery and ancillary equipment. The knitting machine itself was a Relanit E, with a 30-in. diameter, E28, and 90 feeders and operates at 18 rpm. The actual production rate of the machine was 14 kg of fabric per hour. The single process from roving to the knitted fabric is illustrated in Figure 5.38.

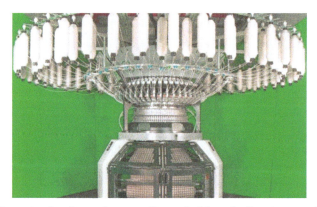

Figure 5.37 Layout and roving packages on the creel of the spinnit systems machine.

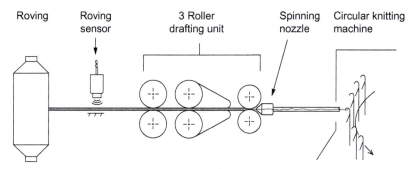

Figure 5.38 Single process from roving to the knitted fabric on the spinnit systems machine.

5.8.2.2   Monarch Company, England

Monarch Knitting Machinery Company exhibited four of its circular machines, two single-jersey and two double-jersey.

OD5-MXC-E3.2RE

This is a high-production, energy-efficient single-jersey machine. It has a 30-in. diameter, E28, with 96 feeders. The machine uses a closed cam system (three race monolithic cams), two race jersey stitch draw amount: 2.0mm. The machine has a central stitch length control mechanism and can be used for knitting either two-race or three-position structures as optional attachment.

OD5-V-SEC7CS

This is superfine gauge single-jersey fully electronic jacquard machine. Its diameter has a 34 in. with E50, and 84 feeders. The machine is equipped with a newly designed two-race SS actuator system. This high-speed individual needle selection system achieves reliable and stable needle selection and is now capable of one-by-one needle selection even at a superfine 50 gauge. The machine is capable of knitting mesh structures that imitate raschel lace structures by using a two-race SS actuator system. The machine is illustrated in Figure 5.39.

V-LEC 6DSIB AND V-LEC 8 BSC

Monarch exhibited two double-jersey electronic jacquard machines with individual needle selection on both cylinder and dial (two-position selection at all feeders), specifically for knitting full-jacquard mattress ticking fabrics. The first machine has a 42-in. diameter, E20, and 80 feeds; the second machine has a 38-in. diameter, E20 with 108 feeders.

Figure 5.39 Monarch model OD5-V-SEC7CS.

5.8.2.3   Beck GmbH, Germany

The Beck group offers an impressive and comprehensive range of both single- and double-jersey large cylinder circular machines. They range from up to four-cam track single-jersey machines for producing basic knitted structures to fully electronic jacquard single- and double-jersey machines with three-way selection of individual needles. The company also offers terry, striper, eight-lock, and many other types of specialist machines.

At ITMA, Beck showcased its BSJEM Select electronic machine. It has a 30-in. diameter, E44 with 60 feeders, single-jersey circular knitting machine with individual needle electronic selection in the three ways (i.e. knit, tuck, and miss) at each feeder. The company also offers the full range of standard and optional equipment to satisfy different customer requirements and solve their problems.

5.8.2.4   Santoni S.p.A, Italy

Atlas and Atlas HS

Among the full complement of the weft knitting equipment exhibited, Santoni in Italy presented its finest gauge machines, which were the talking point of the exhibition. Santoni has produced a 30-in. diameter, single-jersey circular machines with up to 90 needles per inch (E90). The company actually showed an Atlas machine with a 30-in. diameter, E80, and 88 feeders, which was knitting 22/20/1 dtex polyamide yarn and plating yarn of 15 dtex Lycra. These ultrafine gauge machines, up to E90, use a holding-down jack (patented by Santoni), not sinkers, during the loop formation to hold down the fabric when the needle rises to form the loop. The jack is withdrawn when the needle starts to descend to the knock-over position. The sinker loops are thus made with the help of the cylinder tricks (not over the sinkers's bellies). This results in much smaller loops and a tighter fabric. Each holding-down jack has two projections that hold down the fabric. The number of holding-down jacks is 50% of the number of needles, for example, a 30-in. diameter, E90 machine has 8472 needles but requires 4236 holding-down jacks.

Other important features of using holding-down jacks are that they allow an easy restart of the knitting after a fabric press-off. The fabric spirality, which is a feature of multifeeder circular machines, is also much less in these fabrics because the increased tightness of the fabric reduces the tendency of the yarn to untwist and consequently causes less spirality in the finished fabric. Figure 5.40 is a close-up view of Santoni's Atlas machine with the holding-down jacks and the needles.

5.8.2.5   Pilotelli, Italy

This company displayed four large cylinder circular knitting machines: NJ-2.4 TA3, a four-cam track, single-jersey circular, 34-in. diameter, E44 with 82 feeders that is capable of operating at 25 rpm. The machine is capable of producing all four-track structures.

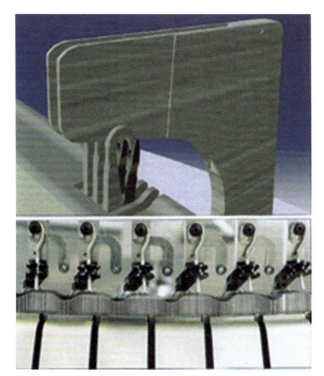

Figure 5.40 Close-up view of the Atlas machine.

JVCE-4 T.A.3
This is also a single-jersey machine equipped with four-cam tracks at each feed; it has an open-width batching device to produce an open-width fabric. The machine on show had a 32-in. diameter with a 96 feeder, and a speed of 28 rpm.

MJE-2.4
This is a single-jersey electronic jacquard circular machine with a 30-in. diameter, E28, and 72 feeders producing a full jacquard pattern in 75 dtex/36-filament polyester yarn and polyamide 6.6, 44 dtex/13-filament yarn.

DJ-3.2
This machine is a double-jersey eight-lock, four tracks, 34-in. diameter, E28, with 108 feeders, producing a fabric in polyester 75/36 and 22 dtex Lycra yarns.

5.8.2.6   Orizio, Italy

JH/V
This is a single-cam track high feeder density machine, with 3.2 feeders per inch, single-jersey, with a needle speed of 2 m/s, used for the production of plain single-jersey rigid and elastic fabrics.

JB2E

This is a single-jersey, two-colour striper machine, 30-in. diameter, E28 with 96 feeders, which can produce striped structures in up to four-cam tracks (up to four wale-wide structures).

MJM/BF2

This is a single-jersey fully electronic jacquard with four-colour stripers. The machine at ITMA had a 30-in. diameter, E32 with 48 feeders and was producing a full jacquard striped fabric for outerwear.

JHFP-LC

This is a sinkerless four-cam track basic single-jersey machine for producing structures with elastomeric yarn. It has a 30-in. diameter, E36, and 90 feeders.

C1/C

It is an eight-lock machine for producing interlock and rib structures. The machine has a 30-in. diameter, E40, and 96 feeders. The machine has four cam tracks in the cylinder and two-cam tracks in the dial at each feeder.

5.8.2.7 Terrot GmbH, Germany

The famous manufacturer of circular knitting machinery exhibited the following six different single- and double-jersey circular knitting machines.

13P284-1

This is an ultrafine eight-lock machine for fine rib and interlock and modified knitted structures. The machine on show was a 30-in. diameter, E40, with 84-feeder machine running at 20 rpm. There are two-cam tracks in both the cylinder and the dial at each feeder.

SCC 6F 548

This is an electronic single-jersey jacquard machine with six stripers at each feeder. It is also a fine-gauge, E36, jacquard machine with three-way technique at each feed. This machine is ideally used for producing outerwear and sportswear fabrics.

S296-2 open width

This is another ultrafine, high-speed, single-jersey machine equipped with an open-width batching frame. It has a 30-in. diameter, E50, and 96 feeders and operates at 27 rpm. The machine has four-cam tracks at each feeder and is used to produce both apparel and technical textiles products.

UCC572T & UCC572

This machine model has a 34-in. diameter, E16, and 84 feeders and operates at 16 rpm. It is an electronic jacquard double-jersey machine with stitch transfer facilities to create holes and openwork in the fabric. The machine has three-way technology in cylinder needle selection that allows the needle to be in the knit-tuck-miss position. The machine can be used to produce jacquard stitch transfer patterns for sports and leisure-wear, underwear, and technical textiles.

The second model is an electronic double-jersey jacquard machine with a three-way technique that uses the latest piezo technology with high-speed actuators for a perfect miss-tuck-knit position. The specific machine exhibited at ITMA was a 30-in. diameter, E40 with 72 feeders and operating at 16 rpm. The machine is ideally suited for the production of sportswear, swimwear, and automotive seating fabric.

UP592M

Terrot showcased its latest minijacquard model with a combination of a high number of feeders and high machine speed that allows a never-before achieved production output for household textiles and mattress ticking. The dial cam system has two needle tracks for knit-, tuck-, and miss-stitches and support. The cylinder needle selection is a two-way technology for achieving knit and miss selection on individual needles. The machine on show was a 38-in. diameter, E20 with 116 feeders and a machine speed of 21 rpm; it is capable of producing 43.3 kg of fabric per hour.

Flat-Knitting Machines

5.8.2.8 Shima Seiki, Japan

This famous company celebrated its 50th anniversary in business in 2012. It displayed 13 models, 11 of which were new.

MACH2X 153 18L and MACH2X 173 15L

These models are the latest addition to the company's whole garment machines that feature four needle beds and its original slide needle. These are high productivity machines that achieve a maximum knitting speed of 1.6 m/s. The following special features make the MACH2X the ideal machine for high-quality, ultrafine gauge whole garment production:

a) These machines use the R2 carriage (rapid response carriage system). With a compact, lightweight carriage, less space is required for the carriage to make turns, allowing more area for the carriage to run at full speed.
b) The split-stitch technique allows efficient knitting by eliminating empty courses.
c) The 18 L gauge capability that uses a special large-hook slide needle and can knit 15-gauge fabrics at 18-gauge pitch to produce tighter fabrics, especially for ribs.
d) The i-digital stitch control system (DSCS) and the dynamic tension control (DTC) system positively control yarn feed through variable electronic controls of yarn tension.
e) The machine has a new computer-controlled takedown system consisting of front and back panels with pins for controlling takedown tension of front and back during whole garment knitting.

First 154 S21

Shima Seiki introduced a 21-gauge shaping machine, a fine V-bed shaping machine that is able to knit rib structures with a 21-gauge appearance and single-jersey fabrics in the 18–26-gauge range.

SSR112–SV

This machine was designed to mass-produce quick, basic knitwear to support the growing demand for knitwear manufacturers in emerging markets where mills operating hand-flat or mechanical machines are looking to upgrade their machines and replace them with higher productivity computerised systems. With a 45-in. knitting width and an ultracompact double-system carriage, this machine is compact, allowing up to 15% more machines to be installed within a given factory floor space. This machine is also said to be up to 25% more energy efficient as compared to the previous machines.

The machine uses a spring-type full sinker system and "v-sinker" on machines with SSR-112 14 and 16 gauges. In addition, the machine uses stitch press technology for effective hold down while knitting rib structures when sinkers are not effective. The v-sinkers used on SSR-14G and 16G machines are shown in action in Figure 5.41.

MACH 2 SIG 123–SC 18G

Shima Seiki exhibited this intarsia machine with the maximum knitting speed of 1.4 m/s. Intarsia knitting is dramatically improved by the use of 40 intarsia carriers. The machine also incorporates all the latest technological achievements that have been described for other models.

The company also demonstrated a computerised coarse-gauge machine, model SCG 122 SN3G, an E3 machine. It also exhibited its most sophisticated and latest apparel design workstation, SDS-ONE APEX3, and its latest version SIP inkjet printing system.

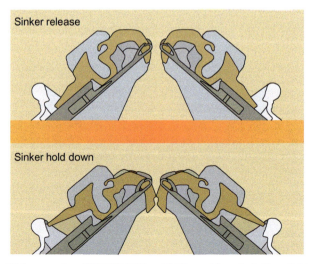

Figure 5.41 v-Sinkers used on SSR112 14G and 16G machines that allow tighter stitches on fine-gauge machines.

5.8.2.9   Stoll, Germany

Stoll's main theme was the evolution of its equipment ranging from the IBO hand flat machine, AJUM mechanical machine, to the state-of-the-art fully computerised machines. Stoll also presented its vintage collection covering 138 years of knitting history for autumn/winter 2011/2012: a combination of craftsman tradition and the most modern industrial possibilities.

Stoll presented its new 4- and 3-L needles at ITMA. They have a bigger needle hook to accommodate higher yarn volume that is said to improve yarn grasping and consequently knitting reliability. The holding-down jack has been reworked, and its control has been adapted to the larger needle hook.

The 3-L needle can be used on the CMS 520C machine, and the 4-L needle is used on the CMS 530 HP, CMS 822 HP, and CMS 502 HP models. Stoll also exhibited its new CMS 502 HP and CMS 502 HP multigauge models. These machines are equipped with more compact knitting systems and shorter carriage reversal times and are said to result in a productivity increase of 10%. The company also demonstrated its new version of M1 plus design and pattern software.

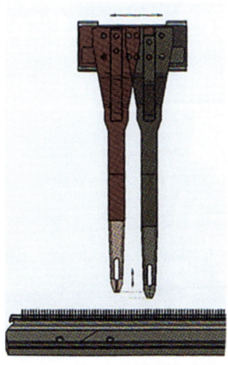

Figure 5.42 Steiger's new programmable pop-up guide said to increase reliability of intarsia knitting.

5.8.2.10   Steiger, Switzerland

The latest Aries flat knitting machine of the Swiss Steiger company is fitted with 32 individually programmable, motorised yarn carriers, which offer increased productivity of up to 40% and almost unlimited knitting possibilities. The dynamic sinker system is suitable for the production of very high-density knitted fabrics with the holding-down sinkers effectively "diving" into the opposite needle bed, much deeper than other systems, resulting in a much wider variety of designs to be knitted. Steiger has also redesigned its carriage system, which is now shorter and has a faster invertion time, resulting in up to 20% higher productivity.

The productivity of the existing Aries 3 or Aries 6 machines can be improved with Steiger's new FAST software.

Aries intarsia machines are now fitted the new pop-up yarn guide, which is simply raised out of the way of the needles. This also knits the inlay patterns efficiently (Figure 5.42).

Acknowledgements

Figures 5.32–5.42 used in this chapter have been taken from the author's paper presentation given at the International Textile Machinery Exhibition (ITMA), Barcelona, Spain, Review Conference, organised by The Textile Institute, Manchester, UK, 1 December 2011. The author has received official permission from The Textile Institute for using these figures in this chapter. The author is grateful to The Textile Institute, Manchester, UK, for their assistance and cooperation in this matter.

References

[1]   Anand SC. ITMA'95 review of circular knitting machines and accessories. Asian Text J, February 1996;49.
[2]   Anand SC. Contributions of knitting to current and future developments in technical textiles. Inaugural Conference of Technical Textiles Group, The Textile Institute, Manchester, 23–24 May 1988.
[3]   Anand SC. Knitting's contribution to developments in medical textiles. Text Technol Int, 1994;219.
[4]   Anand SC. Fine array of warp knitting machines from Karl Mayer. Asian Text J, April 1996;51.
[5]   Anand SC. Warp knitted structures in composites. In: ECCM-7 Proceedings, vol. 2. Woodhead Publishing, Cambridge, UK. 14–16 May 1996. p. 407.
[6]   Anand SC, Burnip MS. Warp–knit construction. Text Asia, September 1981;65.
[7]   Anand SC. Speciality knitting equipment at ITMA'99. Asian Text J, December 1999;49.
[8]   Anand SC. Karl Mayer warp knitting equipment at ITMA'99. Asian Text J, September 1999;49.
[9]   Anand SC. International Textile Machinery Exhibition (ITMA), Barcelona, Spain, Review Conference, The Textile Institute, Manchester, 1 December 2011.

Technical fabric structures – 3. Nonwoven fabrics

S.J. Russell, P.A. Smith
Nonwovens Research Group, School of Design, University of Leeds, Leeds, United Kingdom

Chapter Outline

6.1   Introduction

Defining what a nonwoven is, rather than what it is not, has been a perennial challenge, and numerous definitions have been suggested over the years. A standard definition appears in International Standardization Organization (ISO) 9092 and CEN EN 29092, and more recently an updated version was proposed by EDANA and INDA as follows:

> *A nonwoven is a sheet of fibres, continuous filaments, or chopped yarns of any nature or origin, that have been formed into a web by any means, and bonded together by any means, with the exception of weaving or knitting. Felts obtained by wet milling are not nonwovens. Wetlaid webs are nonwovens provided they contain a minimum of 50% of man-made fibres or other fibres of non vegetable origin with a length to diameter ratio equal or superior to 300, or a minimum of 30% of man-made fibres with a length to diameter ratio equal or superior to 600, and a maximum apparent*

Handbook of Technical Textiles. http://dx.doi.org/10.1016/B978-1-78242-458-1.00006-6

density of 0.40 g/cm³. Composite structures are considered nonwovens provided their mass is constituted of at least 50% of nonwoven as per the above definitions, or if the nonwoven component plays a prevalent role.

Following the emergence of the nonwovens industry in the early 1930s, global trade has grown substantially, driven by the development of new markets and increased production in the European Union, United States, and China. In the last 10 years alone, European production of nonwoven fabric in square metres has almost doubled.

The manufacture of nonwoven fabrics is only part of the overall value chain. Today's industry depends on a diversity of raw materials in various forms, for example, polymer chip, staple fibres, wood pulp, superabsorbent particles, adhesives, lubricants, and many other different auxiliaries provided by a global supplier base. Nonwoven manufacturers purchase these raw materials and produce fabrics that are normally sold in the form of "roll goods". These rolls are then "converted" into final products by various means involving processes such as slitting, cutting, laminating, coating, fluid impregnation, printing, folding, and welding amongst others before being packaged and distributed to the retailer or end user. It is quite common for roll good manufacturing and converting to be carried out by separate companies in the supply chain.

One of the major advantages of nonwoven fabric manufacture is that it is most commonly carried out in a continuously linked process in which raw materials are first made into a web and then into finished fabric, although there are some exceptions to this. This naturally means that labour costs in manufacture are low compared to traditional textile processing because material handling requirements are minimised, and there is a high degree of process automation. In spite of this mass-production approach, the nonwovens industry produces an enormous variety of different fabrics, and therefore, it is nearly always possible to find an exception to the norm. Nonwoven fabrics must address performance requirements in a multitude of different applications, governed by different price points, supply chain structures, and regulatory frameworks. Products are as diverse as a $45 \, \mathrm{g \, m^{-2}}$ water dispersible, single-use wet wipe to a highly durable $1000 \, \mathrm{g \, m^{-2}}$ geomembrane protector capable of resisting point loads when buried beneath $20 \, \mathrm{m}$ of landfill waste for many tens of years. Nonwoven fabrics therefore embrace the gamut of industrial, health, and consumer products that extend from single-use disposable items such as teabags and wet wipes that might be used for just a few minutes to highly durable products such as synthetic leather, automotive components, and roofing membranes designed to last for many years.

The versatility of nonwoven fabrics compared to traditional textile structures reflects distinctive differences in structural architecture. Unlike woven and knitted fabrics, nonwovens consist of fibres rather than yarns and have less periodic, more complex three-dimensional fibre arrangements. Nonwovens are anisotropic in terms of structure and properties and are highly porous and permeable. The relative orientation and spatial separation of fibres as well as bond point positions within the fabric are heavily influenced by the combination of processes as well as their settings used during manufacture. The fibre orientation distribution is particularly important because of its influence on the isotropy of fabric properties. Fibres are usually oriented in many different planar directions in nonwoven fabrics, but there is usually at least one preferential direction.

Consequently, when properties such as tensile strength are measured, it is often found that values are not the same in the machine direction (MD) and the cross direction (CD). This anisotropy is commonly expressed in terms of the MD:CD ratio and has a profound effect on the performance of fabrics during use. Briefly, if the fabric is highly anisotropic in terms of tensile strength so that it is much weaker in one direction than another, it could quickly fail if placed into a product application where multiaxial forces are applied, such as a filter medium. In this particular example, a reinforcing scrim might be incorporated within the fabric to compensate for its inherent anisotropy. Note that many other properties such as liquid spreading, permeability, conductivity, and tear strength are also affected by fibre orientation.

For simplicity, nonwoven fabric production can be conveniently divided into at least two stages: the preparation of a web consisting of a sheet of fibres or filaments, and the bonding of this web to make a coherent fabric structure. Subsequently, the fabric may be converted into a final product, which frequently involves combining it with other materials, which may or may not be in sheet form. There are a number of different ways of forming and bonding webs, each producing its own particular characteristics in the final fabric. The main web formation processes are traditionally referred to as *drylaid* (carding and airlaid), *wetlaid* and *spunmelt* (spunbond and meltblown), and the bonding processes as *mechanical* (hydroentangling, needlepunching, and stitchbonding), *thermal*, and *chemical*. However, it is becoming increasingly difficult for these broad classifications to adequately encompass the full variety of processes used by the industry.

In a nonwoven production line, webs may be produced by one or more separate web formation techniques and then bonded by one or more bonding processes so that the range of different possible manufacturing lines is enormous, allowing for a variety of final properties.

Therefore, the nonwoven fabric production process is essentially an integrated one in which web formation and bonding take place in two separate processes that are linked together. Consequently, it is difficult to describe the combined processes as a whole resulting from the wide number of different web forming and bonding combinations that are possible. Instead, it is convenient in an introductory text such as this to explain the methods of web formation and bonding separately.

6.2   Carded

Carding systems used to make nonwovens rely upon principles similar to those used in yarn production, but there are differences in fibre preparation as well as machine configurations. Unlike yarn manufacture, in nonwoven processing there are limited opportunities to improve the degree of fibre mixing and weight uniformity after carding before the final fabric is produced. It therefore follows that the opening and blending stages before carding must be more intensive in a nonwoven plant and the card should have sufficient opening and mixing power, for instance, by including at least one more cylinder although it must be admitted that many nonwoven manufacturers do not follow this maxim.

Although short-staple flat cards historically have been used to produce some lightweight nonwoven fabrics for products such as wipes, worker-stripper roller cards predominate. Unlike short-staple flat cards, worker-stripper roller cards can be many times wider, up to 5 m in some cases, making them much more productive for nonwoven manufacture.

Hence, a nonwoven carding installation usually consists of an automatic feed hopper linked directly to a wide width, worker-stripper card. Nonwoven cards normally have more than one doffer roller, two doffers is particularly common, and there is the facility to mechanically manipulate the fibre orientation of the fibres in the web by using condensing or scrambling rollers, as required. Typically, this is done to adjust the MD to CD ratio nearer to 1 because there is a tendency for fibres collected by the doffer to be preferentially oriented in the MD. In nonwoven production, the MD:CD ratio is generally important because it reflects the nature of fibre orientation in the fabric and influences the degree of anisotropy of a number of important fabric properties, not least directional stress-strain characteristics.

6.2.1   Parallel laid

In "straight-through" carding installations, the web leaving the card is bonded without lapping. Normally, the straight-through approach enables the highest linear production speeds in carding to be achieved, but the mass per unit area of the carded web that can be delivered is limited, so the process is commonly encountered in the manufacture of lightweight webs for hygiene products such as wipes. In cases when the mass per unit area of a carded web is too low or when the linear output needs to increase, the web may be layered, which also improves the uniformity. Often the only practical option for layering in this case results from combining webs from multiple cards arranged in sequence (Figure 6.1). For example, layering can be achieved by having two cards

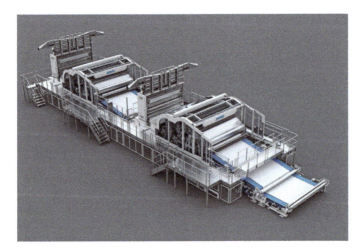

Figure 6.1 Parallel-laid carding installation, showing *from left to right*: hopper feed to first card and first web output; hopper feed to second card and combined single web output. Courtesy of Andritz.

in sequence raised slightly above the floor to allow a long conveyor to pass underneath each of them. The webs from each card are directed onto the conveyor to form a web of the required mass per unit area. If the cards are too long, this method becomes un- wieldy and instead the cards may be placed side-by-side, feeding a common conveyor. In this case, the card webs are turned through a right angle by a guide at 45°, but the composite web produced by this method is identical in all respects to that produced by the previous method. It is important to recognise that this does not produce a crosslaid web despite the similarities between the layouts.

In card webs, leaving the doffer there is a marked tendency for fibres to be pref- erentially oriented in the MD (i.e. along the MD) rather than across it (CD). Because in parallel laying, all the card webs are parallel to each other, it follows that a larger proportion of fibres lies along the web with fewer across it. Typically, this leads to MD:CD ratios much higher than 1.

The advent of cards designed especially for the nonwoven industry have enabled lower MD:CD ratios to be produced and to be adjusted more easily according to the isotropy needed in the final product. Such cards can be operated with a randomising doffer, which, as its name implies, alters the fibre orientation in the web leaving the card. Another approach involves scrambling rollers that condense the card web in length, hav- ing the effect of buckling and displacing fibres altering their orientation. By using these two techniques together, it is possible to bring the MD:CD strength ratio of parallel-laid fabric down from approximately 5:1–20:1 to 1.5:1. However, all parallel-lay processes suffer from a further fundamental problem; the width of the final fabric cannot be wider than the card web, and there are physical limits to how wide a card can actually be built before roller deflection, for example, prevents increases in width.

6.2.2 Crosslapping

When crosslaying, the card (or cards) are placed at right angles to the main conveyor and using a device called a *crosslapper*, the card web traverses backward and forward across the main conveyor, which itself is moving (Figure 6.2). The result is a zig-zag laydown of the web, which is built up to the required thickness based on the relative speeds of the laydown and the main conveyor. This process also results in a change in the MD and permits a large increase in the laydown width depending on the traverse across the main conveyor by the crosslapper.

Usually the main conveyor onto which the web is deposited moves only slowly so that many layers of card web are gradually built up into the form of a batt (a thick web). The thickness of the card web is very small in comparison with the completed batt, so that the zig-zag marks are usually not visible in the final product provided the number of web layers is sufficiently high. There are two major problems with crosslappers; one is that they tend to lay the batt more heavily at the edges than in the middle. This fault can be corrected by running the traversing mechanism rather slowly in the centre and more rapidly at the edges with a very rapid change of direction at the edge. Profiling crosslappers allow the operator to control the relative speeds of the conveyors to achieve the preferred cross-machine weight profile of the laid down batt. The other problem involves matching the input speed of the crosslapper with the card

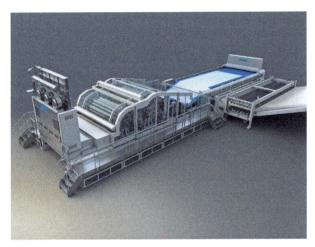

Figure 6.2 Carding and crosslapping installation, showing *from left to right*: hopper feed, worker stripper card, crosslapper, and batt drafter.
Courtesy of Andritz.

web speed. For various reasons, the input speed of the crosslapper is limited and the speed of the card web has to be reduced to match it. Because of economic reasons, the card often needs to run at maximum production; the card web at the lower speed is thicker and crosslaying marks tend to show more. In spite of these problems, crosslappers are used much more frequently than parallel lappers.

Because the main conveyor is continuously moving in the MD, there is a small crosslaying angle, normally less than 10°, so that the fibres in the batt are preferentially oriented in or near the CD. Crosslaid fabrics are consequently strong in the CD and weak in the MD. In many cases, this may not matter because crosslaid fabrics are often quite heavy so that there is adequate tensile strength in all directions after bonding, but in many other cases a more isotropic batt is required. The obvious solution is to combine parallel laying and crosslaying together; this is done very occasionally but it is uncommon because it combines the limitations of both systems, that is, the relatively narrow width of parallel laying and the slow output speed of a crosslapped batt. The common solution is to stretch (or draft) the batt in the MD as it exits from the crosslapper. Various systems are available for doing this; the important criterion is that batt drafting should be uniform and progressive; otherwise, thick and thin places are introduced into the batt. Carding and crosslapping are extremely common in the production of medium- to heavy-weight nonwoven fabrics used to make for example, filtration media, floorcoverings, automotive interiors and headliners, shoe linings, geosynthetics, washable continence management pads, and wound dressings.

6.2.3 Vertical lapping

The fibres in carded and crosslapped structures are mostly oriented in a planar configuration (i.e. fibres lie mostly in the *x*–*y* directions rather than in the *z*-direction through thickness). As a result, the resistance to compression and the compression-recovery properties of bonded fabrics made from such structures may be lower than required in comparison to other porous products such as polyurethane (PU) foam. By corrugating a carded web into a concertina-like structure so that the knuckles are oriented across the MD, a proportion of the fibres becomes preferentially oriented in the vertical or z-direction relative to the plane. After through-air thermal bonding, the corrugations are unable to fall out, and a three-dimensional fabric is produced whose compressive properties can be made to approximate to that of PU foam, at least during initial compression-recovery cycles. Of course, to bond such a structure by means of thermal bonding requires that a proportion of the fibres in the web be composed of thermoplastic fibres, and most commonly these consist of core-sheath bicomponents in which the sheath has a lower melting point than the core. Over the years, various systems have been developed to produce perpendicular-laid webs, probably the best known of which are Struto and V-lap. Such fabrics have found industrial applications in the automotive industry, for example, as lightweight, recyclable foam replacements made from polyester for interior components. They have also been used as part of futon mattresses.

6.3   Airlaid

Airlaying involves the continuous deposition of individual fibres from an airstream onto a moving conveyor to build up a web structure. The way in which this is done varies depending on machine design, and there are so many different variants, it is difficult to generalise. Some airlay processes rely on sifting short fibres through meshes, or perforated/slotted screens, whereas others rely on rotating clothed rollers. Typically, the former is best suited for short-cut fibres (approximately 4–15 mm) and wood pulp and the latter for longer fibres (approximately 15–60 mm), although in some recently introduced systems, there is an overlap. Generally in airlay systems, fibres are dispersed in the air to make a very dilute volumetric suspension. The dilution ratio controls the degree to which dispersed fibres are able to remain separated from each other and free from entanglement as they move around during the airlaying process. Following dispersion in air, the fibres are transported to a conveyor, usually by means of a directed airflow, following which they are deposited onto a moving, permeable conveyor to form a web whereas the air in which they were suspended is removed.

Because fibres suspended in air have the opportunity to move by rotation and translation, thereby changing their geometric position, it is often thought that the fibres are deposited without control and therefore are randomly oriented when the web is formed. This belief is so widespread that airlaid fabrics are frequently called "random-laid" fabrics. However, although airlaid fabrics can have very low MD:CD values, strength ratios are sometimes seen as high as 2.5:1, depending on how the deposition conditions are configured during processing.

The degree of fibre opening available in an airlay system varies from one manufacturer to another, but in all cases it is usually much lower than in a card. Consequently, fibre opening is required prior to airlaying, and the fibres used should be capable of being easily opened; otherwise, the final web contains agglomerations of inadequately opened fibre. In the past, the desire for really good fibre opening, which is needed to make lightweight, uniform airlaid webs, led to a process consisting of carding, crosslaying and then feeding the crosslaid batt to an airlaying machine. The only purpose of the airlaying machine in this particular example is to obtain the desired MD:CD strength ratio, but it is a very expensive solution compared with other approaches. More recently, manufacturers have developed carding systems in which the final doffer section is replaced with an airlayer enabling a lightweight, isotropic web to be produced with a high degree of fibre individualisation.

In traditional roller-based airlay machines, opened fibre from the opening and blending section is fed into the back of a hopper, which delivers a uniform sheet of fibres to the feed rollers. A high-speed toothed or pinned roller then takes the fibre between optional worker and stripper rollers that may be included to increase the fibre opening power. An air stream directed over the surface of the high-speed roller is then used to assist in dislodging fibres from the teeth on the roller surface and to transport them onto an air-permeable conveyor where the web is formed. The size of this formation zone and the airflow dynamics influence the way in which fibres are deposited and their configuration in the web. As the web builds up, incoming fibres are deposited on previously landed fibres and may fall onto an inclined plane. The angle between this plane and the plane of the web depends on the width of the formation zone and the thickness of the batt. It is possible when making thick fabrics to reduce the width so that the fibres lie at a substantial angle to the plane of the batt. This increases the thickness-to-weight ratio of the batt and affects its compression and recovery properties.

In preparation for airlaying using sifting processes and those using slotted or perforated drums, short-cut man-made fibres are first mechanically opened, whereas compressed fluff pulp is defibrated in hammer mills to mechanically separate the fibres into a loose state. To make absorbent single-use hygiene products such as wipes, blends of defibrated pulp, and man-made fibres are then fed into the feed hopper of an airlay system where the fibres and pulp are dispersed in air. Brushes or high-speed rotors are used to assist the passage of individual fibres through meshes or the walls of the slotted/perforated drums and towards the air-permeable conveyor where the web is formed. The size of the mesh or holes in the screen relative to the fibre dimensions influences passage efficiency. Short fibre airlay systems such as these also facilitate the introduction of nonfibrous materials such as superabsorbent particles.

The parallel-laid, crosslaid, and airlaid methods discussed so far are collectively known as *dry-laid processes*.

6.4   Wetlaid

The wetlaid process is derived from the papermaking process, which is capable of very high line speeds while producing a very uniform product in terms of weight per unit area. Amongst all lightweight nonwoven web formation processes, wetlaid fabrics produce some of the most uniform in terms of mass per unit area. Man-made fibres for wet laying are cut very short by textile standards (approximately 4–20 mm), but these are long compared with wood pulp, the usual raw material for paper. The general steps in wet laying are to form a dilute dispersion of fibres in water so that they remain well separated, followed by the deposition of the dispersion on a liquid permeable conveyor on which the web is formed. Fibres are dispersed into water with a dilution ratio sufficient to prevent the fibres agglomerating. The required dilution ratio is roughly 10 times that required for paper, which means that production is best suited to inclined-wire machines in locations where clean water is plentiful. When wood pulp is blended with short-cut fibre and used as the raw material, both the dilution ratio and cost of the raw material can be reduced. High wood pulp contents are common in the manufacture of wipe substrates, including water dispersible wipes and other absorbent hygiene products. Specialty wetlaid webs are produced from a wide variety of thermoplastic and high-performance polymers such as PET as well as aramids, carbon, glass, and metal-coated fibres together with organic binders to stabilise the structure following web formation. Strength modifiers and other additives may also be introduced into the water in an effort to produce satisfactory fabric properties. Squeezing and drying of webs containing large volumes of cellulose can stabilise the structure as a result of hydrogen bonding, but normally for most wetlaid nonwovens, mechanical, thermal, or chemical bonding is required to produce the finished fabric.

6.5 Spunbond

Spunlaid web production involves extrusion of continuous filaments from polymer raw material, drawing the filaments, and depositing them as a web. Web formation and bonding are continuously linked so that spunlaid production provides the shortest route from polymer chip to finished fabric in a single integrated process. Spunlaid web formation starts with extrusion. Although there are some exceptions, thermoplastic polymers and melt extrusion are used by the majority of spunlaid systems, and spunbond fabrics therefore fall into the category of spunmelt nonwoven processes along with meltblown. Polypropylene (PP) and polyester are by far the most common starting materials, but polyamide, polyethylene, and polylactic acid are also important. In addition to homopolymer filament extrusion, some spunlaid installations produce bicomponent (bico) filaments, commonly in the core sheath but also in segmented pie and islands-in-the-sea cross-sectional configurations.

In spunlaid web formation, polymer chips are fed continuously to a heated screw extruder that delivers the liquid polymer to a metering pump and then to a bank of spinnerets or alternatively to a rectangular spinneret plate, each containing a very large number of orifices. The molecular weight distribution (MWD) of the polymer and its melt viscosity are important to enable the molten material to readily be drawn into fine diameter filaments.

The molten polymer pumped through each hole is progressively cooled to the solid state and stretching or drawing is applied to develop the required molecular structure in the filaments and mechanical properties such as modulus. Over the years, various approaches have been adopted for cooling and stretching the filaments during the process. Commonly, cooling takes place sequentially using primary and secondary blow ducts. Some early systems use mechanical (roller) drawing methods similar to those used in standard filament extrusion processes because these have the advantage of enabling greater draws and reduced variation in fibre diameter. Most contemporary spunbond processes use air drawing to attenuate filaments where the high velocity airflow and duct dimensions can be adjusted to ensure sufficient tension is developed in the filaments to cause drawing to take place.

The deposition of the drawn filaments on the conveyor must satisfy two criteria: The web must be as uniform as possible in mass per unit area, and the distribution of filament orientations must be as desired, which may not be isotropic. Taking the regularity criterion first, filaments must be directed onto the conveyor belt in such a way that an even distribution is possible. However, this in itself is not sufficient because the filaments can "twin" together, forming agglomerations that make "strings" or "ropes" that can be clearly seen in the final fabric. Historically, various methods have been suggested to prevent this; examples are charging the spinneret so that the filaments become charged and repel one another and blowing the filaments from the air tubes against a baffle plate, which tends to break up any agglomerations. Another consideration is the filament orientation distribution collected on the conveyor because this influences the isotropy of web and fabric properties. An established approach is to direct the filaments through a Venturi system (high-velocity and low-pressure zone), which encourages fanning (lateral displacement) and entangling of individual filaments prior to their deposition on the conveyor. Despite the mechanisms employed to produce isotropic webs, MD-bias is common because of the relative velocities of filament laydown and of the conveyor.

6.6   Meltblown

The process of melt blowing produces very fine fibres at high production rates. Normally, a molten fibre-forming thermoplastic polymer is extruded through a metal die tip containing from 30 to more than 100 holes per inch, depending on the required production rate and fibre diameter. Most commonly, the polymer is PP, but a variety of other materials – including polyethylene, polyamide, aromatic and aliphatic polyesters, polystyrene, and thermoplastic polyurethane, amongst others – can also be converted into nonwovens using this approach. The hot liquid polymer emerging from the die holes is converged upon by high-velocity airstreams that are heated to a temperature near or above that of the polymer. As a result of the very high initial air velocity (approaching Mach 1) at the die tip and resulting dynamics, the liquid polymer stream attenuates in the airflow but does not do so uniformly, leading to variations in diameter both between and along individual fibres. As the polymer streams travel farther from the die, mixing with ambient air, they begin to cool and solidify. Because of this manner of attenuation, it is usual to observe variations in the diameter of individual meltblown fibres as well as variations in diameter along the length of fibres. Thermal branches and bifurcations of fibres can also be present. In some systems, the curtain of hot, tacky filaments formed during the process is exploited by blowing in wood pulp or other particles to make various coformed assemblies.

The meltblown fibres are deposited into the form of a web on a permeable conveyor assisted by suction.

To promote attenuation in air during meltblowing, the melt viscosity of the polymer should be relatively low, and selection of low molecular weight polymers is therefore common. This combined with the lack of controlled attenuation (rate of stretching and temperature level) of filaments during the process means that meltblown webs are much weaker than spunbonds. However because of the very small fibre diameter and resultant frictional resistance within the assembly, relatively heavy meltblown webs can sometimes be sufficiently strong to be handled without additional bonding other than compression. Mean fibre diameters are typically in the range of 1–5 µm, but fibres smaller than 500 nm have been produced as well as much coarser fibre meltblowns, depending on the intended product application. Meltblown fabrics are frequently combined with other nonwoven fabrics. They are important in many different applications, such as filter media and lightweight single-use hygiene products as well as in heavier weight fabrics ($>100\,\mathrm{g\,m^{-2}}$) as oil sorbents.

6.7   Composite spunmelts

A large proportion of nonwovens containing meltblown webs are manufactured as composite fabrics. The rationale for building fabrics in this way is to exploit the different physical properties of spunmelts produced by different methods to meet the requirements of the end use. For example, SMS fabrics consist of two layers of spunbond (S), which sandwich a layer of meltblown (M). The webs are usually produced in one continuous operation using a common conveyor, and bonded together at the end to produce the final fabric. In SMS constructions, the meltblown inhibits liquid penetration and remains vapour permeable whereas the outer spunbond layers provide mechanical reinforcement and surface abrasion resistance. The performance of barrier fabrics such as these can be engineered by adjusting the basis weight of the meltblown layer, modifying the polymer additives, and the mean fibre diameters in the layers. Composite nonwovens based on spunmelts include SS, SSS, SM, SMMS, SMMMS, SFS (where F = film layer), SNS (where N = nanofibre layer), and others. Applications for nonwoven composites are diverse but include low-cost single-use products found in the medical and hygiene sector such as protective clothing and feminine hygiene and diaper components, packaging and crop cover, and durable products such as roof membranes and vehicle covers.

6.8   Flash spinning

In flash spinning, a polymer is dissolved in a solvent and is extruded at high temperature and pressure. At the point of extrusion, the polymer solution consists of a mass of bubbles with a large surface area and consequently has a very low wall thickness. When the pressure falls on leaving the extruder, the solvent quickly flashes off, leaving behind a network of very fine fibrils that are commonly described as plexifilaments. The mean diameter of the plexifilaments is typically ≤4 μm, and they are delivered on a conveyor in the form of a web. Tyvek® is one example of a nonwoven fabric produced by flash spinning. Commercially, most flash-spun fabrics are manufactured from polyolefins, the majority being polyethylene based.

Flash-spun webs are bonded in two ways. The first method involves melting the fibres under high pressure so that virtually all fibres adhere along the whole of their length and the fabric is almost solid with very little air space. This area bonding method makes a very stiff fabric with high tensile and tear strengths. Traditionally, flash spinning followed by area bonding has been used to make tough waterproof envelopes as well as banknotes. The fact that the constituent plexifilaments are very fine and the surface is smooth and free of fibre ends means that such fabrics are attractive for use as printing or writing substrates. The alternative method of bonding is the same – that is, heat and pressure but is applied only to small areas, leaving the larger areas in between completely unbonded. These point-bonded areas normally form a square or diagonal pattern. Because of the very fine plexifilaments, small pore size, and inherent hydrophobicity of polyethylene, flash-spun fabrics are not only waterproof but also are resistant to the penetration of many other liquids. Tyvek® is used principally for protective clothing in the chemical, electronics, health care, nuclear and oil industries, amongst others. Such garments can be produced so cost-effectively that they are usually regarded as single-use disposables.

6.9   Electrospun webs

The basic electrospinning process, which has been industrially employed for more than 50 years, involves electrically charging a polymer solution, sol–gel, or melt to produce a liquid jet that can be elongated and solidified into very fine, submicron diameter fibres. *Electrospinning* is the process that is most commonly associated with the production of nanofibre webs; however, it is possible to make electrospun fibres in the range of diameters from approximately 0.05–50 μm (most typically with a mean diameter of <1 μm), and strictly speaking, a nanofibre has a diameter of ≤100 nm.

Briefly, in the production of electrospun webs, voltage is applied to a liquid polymer whose viscosity and surface tension are sufficiently high to allow for stretching by electrostatic forces. Initially, a droplet is stretched into a conical form known as a *Taylor cone*. When electrostatic forces exceed the surface tension of the liquid polymer, a liquid jet is initiated at the Taylor cone's tip that elongates as it travels towards the grounded collector, providing a basis for continuous spinning. Elongation into very fine fibres results from bending instabilities in the electrified jet, causing nonaxisymmetric whipping that bends and stretches the liquid jet as it travels towards the collector, and begins to solidify. Solidification of the polymer stream occurs as the solvent evaporates during the process or cools in the case of melt-electrospun materials. A distribution of very fine diameter fibres is deposited on the collector in the form of a web. The mean fibre diameter and the diameter distribution depend on polymer concentration and numerous other parameters related to material and solution properties

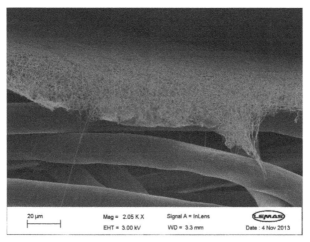

Figure 6.3 Electrospun PVA-based web supported by a polypropylene spunbond fabric. SEL Bulman, University of Leeds.

as well as process settings. Electrospinning is possible using various devices to produce the initial Taylor cone. Traditionally, for the production of webs suitable for manufacturing nonwovens, hollow needles or tubes have been employed, but free surface techniques using lick-rollers, wire cages, and other devices are also commercially used to facilitate wider laydown widths and cost-effective production. Employing multiple electrospinning stations in a series mounted over a common conveyor also enables increased linear production speed. Electrospun webs are typically thin and can be mechanically fragile with basis weights $<1\,\mathrm{g\,m^{-2}}$. Consequently, they are normally deposited onto supporting nonwoven or paper substrates or sandwiched between other fabric layers to facilitate their conversion into finished products (Figure 6.3). The production of composite fabrics in which a very thin layer or electrospun web is applied to one side can have a remarkable effect on many bulk properties, two of which include sound absorption and filter efficiency.

6.10   Centrifugally spun webs

The basic principles of centrifugal spinning, encompassing force spinning, and rotary jet spinning as a means to produce submicron diameter fibres have been known for years. The process takes various forms but essentially involves high-speed rotation of a radial spinneret containing liquid polymer in the form of a melt or solution. The spinneret can contain multiple orifices. The viscoelastic liquid polymer jet emerging from each orifice in the rotating spinneret is subject to tension and air drag, which stretches the jet and reduces its diameter. Depending on the prevailing extensional and viscoelastic forces, the jet elongates and finally solidifies in the ambient air into the form of a filament before being deposited onto a collector in the form of a thin web. Centrifugal spinning is an alternative to electrospinning in the production of webs containing submicron fibres although the fibre diameter distribution in centrifugally spun webs is often narrower. The process has the advantage of not requiring high voltage, and spinning from polymer melts or solutions is feasible using the same basic process equipment.

6.11   Extruded split or fibrillated films

Films of thermoplastic extruded polymer can be fibrillated or split using various means to form netlike structures similar to a fibrous web. These films are manufactured mostly from polyolefins but also other thermoplastics based on PET, PA, and copolymers. The polymer is normally extruded using a slit die to form a film in sheet format and then split in different ways to develop the required structure. There are at least two basic approaches. The first involves drawing an extruded film to increase molecular alignment using an appropriate draw ratio and temperature before fibrillating or splitting it under tension to form a netlike structure. In some cases, the film can be embossed prior to stretching. The second approach involves mechanically splitting an undrawn film and then tensioning it to develop a netlike structure. Tension can be applied uniaxially or biaxially to the film. A variety of techniques have been developed to split and fibrillate drawn film; the most common of which is the use of a pinned roller or grooved rollers that introduce perforations and enable a net like structure to be developed by applying tension to the film. Modifying the pin or groove density of the rollers, the relative speeds of the rollers and film can produce different net structure, as well as the degree of uniaxial or biaxial tension applied during manufacture. Depending on the polymer composition and basis weight, split or fibrillated films find uses in areas such as filtration and packaging and as layers in composite nonwovens.

6.12   Chemical bonding

Chemical bonding involves treating either the entire web or isolated portions of it with an adhesive with the intention of sticking the fibres together. Fabric properties therefore depend on the chemical, thermal, and mechanical properties of both the fibres and the binder in the structure as well as their relative proportion. Although many different bonding agents can be used, the modern industry predominantly uses synthetic polymer latexes produced by emulsion polymerisation of which acrylic, styrene–butadiene, and vinyl acetate-based compounds represent a large proportion. Chemical bonding is extremely versatile because of its ability to select from a large number of different polymer and copolymer latex formulations, additives, and methods of applying the binder to the web. Clearly, the chemical composition of the binder – its wet and dry mechanical properties when cured – and its attritional and thermal properties are important considerations. In relation to the latter, the glass transition temperature (T_g) can influence the softness of the final fabric. The binder is most commonly applied to webs in the form of an aqueous dispersion rather than a polymer solution, which necessitates drying and binder cross-linking to produce the final bonded nonwoven. Note that in some cases the liquid phase may be a solvent other than water. The properties of chemically bonded fabrics depend on the fibre composition, binder composition, adhesion of the binder to the fibres, binder cohesion, and binder distribution. When the latex binder is applied, it must be capable of wetting the fibre surfaces; otherwise, there is poor adhesion between the two phases after drying. Most lattices already contain a surfactant to disperse the polymer particles, but in some cases, additional surfactant may be needed to aid wetting. The next stage is to dry the latex by evaporating the aqueous component and coalescing the polymer particles in the latex to form a film over the fibre surfaces. During this stage, the surface tension of the water and capillary pressure coalesces the binder particles to form a film over the fibres, particularly around fibre intersections. Smaller binder particles form a more effective film than larger particles, other things being equal. During high temperature drying, care is required to minimise binder migration to the exterior surfaces of the web because this alters the intended binder distribution throughout the thickness. The final stage is curing, which involves increasing the temperature of the web beyond that used for drying. The purpose of curing is to develop cross-links both inside and between the polymer particles to develop good cohesive strength in the binder film. Although not all binders used for chemical bonding are designed to be cross-linked, it is usually necessary to do so to ensure adequate mechanical stability.

6.12.1 Saturation bonding

Although the principles just discussed apply to all forms of chemical bonding whatever the method of binder application, saturation bonding has a profound influence on the properties of the nonwoven fabric. As implied by the name, it wets the entire web structure with bonding agents so that all fibres are covered in a film of binder.

To saturate or impregnate the web, it is carried under the surface of the bonding agent. In most cases, webs are very open and weak so that care is needed to avoid distortion. Consequently, webs may be prebonded by other methods prior to saturation bonding to increase the mechanical stability. The penetration of the liquid binder into the web structure reduces the web thickness, and it reduces further as a result of the squeeze rollers that follow impregnation. Hence, saturation-bonded fabrics are generally compact and relatively thin.

Drying is often done with the web supported on a permeable conveyor to transport it through the process. Various approaches to drying are employed, depending on the design of the system. In some hot air drying systems, air is blown against the top and bottom surfaces to facilitate drying in which case the top air pressure is slightly greater than the bottom pressure to press and control the nonwoven against the conveyor. In such systems, if the air penetrates only the immediate surfaces of the nonwoven, drying is confined to these areas and the central layers of the nonwoven remain wet. The result is that capillary pressure drives the liquid from the wet areas to the dry ones, carrying suspended binder particles with the water. This leads to excessive binder migration and under extreme conditions, a large proportion of the binder content can be found in the surface layers, leaving the central section very weak. Such a fabric can easily split into two layers, producing delamination. Through-air drying is one of the ways that binder migration can be successfully minimised. In through-air drying, a hot air stream is drawn through the entire web from one side at a time. Drying conditions are then almost the same in all parts of the fabric and little or no binder migration takes place. However, the air pressure may exert a significant force on the nonwoven, pressing it against the conveyor so hard that it may be imprinted with the pattern of the conveyor.

Following drying, curing is ideally carried out in a separate compartment in order to achieve the correct temperature. However, it is quite common for curing to be done in the final part of the dryer in order to keep machinery costs down.

Many of the physical properties of saturation-bonded fabric derive from the fact that all fibre surfaces are covered with a film of binder. The surface chemistry that is presented to incoming liquids is dominated by the film coating on the fibre surfaces, which influences wetting behaviour. The binder also influences frictional and mechanical properties as well as the handle of the fabric.

The mechanical properties can be explained from a model of a network of fibres bonded together at close intervals. The fabric cannot stretch without the fibres also stretching by a similar amount. Hence, the fabric modulus is of the order of the fibre modulus, that is, extremely high. A high modulus in a spatially uniform material means that it will be stiff, which explains why saturation-bonded fabrics are very stiff relative to conventional textiles. At the same time, tensile strength is low because the bonds tend to break before most fibres do. Fabrics can be made more flexible by selecting extensible or elastomeric binders and applying a lower proportion of binder to the web, but both can reduce the tensile strength. Interestingly, the ratio of fabric modulus to tensile strength can remain remarkably constant.

One of the major uses of saturation-bonded fabric turns the apparent disadvantage of stiffness into an advantage. Interlining fabric for textile clothing is required to be stiff and to have a high modulus.

6.12.2   Foam bonding

One of the problems of saturation bonding is that a large volume of water is required. This increases not only the cost of drying but also the risk of binder migration. Application of chemicals from foams rather than liquids was developed not only for nonwovens but also for the dyeing and finishing industry as a means of reducing water use. The binder and a measured volume of air are passed continuously through a driven turbine, which beats the two components into consistent foam. The foam can then be delivered in various forms, but commonly it is directed to the horizontal nip of a set of impregnating rollers. The foam delivery has to be traversed because the foam does not flow easily. End plates prevent the foam from running out of the gap at the end of the rollers. The rollers serve the dual purpose of metering the amount of foam applied and of squeezing the foam into the web. If foam penetration needs to be increased, then the web can be entered vertically and the foam can be applied from both sides. By adjusting the roller pressure, the extent to which the foam penetrates the web and its eventual distribution through the cross-section of the fabric can be graduated or localised. As in other chemical bonding processes, prebonding using processes such as needling or hydroentangling may be used to increase the strength of the substrate before it enters the foam bonding system. Foam application can be thought of as an alternative method of saturation bonding because the properties and uses of the fabrics are similar.

6.12.3   Print bonding

Print bonding involves applying the same types of binder to the web, but the application is restricted to limited areas in a predefined pattern. The binder does not penetrate well into the dry web, so it is first saturated with water and then printed with either a printing roller or a rotary screen printer. The final properties of the fabric depend vitally on the ratio of printed to unprinted area, which can change significantly if the binder migrates sideway from the printed area. To prevent migration, the binder formulation must contain a viscosity modifier or thickening agent.

Print-bonded fabrics are usually much softer in handle and much more flexible because of the unbonded fibre segments between bond points. They are also significantly weaker than saturation-bonded fabrics because of fibre slippage in the unbonded areas, but given the prevailing fibre length and fibre orientation distribution, it is possible to design a print pattern that minimises the strength loss.

Print-bonded fabrics tend to be used in applications in which the textile-like handle is an advantage. Examples are single-use protective clothing, cover stock and wipes, particularly domestic dishcloths and dusters.

6.12.4   Spray bonding

Latex binders can also be applied by spraying the web supported on a perforated conveyor by means of compressed air or airless spray systems. To aid penetration into the structure, the web is often sprayed from both sides sequentially. On the first passage, spray droplets, which can average about 20–200 µm in size, penetrate about 5 mm into the top surface, and then the web is turned over for a spray application on the lower surface. Suction can be applied from beneath to aid through-thickness penetration. Each spray application reduces the thickness of the batt slightly, but it is still left substantially lofty; the drying and curing stage also causes some small dimensional changes. The final product is usually a thick, open, and lofty fabric used widely as the filling in quilted fabrics, for duvets and for some upholstery; for abrasive pads for polishing and cleaning operations; and for some types of filter media.

6.13   Thermal bonding

Thermal bonding exploits the characteristic ability of thermoplastic materials to soften and melt at an elevated temperature and then solidify on cooling. When adjacent fibres of similar composition are in a molten state and both are in direct contact, segments of long-chain molecules from each can migrate into each other so that when the polymer cools, these chains become embedded, producing a cohesive bond. However, not all the bonds created in thermal bonding are of this type because the molten polymer may not always be in contact with fibres made of similar material. In such cases, adhesive rather than cohesive bonds are usually formed. Thermal bonding in nonwovens can use three types of fibrous raw material, each of which may be suitable in some applications but not in others. First, the fibres may be homogeneous with the same melting point. This is satisfactory if the heat is applied at specific locations (point bonding), but if applied to the full area of the web (area bonding), it is possible that all the fibres will melt into a plastic sheet with little or no value. Second, a blend of thermoplastic fibre with either a fibre with a higher melting point or a nonthermoplastic fibre can be used. This is satisfactory in most conditions except when the thermoplastic fibre melts completely, losing its fibrous nature, and causing the web to collapse in thickness. Finally, the thermoplastic fibre may be a bicomponent (bico) that for thermal bonding usually consists of a core of high melting point polymer surrounded by a sheath of lower melting point polymer (core–sheath bico). Such bico fibres are available with a variety of thermoplastic polymer combinations and different core–sheath ratios, and the concentricity of the core and sheath can also be varied. As the sheath polymer begins to melt, the core of the fibre remains in its fibrous state, reinforcing the sheath. Thermal bonding is applicable to webs made by virtually all formation methods, but there are variations in the heat transfer mechanisms that are used to bond the fibres.

6.13.1   Through-air bonding

Webs pass through a hot air oven supported on an air-permeable conveyor in the form of a belt or drum. The aim is to rapidly heat the web to the melting point of the constituent fibres by drawing hot air through the web so that all the constituent fibres are heated uniformly. The air temperature, air velocity, and dwell time of the web in the oven all influence the bond strength developed in the fabric during thermal bonding.

In some through-air processes, an upper perforated surface is provided in the oven to slightly compress the web to the required thickness as well as to control shrinkage. As the web leaves the oven, its final thickness can also be controlled by passing it between two calender rollers, which can be set to a specific gauge as required to bring the web to the required thickness. Chill rollers are also employed to cool the thermally bonded fabric before winding or further processing to prevent layers from sticking. Through-air bonding is particularly suitable for the production of high-loft or low-density fabrics because there is minimal compression of the web during the process. It is also an effective means of bonding heavyweight webs uniformly through their thickness, which can be difficult to achieve using direct contact methods such as calender bonding.

6.13.2   Calender bonding

The web passes between the nip of two large heated (calender) rollers that operate under pressure, compressing the fibrous assembly and conducting heat into the fibres, causing them to soften and melt (Figure 6.4). Provided that the web is not too heavy in mass per unit area, the heating is very rapid and the process can be carried out at

Figure 6.4 Calender for thermal bonding.
Courtesy of Andritz.

high linear speed ($>350\,\mathrm{m\,min^{-1}}$). The design of calender rollers for this purpose has become highly developed; they can extend to more than 5 m width and can be heated to produce less than 1 °C temperature variation across the rollers. Also systems have been developed to ensure that uniform pressure is applied all the way across the rollers because rollers of this width have a tendency to deflect. Most commonly, at least one engraved roller is used in calender bonding systems because the area bonding resulting from the use of two plain rollers produces fabrics that are too stiff and impermeable for practical use.

Calender rollers for point bonding are engraved with a pattern that limits the degree of contact between the rollers to roughly 5–25% of the total area to maintain the permeability and flexibility of the fabric. The size, shape, and geometry of the pattern influence not only the appearance of the resulting fabric but also its physical properties. Thermal bonding is mostly confined to those raised or embossed points on the rollers where they touch and compress the web and leaves the rest effectively unbonded. Fabrics made in this way are flexible and relatively soft because of the unbonded areas. At the same time, fabrics maintain reasonable strength, especially in the case of spunlaid fabrics. These fabrics have many uses, for example, as a substrate for tufted carpets, in geosynthetics, in filtration media, in protective/disposable clothing, as coating substrates and as hygiene cover stock. Another use of calendering is to melt fibres on just one side of a fabric in a process known as "skinning" the surface to increase mechanical stability. This can be done by passing the fabric between a set of plain calender rollers, only one of which is heated to a temperature near the melting point of the polymer.

6.13.3 Powder bonding

In some applications, thermoplastic powders are applied to webs for the purpose of bonding. The powder can be mingled with the fibre during web formation, which is particularly convenient in some airlaying processes, after web formation, or can be applied to the surface of prebonded nonwoven fabrics. One of the challenges is to uniformly apply the powder to the web. Particularly when the powder is applied after web formation penetration into the internal structure can be quite limited. Products made by powder bonding are characterised by softness, low density, and flexibility and in general they have relatively low strength. Again there is a very wide range of uses covering particularly interlinings, shoe fabric components, and floorcoverings.

6.14   Solvent bonding

Although solvent bonding is in limited use, it provides a valuable means of producing cohesive bonds in nonwoven fabrics. Solvents used in the process can be recycled, although the costs of solvent recovery must be considered. In one embodiment of the process, a spunlaid polyamide web is carried through an enclosure containing the solvent gas nitrogen dioxide (NO_2), which softens the filament surfaces. Subsequently, the web is passed between cold calender rolls and the solvent is washed off the fabric. Another example uses a so-called latent solvent, which means one that is not a solvent at room temperature but becomes a solvent at higher temperatures. This latent solvent is used in conjunction with carding and crosslapping and is applied as a liquid before carding. The action of carding spreads the solvent and at the same time, the solvent lubricates the fibres during the process. The web is passed into a hot air oven that first activates the solvent and later evaporates it. The resulting fabric is normally low in density, but if it needs to be increased, compression rollers can be used.

6.15   Needlepunching

Mechanical bonding of fibre webs relies on frictional forces and fibre entanglements to produce strong bonds. Needlepunching, also referred to as *needlefelting*, is one of three main mechanical bonding methods along with stitchbonding and hydroentanglement.

Needlepunching is commonly used to bond carded and crosslapped batts but can also be used in combination with long-fibre airlaid or heavyweight spunlaid webs. The web is led between upper and lower stationary plates, known as the *stripper* and *bedplate*, respectively, that are intended to control the position of the web relative to the needles used in the process. Both plates are perforated or slotted to allow the passage of a barbed needle. While between the plates, the web is penetrated by a large number of oscillating barbed needles that are held in a needle board. Depending on the configuration of the process line, the number of needles can range from 1500 to 20,000 m^{-1} of working width and the number of strokes from about 500 to >3000 min^{-1}. Introducing additional needle boards in the same machine or additional needlepunching stations in sequence increases the needling capacity of the line (Figure 6.5). The first machine is normally referred to as a *preneedler* and the final ones as finishing looms. The web can be needlepunched from above (downstroke) or below (upstroke) to control the degree to which fibre entanglement is developed in the fabric as well as the surface uniformity

Figure 6.5 Needlepunching installation, showing *from left to right*: hopper fed carding and crosslapping; batt drafting; pre needling unit; second needlepunching unit; third needlepunching unit; winding up of final roll. Felt drafting units are employed after the first and second needle punching units to adjust the MD:CD ratio.
Courtesy of Andritz.

of the fabric. Some looms incorporate a double-punch, that is, one board that punches down and one board that punches up. Double-punch looms in which needles in both boards penetrate through the same bed and stripper plates are also available. The normal operation is for the needles in the boards to penetrate the fabric alternately, one from above and the other from below.

Fundamental to the operation of a needlepunching system is the needle itself. Needles intended for bonding usually have a triangular cross-section and have three barbs on each of the three edges of the working blade. When the needles descend into the web, they catch some fibre segments and pull them past other fibres, which are mostly aligned in-plane, to form "pillars" or "pegs" within the fabric cross-section. The number of fibre segments that can be transported in this way depends on various factors, including the geometry of the barb, particularly the angle of kickup and throat depth, fibre to metal friction, and fibre diameter relative to the barb dimensions. When the needles return upward, the loops of fibre formed on the downstroke remain essentially in position as the barbs release them. The repeated oscillation of the needles gradually increases fibre entanglement and frictional forces as well as fabric density, resulting in a strengthening of the structure. Gradual wear of the needle, particularly the leading edge of the barbs as well as occasional needle breakage, means that needles are consumable items and must be periodically replaced to ensure that fabric properties are not adversely affected.

Needlepunching is more complex that it seems at first. The needles migrate only a relatively small number of the fibres through the web thickness to increase entanglement. Although there are exceptions, in most systems, the needle penetrates the web perpendicular to the surface, leaving a vertically oriented pillar in the fabric cross-section. This alone does not form a strong fabric unless the vertical pillars pass through loops already present in the horizontal plane of the structure. It follows from this that parallel-laid fabric is not very suitable for needling because few fibre loops are created, so most needling processes are carried out with crosslapped, long-fibre airlaid, and spunlaid webs. The amount of needling is determined partly by the distance the drawing rollers move between each oscillation of the needle board, the "advance", and partly by the number of needles per metre across the loom. The advance per stroke cannot be too large because of the risk of needle breakage.

If the chosen advance happens to be equal to, or even near the distance between needle rows, then the next row of needles come down in exactly the same position as the previous row, and so on for all the rows of needles. The result is a severe needle patterning of the fabric. To avoid this, the distance between each row of needles must be different. Computer software can be used to calculate the best set of row spacings.

To decide how to make a particular needled fabric, it is necessary to choose the number of needle penetrations per unit area, the depth of needle penetration, and what type of needles should be used from many different types. There are so many variations that optimisation becomes very difficult.

To increase the modulus and recovery from extension of needled fabrics, reinforcements may be incorporated. These can consist of yarns or scrims made of open weave fabrics or meshes that are combined with the batt after web formation and needled together to make a fully integrated fabric.

Another form of needlepunching is referred to as "structuring"; it is responsible for introducing a pile surface in the form of structural patterns or a velour-like surface on preneedled fabrics. To make structured fabrics, the perforated bedplate that supports the substrate during needling is replaced with either a series of lamellae plates between which the needles pass (loop pile patterned surface) or a brush bed (velour surface) into which the needle points can penetrate. For structuring, different types of needles designed to release fibres in the same position on the upstroke are used. These include fork and crown needles, which differ from the normal needles in that there is either a fork at the needle tip instead of barbs (fork needle) or there is one barb on the working edges of the blade each spaced near to, and at the same distance from, the needle tip (crown needle).

Needlepunched fabrics are used in a large variety of applications, including filtration, geosynthetics, papermakers' felts, synthetic leather, floorcoverings, automotive headliners, and wound dressings.

6.16 Stitchbonding

Stitchbonding refers to a family of processes and fabric types in which mechanical strengthening is accomplished either by knitting yarns into preformed webs or laid-in yarns or by stitching through preformed webs, locking the fibres into the structure, increasing their frictional resistance. Some of the available systems are briefly discussed.

6.16.1 Batt bonded by threads

Stitchbonding uses mainly crosslapped and long-fibre airlaid batts. The batt is taken into a modification of a warp knitting machine and passes between the needles and the guide bar(s). The needles are strengthened and are specially designed to penetrate the batt on each cycle of the machine. The needles are of the compound type, having a tongue controlled by a separate bar. After the needles pass through the batt, the needle hooks open and the guide bar laps thread into the hooks of the needles. As the needles withdraw, the needle hooks are closed by the tongues, the old loops knock over the needles and new loops are formed. In this way, a form of warp knitting action is carried out with the overlaps on one side of the batt and the underlaps on the other. Generally, as in most warp knitted fabrics, continuous filament yarns are used to avoid yarn breakages and stoppages on the machine. Two structures are normally knitted on these machines, pillar (or chain) stitch and tricot. Knitting chain stitch laps the same guide around the same needle continuously, producing a large number of isolated chains of loops. Knitting tricot structure results when the guide bar shogs one needle space to the left and then one to the right. Single-guide bar structure is called *tricot*, and the two-guide bar structure is often referred to as *full tricot*.

Fabrics produced in this way are textile-like, soft, and flexible. At one time, the process was widely used for curtaining but is now used as a backing fabric for lamination, ticking for mattresses and beds, and fabrics used in training shoes amongst other end uses. As to whether to use pillar or tricot stitch, both have a similar strength in the MD, but in the CD, tricot produces stronger fabrics because the underlaps lie in that direction. A crosslaid web is already stronger in that direction, so the advantage is relatively small. The abrasion resistance is the same on the loop or overlap side, but on the underlap side, the tricot fabric has markedly better resistance because of the longer underlaps. However, continuous filament yarn is relatively expensive, so tricot fabrics tend to cost more.

6.16.2 Stitchbonding without threads

It is possible to bond fibres in a batt by knitting through the structure to mechanically bind the fibres together without the use of threads. In this case, the machine is basically the same as described in the previous section, but the guide bar(s) is(are) not used. The needle bar moves backward and forward as before, pushing the needles through the batt. The main difference is that the timing of the hook closing by the tongues is somewhat delayed so that the hook of the needle picks up some of the fibre from the batt. These fibres are formed into a loop on the first cycle; on subsequent cycles, the newly formed loops are pulled through the previous loops as in normal knitting. The final structure appears nonwoven on one side and similar in appearance to a knitted fabric on the other. Traditionally, such fabrics have been used for insulation and as decorative materials.

6.16.3 Stitchbonding to produce a pile fabric

Forming a pile fabric normally requires two guide bars, two types of warp yarns (pile yarn and sewing yarn), and a set of pile sinkers that are narrow strips of metal over which the pile yarn is passed and whose height determines the height of the pile. The pile yarn is not fed into the needle's hook and so does not form a loop; it is held in place between the underlap of the sewing yarn and the batt itself. It is clear that this is the most efficient way to treat the pile yarn because any pile yarn in a loop is effectively wasted. This structure has been used for making towelling with single-sided pile and loop-pile carpeting, but such fabrics compete with those produced by double-sided terry towelling and tufted carpets.

6.16.4   Batt looped through a supporting structure

The needles pass through a supporting fabric and pick up as much fibre from the batt as possible. Special sinkers are used to push fibre into the needle's hook to increase the pickup efficiency. The fibre pulled through the fabric forms a chain of loops with loose fibre from the batt on the reverse surface of the fabric. The fabric is finished by raising to give a thicker pile. This structure was widely used in the manufacture of artificial fur, but such fabrics compete with those produced by sliver knitting, which gives a fabric with similar properties.

6.16.5   Laid yarns sewn together with binding threads

Two distinct types of fabric can be made by using the same principle. The first is a simulated woven fabric in which the weft direction yarns are laid many at a time in a process a bit-like crosslaying. The MD yarns, if any are used, are simply unwound into the machine. These two sets of yarns are sewn together using chain stitch if there are only cross-direction threads and tricot stitch if machine-directed threads are present with the underlaps holding the threads down. Fabric can be made rapidly by this system. Unlike a woven fabric, the weft threads do not interlace but lie straight in the fabric. Consequently, the initial modulus of the fabric is very high compared to a woven fabric because of the lack of yarn crimp. These fabrics have been used to make fibre-reinforced plastics (FRP) using, for example, continuous filament glass yarns or other high modulus yarns.

The alternative system makes a multidirectional fabric. Again, sets of yarns are laid across the width, but in this case, not in the CD but at angles such as $45°$ or $60°$ to the CD. Two sets of yarns at, say, $+45°$ and $-45°$ to the CD, plus another layer of yarns in the MD can be sewn together in the usual way. Again, high modulus yarns are used with the advantage that the directional properties of the fabric can be designed to satisfy the stresses in the component being made.

6.17   Hydroentanglement

The manufacture of hydroentangled or spunlace fabric relies on columnar high-velocity water jets to mechanically bond the fibres or filaments in a web while it is supported on a moving conveyor. Hydroentangling is compatible with drylaid, wet-laid, and spunlaid webs, although carded webs are most commonly encountered in currently installed production lines (Figure 6.6). During hydroentangling, pressurised water (approximately 10–400 bar) is forced into an injector (or manifold) and then through capillary cone-shaped nozzles in a thin jet strip to produce one or more rows of closely spaced water jets that extend the width of the machine. The nozzle diameters in the strip are very small, approximately 80–150 μm, and it is important that these fine jets remain intact and do not break up into droplets to ensure efficient kinetic energy transfer to the web. Typically, multiple injectors are employed in series, and the process is usually configured so that the web can be hydroentangled on both sides to

Figure 6.6 Hydroentangling installation, showing *from left to right*: two hopper-fed cards in series; hydroentangling unit; through-air dryer; winder.
Courtesy of Andritz.

gradually build up fibre entanglement and the required fabric structure. The number of injectors fitted to hydroentanglement installations varies, but typically 5–8 are required to manufacture fabrics with adequate bonding and surface uniformity, assuming that no additional bonding processes precede or follow. Some installations operate with only 2–4 injectors because additional chemical or thermal bonding to complete fabric production follows them. Normally, the water pressure in the first injector is low and increases as the web moves through the system. Excess water is removed from the web by suction applied from below the conveyor. Hydroentangling uses large volumes of water, and the water quality both in terms of solid particle content and chemical composition affects process efficiency so that the filtration system is normally a major cost in a hydroentanglement installation.

The formation and structure of the fabric is influenced by the interactions between the fibres, the energised water, and the conveyor surface during the process. For fibres to entangle efficiently, they need to be flexible and capable of entwining around others, which highlights the importance of fibre's mechanical properties. Many process parameters, including the water pressure from each injector and the total specific energy consumed by the web, influence fabric properties. Although there are ways to minimise them, jet marks running in the MD are a characteristic feature of hydroentangled fabrics. In addition to water jet conditions, the design of the conveyor surface can have a profound effect on fabric structure because of its influence on the rearrangement of fibres during the process. For example, apertures are formed in fabrics if fibres are displaced and moved away from the raised sections in the conveyor by the impinging water jets. Other structural effects such as ribs and embossed patterns can be introduced by engineering variations in the surface profile of the conveyor. Hydroentanglement also provides a convenient means of mechanically combining two or more different webs to produce multilayer fabrics. Additionally, preformed nets or scrims can be hydroentangled with webs to increase the strength of resulting fabrics. Commercially, segmented pie bicomponent fibres can be split during hydroentangling to manufacture microfibre fabrics. Hydroentangled fabrics are produced for a large variety of applications including, dry and premoistened wipes, medical gauze, surgical gowns, scrub suits, sheets and drapes, protective clothing liners and moisture barriers, automotive components and filtration.

Bibliography

Albrecht, W., Fuchs, H., Kittelmann, W., 2003. Nonwoven Fabrics—Raw Materials, Manufacture, Applications, Characteristics, Testing Processes. Wiley-VCH Verlag GmbH & Co., Weinheim, Germany.

Batra, S.K., Pourdeyhimi, B., 2012. Introduction to Nonwovens Technology. DEStech Publications, Inc., Lancaster, PA, USA.

Chapman, R. (Ed.), 2010. Applications of Nonwovens in Technical Textiles. Woodhead Publishing Ltd, Cambridge, United Kingdom.

Das, D., Pourdeyhimi, B., 2014. Composite Nonwoven Materials: Structure, Properties and Applications. Woodhead Publishing Ltd, United Kingdom.

Russell, S.J. (Ed.), 2006. Handbook of Nonwovens. Woodhead Publishing Ltd, Cambridge, United Kingdom.

Turbak, A.F. (Ed.), 1993. Nonwovens—Theory, Process, Performance and Testing. 1993. TAPPI Press, Atlanta, GA, USA.

Technical textile finishing

7

Roy Conway
Textiles Division, School of Materials, University of Manchester, Manchester,
United Kingdom

Chapter Outline

7.1   Introduction

Textile finishing refers to the treatment of a base fabric by either mechanical or chemical finishing process to enhance the appearance, aesthetics, properties, and performance to meet the functional requirements in the final product. Textile finishing is carried out on both woven and knitted fabrics as well as on some non-woven materials to manufacture a product with the final overall performance characteristics. In technical textiles, the base fibres may be either natural, manufactured, or a natural/manufactured fibre blend. Whereas all commercially acceptable fabrics must undergo one or more finishing treatments, this chapter focuses on those that have significance mainly within the technical textile sector. The reader should consult the wider literature on finishing technologies for textiles designed for non-technical applications [1]. This chapter builds on that by Hall [2] in the first edition of the *Handbook of Technical Textiles*.

7.2   Pre-treatment of fabrics for finishing

Before any mechanical or chemical finishing process is undertaken to add functional properties to the substrate, the removal of impurities is essential for the following:

- Natural fibre impurities if fibres like cotton, flax, jute, or wool are present because they may include natural colorants.
- Fibre spin finishes and yarn finishing/lubricating oils.
- Warp size presents only in woven fabrics.
- Adventitious impurities such as oil stains, particulate dirt, and so on.

Handbook of Technical Textiles. http://dx.doi.org/10.1016/B978-1-78242-458-1.00007-8

Removing such impurities may require one or more pre-treatment processes such as *desizing*, *scouring*, and *bleaching*, which may be undertaken either in sequence or in a combined process depending on the severity of treatment required. In conventional textiles containing natural fibres like cotton, chemical bleaching is almost always undertaken because the textile may be sold in the white state (e.g. bed and table linen). Fibres, yarns, and fabrics may also be scoured and bleached in order to promote good absorbency for subsequent dyeing or printing. In technical fabrics, these factors are less important, so bleaching is not often undertaken unless some additional feature such as high absorbance is required as in medical textiles, for instance. The manufactured fibres/filaments do not normally require chemical bleaching because they are already produced in a white or spun dyed state.

Once the textile has been pre-treated and dyed or printed if required (see Chapter 9), it is ready for mechanical finishing and/or chemical finishing by the application of appropriate chemical treatments.

The following pre-treatment processes are briefly described next.

7.2.1 Desizing

Sizes are normally applied to the warp yarns prior to weaving to protect and lubricate the warp and, hence, increase the weaving efficiency. The removal of the size is accomplished by desizing.

Traditionally, starch is the main sizing agent used for cotton and cellulosic yarns (including blends with manufactured fibres). However, it is not soluble in water and so must be chemically broken down into water-soluble products. To describe the different desizing methods used in detail is beyond the scope of this chapter; suffice it to say that a number of processes exist to break down and/or solubilise the sizes present including:

* Overnight steeping in warm water with or without enzyme.
* Enzyme desizing using bacterial alpha amylase or malt enzymes.
* Mineral acid desizing to hydrolyse and break down the starch.
* Oxidative desizing using alkaline hydrogen peroxide, a treatment that can be boosted by the addition of potassium peroxodisulphate.

Each of these treatments is normally followed by a thorough washing off process to remove solubilised size and desizing agents, which if left on the fabric for a significant time, may damage the cellulosic fibres themselves.

Wool-containing yarns may be traditionally sized with natural proteins such as gelatin mixed with waxes that are more easily removed by simple scouring using a synthetic detergent. With the advent of manufactured fibres, however, synthetic sizes have been developed for removal by a steeping and washing process or by aqueous scouring. These synthetic sizes are more expensive than starch-based sizes but are easier and cheaper to remove and, in some cases, can be recycled by an ultrafiltration treatment combined with heat recovery from the hot water used for desizing. Obviously, desizing and scouring of warps containing soluble synthetic sizes may be combined into a single process.

7.2.2　Scouring

Scouring is a process that depends on the fibres present in the technical textile. In the case of cotton textiles, the natural impurities such as waxes, proteins, dirt, and seed coat fragments as well as any residual size remaining may be removed by alkaline scouring. Cotton waxes are difficult to remove simply by "washing off"; efficient removal involves the scouring process at a high temperature treatment, often under pressure, using an alkaline solution, usually sodium hydroxide, to hydrolyse and thus solubilise the waxy components. High temperature in pressurised machinery is essential if a fast process is required and for the prevention of cellulosic degradation from any oxygen present. For other non-cellulosic fabrics, scouring is a simpler process involving the use of synthetic detergents to disperse the removed, water-soluble impurities in the scour liquor and to prevent redeposition on the fabric.

Scouring has traditionally been undertaken as a batch rope process in a kier, but more recent developments include open-width high-pressure batch-scour systems and continuous rope J-box treatments as well as pad/steam/wash-off treatments.

7.2.3　Bleaching

Although desizing and scouring pre-treatments remove impurities and unwanted residues in the textile, there is a tendency for the textile (e.g. cotton) to have a natural off-white colour that may become more yellow during alkaline scouring; consequently, there is usually a requirement for the fabric to be bleached in order to remove these colorant species present. Furthermore, the oxidative nature of modern bleaching processes removes all final traces of remaining impurities and increases the absorbency of the fabric and its ability to receive subsequent coloration and finishing processes more effectively and uniformly both across its width and along its length.

Technical fabrics are not normally chemically bleached but those that are most likely are to be work wear and corporate wear composed of natural fibre contents that require dyeing to standardised shades. Some contract furnishing fabrics may also require a moderate bleach if the subsequent dyed colour is a pale or pastel shade. For these applications, the use of a hydrogen peroxide bleach formulation is more environmentally friendly than the chlorine-based bleaching agents such as sodium hypochlorite and sodium chlorite. Hydrogen peroxide bleaching is conducted under alkaline conditions (pH 12) at elevated temperatures (typically 75–100 °C or higher). Hydrogen peroxide treatments can form part of a combined desize/scour/bleach process. A more comprehensive review of hydrogen peroxide bleaching treatments is described elsewhere [3].

7.2.4　Mercerising

Mercerising is a chemical finishing process that is carried out during or just after the pre-treatment process on cotton-containing fabrics to improve a number of the fibre properties. Mercerising can be carried out on cotton yarns in hank or warp form as well as on woven and weft knitted cotton fabrics. Mercerising improves a number of the cotton fibre properties and is carried out to improve the lustre, strength, and affinity for dyestuffs for dyeing and printing should these be required in a subsequent process for the final product. The process is carried out with the fabric (or yarn) under tension when it is treated with a concentrated (≥ 18 wt%) sodium hydroxide solution that causes the cotton fibres to swell. It is important to wash out the sodium hydroxide and then neutralise it with acid while the material is still under tension; otherwise, the textile will shrink. However, the balance of strength, lustre, and dye affinity may be altered by allowing some degree of controlled shrinkage. The highest strength and lowest improvement in dye affinity occurs under the highest tension and vice versa. It is likely that the mercerising process would be used in the technical textile sector only for contract furnishings and wall décor (e.g. in marine transport), where the improved lustre is a premium requirement.

7.3   Mechanical finishing processes

7.3.1   Calendering

A *calender* is a machine in which fabric is passed at open width through the nip (pressure zone) between rollers under pressure. There are four basic types of finishing calender, each one producing a different effect on the textile. Normally, the pressure between the calender bowls is controlled hydraulically. A *rolling* or *swizzing* calender is one that runs all rollers at the same speed. A *frictioning* calender is one in which one roll runs faster than the others, and preferably three bowls are used. A *chasing* calender is used to provide a linen effect surface on the textile, and normally five bowls are used where the textile is passed several times through the calender with each pass on top of the material in the previous pass.

In *rolling* mode, a matt finish can be produced at low temperatures and different pressures to close the threads and improve the handle or feel of the fabric. A glazed finish can be achieved on one side of the substrate with a heated steel roll, depending on the temperature, pressure, and speed of operation. For a very high glaze on 100% cotton materials, the calender would be operated in friction mode with a heated steel roller but for polyester, cotton blends, and synthetic fibre fabrics, chintz (glazed) effects can be achieved by temperature and pressure without the use of friction. Chintz effects on printed plain weave cotton fabric can either be "semi-glazed" or "half-glazed" through the stiffening action of friction calendering alone whereas cotton fabric that has been stiffened by starch or another substance and friction calendered generates a fully glazed appearance.

A two-bowl calender (see Figure 7.1) is a compact and versatile system that can cater to a wide variety of finishes. The machine is fitted with a heated steel bowl and either a compressed cotton bowl (roll) or a roll fitted with a polyamide sleeve. The polyamide sleeve allows for higher production speeds and lasts longer than a compressed cotton bowl, because it is less susceptible to damage and marks from sewn joints passing through the nip zone and so on. Gloss effects on synthetic or polyester cotton-blended textiles are possible with increased temperature and pressure without frictioning, caused when the driven rolls rotate at different speeds and hence friction, where the steel roll runs faster than the back-up bowl.

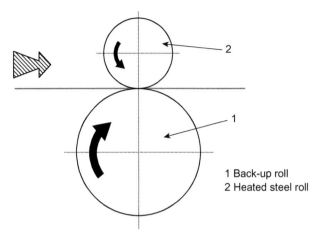

Figure 7.1 Two-bowl calender.

A three-bowl calender is normally used when higher speed or friction is required because the textile would pass through a double nip that consequently doubles the effect for a given speed (see Figure 7.2). It is also possible to use the three-bowl calender as a two-bowl machine should this be required or thought to be an advantage for a particular technical textile.

A five-bowl calender, shown in Figure 7.3, allows the possibility for the production of a high gloss and soft handle in one pass through the machine. The larger nip width and lower nip pressure between two cotton bowls result in a softer nip zone that creates a more bulky fabric with a softer handle than that possible when using a steel/cotton bowl combination. Because of the configuration of the bowls, it is possible to use this type of equipment as a two-bowl or three-bowl calender, depending on the requirements of the final product. It is therefore the most versatile machine and consequently the most expensive.

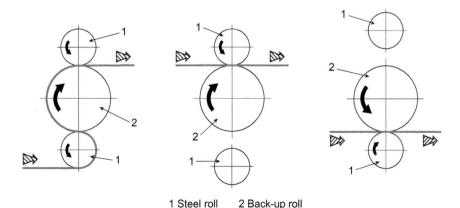

Figure 7.2 Three-bowl calender.

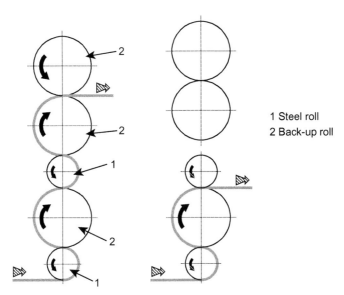

Figure 7.3 Five-bowl calender.

7.3.2   Raising and cropping

Raising (also termed *napping* in the United States) is a process that lifts the fibres in a textile from the material surface to give a layer of protruding fibres either by brushing, teazling, or rubbing in order to give an altered surface appearance and a softer handing fabric. The use of natural teazles set in frames on a raising cylinder can be used to raise surface fibres to produce a drawn or laid pile when the fabric is processed under damp conditions. More commonly, the fabric is raised under air-dry conditions in a blanket style finish by passing the textile fabric at open width under tension over a series of rollers that are covered in a card clothing material. This process increases the amount of air trapped in the fabric, thereby increasing the thermal insulation and providing a softer handle. The type of card clothing, the height of the wire, and the flexibility of the wire all exert an influence on the effect obtained. The process of card wire raising for a single action machine is shown schematically in Figure 7.4.

In the single-action machine, the fabric revolves in the opposite direction to the raising cylinder (or drum). In the double-action raising machine, the fabric passes through the machine in the same direction as the raising cylinder revolves. The card wire rollers in the raising cylinder rotate in the opposite direction. Two sets of rollers arranged alternately work against the fabric surface. In one set (pile rollers), the card wire teeth point in the direction of revolution whereas in the other set (termed *counter-pile rollers*), the teeth point the other way. The maximum amount of raising action is obtained when the pile rollers rotate as slowly as possible and the counter-pile rollers are at maximum speed.

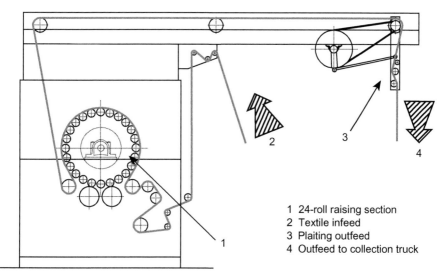

Figure 7.4 Raising machine.

1 24-roll raising section
2 Textile infeed
3 Plaiting outfeed
4 Outfeed to collection truck

Modern raising machines are micro-processor controlled, and the zero point raising condition occurs when the fabric passes through the raising system without any raising action taking place. A finite amount of torque is then applied to produce a defined level of raising action set by the micro-processor control system to produce a reproducible raising action and raised fabric appearance on a particular type of substrate. Raising machines may contain more than one raising drum and may be arranged in sequence for a continuous raising line.

In order to give a completely uniform surface finish, a textile that has passed through a raising machine would then be passed through a rotary cutting (cropping or shearing) machine to ensure that the raised fibre lengths are uniformly similar. A rotary cutting machine consists of a series of spirally oriented blades extending across the width of the cutting roller. The distance of this cutting roller from the open-width fabric passing over a fixed bed can be adjusted to determine the depth of cut. The rotary cutting machine consists of a series of cutting heads that progressively cut closer and closer to the textile surface to produce a consistent and uniform surface appearance and handle. The raising and cropping processes are used to manufacture such things as natural velvet products and some industrial filtration cloths to improve their particle collection effectiveness (see Chapter 4, Volume 2).

Another finishing process that alters the surface of a technical textile is *emerising* (also known as *sueding* or *sanding* in the United States). This process consists of a machine in which fabric at open width is passed over one or more rotating emery-covered rollers to produce a suede-like finish. Woven, knitted, and laminated technical textile fabrics may be emerised, altering the surface appearance, texture, and handle of the emerised fabric according to the emerising conditions. The major change in the fabric after emerising is the production of very short fibres protruding from the fabric surface. Fabric handle is much softer after emerising, and the softness can be greatly enhanced by using micro-fibres (less than 1 dtex per filament) together with chemical softening agents to give a peach skin finish. The most versatile and common form of emerising machine is the multi-roller type that typically may have 4–8 rollers. Each roller is independently driven and may be rotated with or against the direction of the fabric run. The grade (grain or grit) size on the surface of the abrasive emery-covered cylindrical rollers may be varied according to the technical application requirements.

7.3.3 Singeing

Singeing is a process that burns any loose or raised fibres on the surface of a textile fabric using a gas-fuelled flame or by infrared radiation to give a smooth surface finish. The height and size of the gas flame coupled with the machine speed regulates the extent of the burning effect on the fabric surface. The process can be carried out using a single-burner machine (see Figure 7.5) or a double-burner machine, depending on the requirements of the final product. In single-burner machines (see Figure 7.6), it is usually possible to adjust the position of the burner to singe either the top or the underside of the substrate. Double-burner machines are normally used to allow for singeing both textile surfaces in a one-machine pass operation.

Figure 7.5 shows how the adjustment of the burner position can determine the surface that is to be singed.

Singeing technical fabrics increase the surface smoothness, and non-woven fabrics, such as needle-punched fabrics in particular, benefit from this treatment. For instance, the release of collected particle cakes in a fabric filter will be aided if the original fabrics are singed (see Chapter 4, Volume 2).

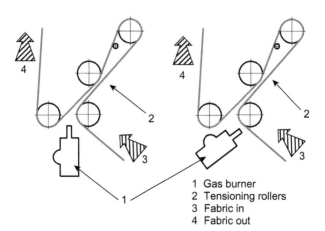

1 Gas burner
2 Tensioning rollers
3 Fabric in
4 Fabric out

Figure 7.5 Single-burner singeing machine.

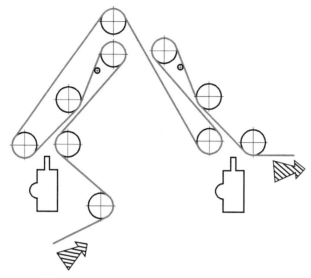

Figure 7.6 Double-sided singeing machine.

7.3.4 Embossing

Embossing a textile fabric produces a pattern in relief by passing the fabric through a calender in which a heated metal bowl engraved with the pattern works against a relatively soft bowl built of compressed paper or cotton on a metal centre. The pressure of the calender nip in the two-bowl calender imparts this pattern onto the surface of the fabric. Usually a compressed paper or cotton bowl would be used underneath the engraved roll so that a mirror of the engraved roll that would allow for a better definition of an embossed effect on the textile is made. The design of the pattern can be wide and varied, usually depending only on the imagination of the product designer. Many fabrics have a Schreiner-embossed finish that is achieved by using a trihelically engraved steel roller, which looks similar to a very fine screw thread. The angle of the engraving must be appropriate for the type of textile being processed in order to obtain the maximum lustre from schreinering. Schreiner fabrics are mainly used for linings in garments. The pressure of the embossing nip is hydraulically controlled and can reach up to 40 tonnes across the width of the textile substrate.

7.3.5 Brushing

Brushing is a process designed to make a textile have a softer handle. The surface of the textile is subjected to the action of a brush made from horsehair or nylon being rotated in contact with the fabric surface and in the opposite direction to the direction of travel of the fabric as it passes through the machine. This action raises individual fibres from the surface to give a "nap" to the material. This process does not lift the fibres to the same extent as the raising process described in Section 7.3.2 and is not usually followed by any shearing process. The substrate is normally agitated by the action of beater bars that consist of rollers running at a high speed fixed on an eccentric centre. The process illustrated in Figure 7.7 reduces any lustre in the textile.

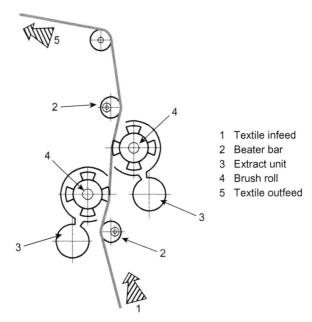

1 Textile infeed
2 Beater bar
3 Extract unit
4 Brush roll
5 Textile outfeed

Figure 7.7 Typical brushing arrangement.

7.3.6   Sanforising

Sanforising is a controlled compressive shrinkage process for which the trademark Sanforized™ (Cluett-Peabody, Inc.) can be applied to cellulosic fabrics that meet defined and approved standards of washing shrinkage. Thus, sanforising minimises the fabric shrinkage, especially in cotton-based textiles in garments after domestic washing or laundering. This is a mechanical process that involves the compression of textile between a heated steel cylinder and an extensible thick rubber blanket. This process is carried out after the fabric has been dampened with water or steam at an elevated temperature. Although this process does reduce shrinkage in the final product, it does not eliminate it completely. Although this process is used mainly on non-technical fabrics, work wear that is often made of 100% cotton or cotton-rich blends with polyester is sanforised in order to minimise shrinkage through the many service-life wash cycles that will be experienced.

7.3.7   Decatising

Decatising (also termed *decating* in the United States) is a finishing process used chiefly to improve fabric handle and appearance in wool-rich fabrics. It also helps to stabilise the fabric dimensions and to minimise shrinkage in the final product. In addition to wool fabrics, decatising can be used for other materials such as polyester, linen, and cotton. In batch decatising, the fabric is interleaved with a smooth cotton wrapper and wound tightly on a perforated roller through which steam is blown. In continuous decatising, the steam is blown through fabric that is passed continuously between one or more perforated drum rollers and a smooth cotton wrapper that forms an endless belt. The substrate would be kept under tension until it is cool in order to minimise the shrinkage in the final product.

It should be noted that both sanforising and decatising only minimise any shrinkage in the final product but does not prevent total shrinkage. This process is often used in formal work wear (e.g. uniforms) and contract and transport upholstery fabrics containing wool as a significant component.

7.3.8 Heat setting

Heat setting is an important finishing process for technical fabrics containing synthetic fibres or filaments. Heat setting may be carried out in hot air, saturated steam, superheated steam, or hot water (hydrosetting). The objective of heat setting is to stabilise the construction and dimensions of the material. All synthetic fibres are subjected to a drawing process during filament extrusion, and heat treatment above the glass transition temperature (T_g) gives rise to molecular relaxation of the polymer chains leading to fibre, yarn, and fabric shrinkage. The second-order or glass transition temperature is the temperature at which molecular movement starts in the amorphous (disordered) regions of the polymer and at which the polymer changes from a glassy solid to a rubbery solid. The glass transition temperature is the temperature at which the polymer segmental motion opens up the fibre structure; hence, at and above T_g, dye molecules and some chemical finishes can diffuse inside the fibre. Thus, if the synthetic filament, yarn, or fabric is held under tension at the requisite dimensions for a relatively short period of time, molecular relaxation takes place within the fibres, and by cooling down below T_g, the structure and dimensions are heat set. Of course, if the heat-setting conditions are subsequently exceeded, then further molecular relaxation can take place and the material will shrink. The general aim of heat setting is to dimensionally stabilise the technical textile to all subsequent processing and end-use application requirements.

The dry heat-setting temperature is clearly lower than the fibre melting point, and, in the presence of steam or water, molecular relaxation is facilitated; hence, the heat-setting temperature under such conditions is lower than the dry heat-setting temperature in hot air.

Care is required during the heat-setting process that temperatures are not too high for the fibres; otherwise, there is a risk of partial decomposition and yellowing and even melting of the textile. In general, the temperature is approximately 20 °C below the melting point of the polymeric constituent in the fibres. Table 7.1 lists typical heat-setting temperatures according to Miles for a selection of fibres and blends [4].

Most technical textile fabrics are heat set in hot air on a stenter (tenter, or frame in the United States) in which the selvedges of the fabric are held by a pair of endless travelling chains maintaining fabric tension through the drying/heat-setting zones. Synthetic fibre fabrics are normally transported on pin plates whereas natural fibre fabrics are transported on clips. Arrangements for adjustment of fabric width and length as well as automatic weft straightening are essential for efficient heat setting.

Table 7.1 **Recommended heat-setting temperatures**

Fabric	Temperature (°C)
Nylon 6,6 (filament) woven	200–210
Nylon 6,6 (filament) knitted	220–225
Nylon 6 (filament)	180–190
Polyester (bulked yarn)	150
Wool blends	170
PE cellulosic blends	180–210
Acrylics	140
Tricel	190–210

It is preferable to remove any processing oils from fibre extrusion/yarn spinning or sizing agents in woven fabric prior to heat setting. Heat setting may be carried out on grey state fabric as an intermediate setting process or as a final post-setting process after coloration. The conditions for post-heat setting must not lead to fabric yellowing or to changes in shade of dyed or printed fabrics.

7.4 Chemical finishing processes

By far the vast majority of chemical finishes are applied using either a pad-dry or a pad-dry-cure process using a pad mangle system by which the textile is passed into a bath of the relevant chemical and, typically, around two small diameter rollers. The upper roller is in a fixed position with the bottom roller being capable of vertical movement up to the upper roller to form a nip. The pressure on the fabric in this nip is regulated by the overall tension in the material. The higher the tension in the substrate gives higher pressure in the nip. Once the material passes through this nip, it passes into a driven nip, often hydraulically controlled but sometimes pneumatically controlled, which removes any excess chemical from the substrate. The second nip is driven at an overall machine line speed but has a speed trim facility to maintain a constant tension in the overall substrate. After the wet textile has passed from the mangle, it passes into a drying oven, or more often into a stenter (tenter, or frame in the United States). The type of stenter depends on the nature of the textile being processed, but generally this is a clip stenter for natural materials and a pin stenter for synthetic products. Typical widths of the fabrics are generally 180–250 cm, but many installations have widths 300–350 cm. Typical production speeds can be up to 80 m/min, depending on the number of drying bays/modules in the stenter (Figure 7.8).

Although many of the chemical finishes are applied to substrates using this type of process, it is also possible to apply the chemicals using alternative processes such as a three-bowl pad mangle or traditional coating techniques. For many functional surface finishes, it is often preferable to apply the finish to one side of the fabric by nip padding (e.g. lick roller or foam application system). This has the additional advantage of decreasing the wet pickup on the fabric, making the fabric easier to dry, and saving

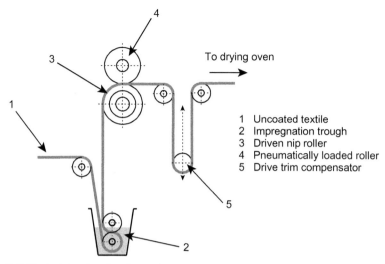

Figure 7.8 Typical two-bowl pad mangle arrangement.

1 Uncoated textile
2 Impregnation trough
3 Driven nip roller
4 Pneumatically loaded roller
5 Drive trim compensator

raw material and heat energy costs. The choice is generally based on the type of chemical finish to be applied and machine availability.

7.4.1 Antistatic agents

Antistatic agents are required for some products to prevent loose dust particles from becoming attached to the surface of a textile. Static electrification arises from friction of the technical textile with a secondary material that can create an electrical charge on the fibre surface. Generation of static may also cause textile filaments, yarns, and fabrics to cling to adjacent surfaces. Voltages generated may be significant in the order of several thousand volts that, if earthed via a human being or animal, can give a nasty shock. Furthermore, the sparks generated can ignite any flammable solvent vapours present that could lead to a fire, explosion, or both. Clearly, therefore, the minimisation and possible removal of static from a textile is crucial to ensure safety in processing and use. Whether or not the charge is positive or negative depends on the position of the fibre place in the so-called triboelectric series shown in Table 7.2.

The static generating property of a textile may be reduced by increasing its electrical conductivity. This may be done physically by including metallic (e.g. stainless steel fibres) or carbon-containing fibres or filaments in yarns; this technique is often used for gas filtration fabrics in which high temperatures and corrosive gases may be present.

On the other hand, antistatic chemicals, commonly polyether-type materials, may be applied using a pad mangle to increase the electrical conductivity of the fabric and consequently very significantly decrease or eliminate the generation of static electricity at the textile surface. Polyether- (or polyglycol)-based finishes are usually hydrophilic, so the water they attract from the atmosphere causes any static electricity to

Table 7.2 **Triboelectric series**

	Reference [5]	Reference [6]	Reference [7]
Positive (+)	Wool	Glass	Wool
	Hercosett wool	Nylon 6,6	Nylon
	Nylon 6,6	Nylon 6,6	Viscose
	Nylon 6,6	Wool	Cotton
	Silk	Silk	Silk
	Regenerated cellulose	Viscose	Acetate
	Cotton	Vinylon (PVAc)	Lucite (PMMA)
	Poly(vinyl alcohol) (PVAlc)	Acrilan (acrylic)	PV Alc
	Chlorinated wool	Steel	Dacron (Polyester)
	Cellulose triacetate	Cotton	Orlon (acrylic)
	Calcium alginate	Orlon (acrylic)	PVC
	Acrylic	Acetate	Dynel (VC/AN)
	Cellulose acetate	Dynel (VC/AN)	Velon (VDC/VC)
	Polytetrafluoroethylene	Saran (PVDC)	Polyethylene
	(PTFE)		
	Polyethylene	Rhovyl (PVC)	Teflon (PTFE)
	Polypropylene	Rubber	
	Poly(ethylene terephthalate)		
	Poly(1,4-butylene		
	terephthalate)		
	Modacrylic		
Negative (−)			

dissipate. In dry environments, however, such antistatic agents are not very efficient. The generation of static electricity is more predominant in synthetic fibres because of their more hydrophobic character, whereas a majority of natural fibres normally contain a small percentage of regain water that would act as an electrical conductor. Other antistatic chemicals may be composed of ionic species that conduct away any generated static. These chemicals include polar compounds such as amine oxides, ammonium salts and quaternary ethoxylated amines, quaternary fatty amide amines, phosphate esters, and pyridinium salts with fatty acids [6].

7.4.2 *Flame retardants*

Many textiles, both natural and synthetic, are susceptible to burning, which is totally unacceptable in a majority of their final applications, especially technical applications where regulatory fire standards prevail. The primary requirement is to reduce the ease of ignition and, if ignited, increase the likelihood of self-extinguishing any fire and consequently preventing its spread. A full description of the means generating heat and flame resistance in technical textiles is discussed in Chapter 8, Volume 2, but here it will suffice to give an overview of the principles and techniques involved in the finishing techniques used.

The chemistry of these processes, mainly for natural fibres and their blends, has been well documented over the last 50 years with a number of classic reviews [8,9] and more recent ones [10,11].

Cellulosic (e.g. cotton) fibre technical fabrics are generally used in applications such as work wear, barrier fabrics, and some contract furnishings whose high level of durable flame retardancy, usually dictated by regulatory standards, determines whether flame-retardant finishes are acceptable or not. In work wear, durability after multiple, high temperature ($\geq 75\,°C$) washings is essential although this is less an issue for barrier and furnishing fabrics, including mattress covers and tickings, which usually require less stringent cleansing processes. The majority of acceptable flame-retardant treatments involve one of the following chemical processes:

(i) Durable finishes based on N-methylol dialkylphosphonopropionamides and typified by the Pyrovatex® (formerly Ciba, now Huntsman) chemistry range of products and the many equivalents on the market [11].

(ii) Durable finishes based on tetrakis (hydroxymethyl) phosphonium chloride (THPC)/urea chemistry [9], which in the Proban® process (formerly Rhodia, now Solvay) involves a final ammonia cure and oxidation.

(iii) Semi-durable treatments based on ammonium polyphosphate, pad, dry, and cure processing.

(iv) Back-coating treatments based on organobromine/antimony trioxide formulations in the main and applied in a resin matrix [11,12].

For work wear, the preferred treatment is usually one based on Proban® processing because of the minimal effect it has on abrasion resistance and tensile strength of fabrics although dyestuffs must be chosen carefully because of the potentially reactive environment present during the ammonia cure process. For 100% cotton and high cotton content contract furnishings, the Pyrovatex®-type finishes are preferred may be applied because these treatments have little effect on dyestuffs, so printed fabrics may be safely flame retarded. This system also uses conventional pad-dry-cure processing technology.

For barrier fabrics at the lower end of the contract market, cured ammonium polyphosphate-based finishes may be used because durability in dry cleaning and/or warm water soaking may be the cleansing requirement.

Flame-retardant finishes are rarely applied in isolation, and there may be other finishes applied either before, during, or after their application to improve properties such as handle and soil resistance. This is not the case with coatings, including back coatings that are usually applied to the reverse face of furnishing fabrics, whether UK domestic or contract. This technology has proved to be very successful and cost effective because it has little if any effect on the aesthetics of the fabric. Typical back coatings comprise an organobromine (or chlorine) compound, antimony trioxide as synergist, and a binding resin.

Coated (see Chapter 8), flame-retarded fabrics, often combining waterproof characteristics and flame retardancy, include some of the traditional tarpaulins, tentage, awning, and similar materials in which cotton was the traditional textile substrate. Currently, most of such products have a synthetic fibre-containing or even glass fabric as the base fabric if high temperature resistance is also required. The whole area of flame-retardant coatings has been reviewed recently [13], and, of course, where waterproof properties are of prime importance, inherently flame-resistant poly(vinyl chloride) is still an important choice for achieving acceptable levels of flame retardancy on many textile substrates although such formulations require the presence of antimony trioxide as a synergist if they are to be really effective.

Although important generally within technical textiles, polyester has a limited role within the heat and flame-resistant product area because of its melting tendency. However, it is used in webbings and other fabrics that require both high tensile strength and flame resistance. However, both here and in the contract furnishing sector, the use of durable flame-retardant finishes competes with inherently low flammable polyesters such as Trevira CS® (see Chapter 8, Volume 2).

Most flame-retardant treatments for thermoplastic fibres are limited to reducing afterflame and self-extinguishment of melt drips that do not allow them to be used in protective end uses such as work wear. However, simultaneous dyeing and flame retardation of 100% polyester fabric can be achieved with reasonable levels of durability either in the dye bath or by a thermosol process. Currently in the United Kingdom, for example, Solvay (formerly Rhodia) markets Amgard CU that is the same as the former Antiblaze® 19 for finishing polyester textiles based on the cyclic phosphonate structure below where $n=1$:

$$(CH_3O)_{2-n}.P \begin{array}{c} O \\ \| \\ | \\ CH_3 \end{array} \left[OCH_2.C \begin{array}{c} CH_2CH_3 \\ | \\ CH_2.O \\ CH_2.O \end{array} \begin{array}{c} O \\ \| \\ P.CH_3 \end{array} \right]_n$$

The dimer form ($n=2$), Amgard 1045 (formerly Antiblaze 1045, Albright and Wilson), has a lower volatility than Amgard CU and so has been used in thermosol operations and as a melt additive for polyester, although the only current suppliers of this system appear to be Chinese. Thor Chemicals produces an Aflammit PE product believed to be a similar species and claimed to be applicable from the dye bath followed by optional thermosol curing. Avocet Chemicals in the United Kingdom promotes its Cetaflam® DB series of flame retardants of which Cetaflam® DB9 is a halogen-free formulation for polyester textiles used in the automobile, clothing, work wear, furnishings, curtains and decoration, and public transport textile markets.

Effectively flame-retarding polyester/cotton blends is not simple, and the problems associated with it have been discussed elsewhere [9]. Within the technical textile sector, however, the uses of these blends probably compete with flame-retarded cotton in the work wear sector; it is generally known that so long as the blend comprises $\geq 50\%$ cotton, durable treatments such as the Proban® system may be used successfully. Otherwise, coating processes based on organobromine/antimony oxide/resin binder formulations can be used.

Flame-retardant finishes for wool and wool blends are used primarily in the industrial, defence, and civil emergency protective clothing sectors as well as seating fabrics for transport (see Chapters 9 and 11, Volume 2). During the last 40 years, the level of thermal and flame protection has most successfully been enhanced using the well-established Zirpro® process finish developed by Benisek in the 1970s and reviewed recently [11]. Based on hexafluorotitanate and hexafluorozirconate chemistry, the treatment can be carried out in the dye-bath and yields fabrics that, when exposed to heat, produce an intumescent char most beneficial for protective clothing including aprons, trousers, and gloves for furnace workers, especially where molten metal splash hazards are present. The formation of the intumescent char while giving protection also allows incident metal droplets to break away from the surface before they can penetrate into the fabric structure and endanger the wearer. The addition of tetrabromophthalic acid (TBPA) to the basic Zirpro® treatment produces a finish suitable for end uses in which low after flame times are required, although it does increase smoke generation. To attempt to address the smoke issue, low-smoke Zirpro® treatments based on a fluorocitrate–zirconate complex are available.

When wool is to be used in protective clothing as well as transport upholstery applications, the use of multi-purpose finishes is important, and their compatibility with the Zirpro® present is essential. Such finishes may include [9,11]:

- Oxidative shrink-resistant treatments that should be applied before the Zirpro® treatment.
- Insect-resistant treatments that should be added to the Zirpro® bath first.
- Resin-based, shrink-resistant treatments that can promote flammability unless, like the Hercosett (Hercules) resin, they contain elements such as chlorine and nitrogen, resins that should be applied after Zirpro® treatment.
- Co-application of water-repellent (e.g. resin-wax dispersions) and oil-repellent (e.g. fluorocarbon) finishes that should follow Zirpro® treatment, for example, by an additional pad-dry-cure-rinse-dry process.

7.4.3   Oil and water repellents

Oil- and water-repellent properties are important on substrates used for outdoor and leisure garments and military textiles along with umbrellas and some work wear. The term *waterproof* normally applies to textile materials that can prevent the absorption of water and the penetration of water into the structure. Thus, a waterproof surface provides a barrier to water under all practical end-use conditions. The most widely used method of producing a waterproof fabric is by coating it with a solid polymeric coating (e.g. neoprene [synthetic rubber]), polyvinyl chloride, or polyurethanes. A waterproof fabric is thus impermeable to both the passage of air and water vapour because such coatings are non-porous.

For many waterproof technical textiles, industrial fabrics and textiles destined for outdoor use, for example, tarpaulins or awnings this does not create a problem. However, in technical fabrics used in apparel, transpiration of air and moisture is required to pass through the fabric to maintain high levels of thermophysical and thermophysiological comfort for the wearer. These conditions can be achieved using the application of water-repellent finishes to porous textile fabrics.

Water is a hydrogen-bonded liquid with a high surface tension ($72.75\,mN/m^{-1}$ at $20\,°C$), whereas hydrocarbon-based oils have a much lower surface tension (around $20–31\,mN/m^{-1}$ at $20\,°C$). Thus, some water-repellent finishes are satisfactory for repelling water but are inadequate for repelling oils. The major finishes that have been applied for water repellency are silicones, but to achieve water and oil repellency, fluorocarbon-based finishes have been widely used. However, fluorocarbon finishes are currently under pressure because they are difficult to biodegrade and hence can bioaccumulate in the environment.

Many types of water-repellent finishes have been applied to technical textiles, the properties and performance of which have been selected for the particular end use. The application of soaps and fatty acids were originally used to impart the required properties, but these have proven to have poor longevity after washing as well as a poor water-repellent performance. Currently, a vast majority, if not all, water and oil repellence is achieved by the use of emulsions of fluorocarbon compounds that are applied using dip or nip padding techniques followed by curing at a temperature in the range 150–180 °C. Silicone compounds, applied together with a catalyst followed by curing, do not give good oil repellency and hence are used principally as water-repellent finishes. The repellency of silicones and fluorocarbons can be increased by the use of extenders that are chemical cross-linking agents in order to achieve maximum efficiency.

7.4.4 Softeners

Different fibres and manufacturing processes produce textiles with varying degrees of stiffness that also affect the softness of handle. A closely woven fabric produces a stiffer material than a more loosely woven one. Depending on the end use of the final product, a certain type of fibre with a specific fabric construction could be required, but the softness or handle may be lower than desired. The application of emulsions of waxes and oils has been used, but developments have shown that emulsions of organic silicones and polyacrylates can improve the speed of the effects and give a softer characteristic to the textile. The most effective of the many chemical softening agents available are amino-functional polysiloxanes. Most softening agents are co-applied with other chemical finishes to give the required overall level of properties and performance. As with many of the other processes in textile finishing, it is normal to apply these emulsions with a pad mangle system followed by an appropriate curing treatment.

7.4.5 Photoprotective agents

Photoprotective agents are required for some technical end-use applications because of the combined action of ultraviolet radiation and atmospheric oxygen, which can lead to fibre degradation, thereby decreasing the useful life of the technical textile. Application of ultraviolet absorbers is essential for materials in outdoor end uses, especially where these are exposed to strong sunlight. To prevent fibre degradation, the materials may be treated with an antioxidant and an ultraviolet absorber that together inhibit the deterioration process, prolonging the useful service life of the fabric. Both organic and inorganic compounds are used to protect the substrate where there is an interaction with ultraviolet light that scatters the light and emits small amounts of heat as any absorbed energy is released. The photoprotective agents used can consist of metal oxides (e.g. titanium dioxide, zinc oxide) often in nanoparticulate form coupled with organic-based ultraviolet absorbers (e.g. derivatives of benzophenone, benzotriazole, phenyl salicylate, and cyanoacrylates). Ultraviolet absorbers can be co-applied during dyeing.

7.4.6 Soil release finishes

Soil release finishing is more commonly required in technical textiles used for garments, upholstery, and military apparel made from synthetic fibres. The main reason for this is the hydrophobic nature of the materials that can result in soils becoming more attached to the synthetic fibres. The use of hybrid fluorocarbons based on dual-action block co-polymers or other hydrophilic polymers based on carboxy-, hydroxy-, or ethoxy-based finishes are generally used. This makes any soiling more accessible to water and consequently easier to remove during laundering. The polymeric finishes are applied using a padding technique and are subsequently dried and cured in a stenter drying system. Even though a soil release finish applied to a synthetic substrate does improve fibre wetting, moisture dissipation, and soil removal, the effect is not as good as the soil release properties from a natural material. Natural fibre textiles such as cotton or wool are inherently water absorbent and easily washable, allowing easy removal of any soiling in the laundry process.

7.4.7 Antimicrobial and fungal treatment

Many technical textiles, particularly those containing natural fibres, can become wet during the end use, which renders them susceptible to microbial and fungal attack that can lead to fibre degradation (rotting) and the growth of moulds or mildew. It is necessary, therefore, to treat the substrates with an antimicrobial compound to eliminate the formation and growth of bacteria, fungi, and algae. This is particularly important in clothing and household linens whose microbial and fungal attack is easily recognised by malodour formation, discoloration, and a slick, slimy handle.

Antimicrobial compounds are used in hygiene finishes to protect the textile user against pathogenic or odour-causing micro-organisms. They are also used to protect the textile material from damage that can adversely affect service life. This is particularly important for end uses such as awnings, screens, tarpaulins, tents, and ropes when the physical and mechanical properties of the fabric must be maintained.

The most widely used antimicrobial finishes are based on the antibacterial agents of two main types. When the applied antimicrobial finish is not chemically bound to the substrate, it migrates from the substrate to enter the microbes, acting as a poison to kill them. This type of antimicrobial is subject to leaching under wet conditions that can result in loss of the finish and, hence, in a decreased treatment efficiency. The second type of antimicrobial finish is chemically bound to the substrate and remains fixed to the surface, thereby imparting durability to the treatment. Microbes entering the substrate are poisoned by the chemical treatment. The second method provides a longer-lasting resistance to attack, but its efficiency can be reduced by abrasion to the substrate during service.

The major types of antimicrobial agents that have been used on textiles include:

- Metals and metal salts (e.g. silver, titanium dioxide, and zinc oxide).
- Quaternary ammonium compounds (e.g. 3-trimethyoxysilylpropyl-dimethyloctadecyl ammonium chloride).
- Polyhexamethylene biguanides.

- Triclosan (2,4,4′-trichloro-2′-hydroxydiphenyl ether).
- Chitosan (poly-(1,4)-2-amido-deoxy-D-glucose).
- Regenerable *N*-halamine and peroxyacid.

Manufacturers and importers of antimicrobial-treated textiles manufactured in the European Union or imported to the European Union are allowed only if the biocides (active substances) used to impart antimicrobial properties have been approved in accordance with the European Union Biocidal Products Regulation (BPR) 528/2012 that repeals and replaces the former Biocidal Products Directive EU 98/8. The BPR has been in force since September 1, 2013.

7.5   Finishing for the future

7.5.1   Flame retardants

Many of the conventional treatments for flame retardance are based on halogen compounds or formaldehyde-based chemicals that suffer from both real toxicological (formaldehyde) and claimed ecotoxicological (for antimony-halogen-based finishes) properties [14]. With regard to the latter, the latter may be replaced with compounds containing phosphorous although they tend not to be as effective. This is especially a problem in the contract furnishing sector where antimony/bromine-based, back-coating formulations are effective on all fibre types and blends whereas possible replacements are substrate specific and often more expensive. Developments have been made to incorporate the use of nanotechnology and the phenomenon of intumescence to provide a flame and thermal barrier on the textiles, but few effective commercial examples exist as yet [15]. The incorporation of nanoparticles in traditional back-coating technology has been studied but has yet to be proved to be successful along with plasma technology, which modifies the surface of the textile to prevent ignition and the spread of flame. However, it is expected that some of these new approaches will prove to be commercially successful in the future.

7.5.2   Plasma technology

Plasma technology has been shown to have the ability to improve numerous properties of textiles [16,17]. For example, it can be used to enhance the hydrophilic/hydrophobic properties of fibres, promote good adhesion in coatings and laminates, confer antistatic properties on manufactured fibres, improve the dyeability or printability of textiles, and functionalise fibre surfaces for a wide variety of end uses. The effectiveness of the plasma treatments varies with the actual plasma used, however. The most commonly used plasma treatments are oxygen/helium or air/helium. Because there are differing effects with the two types of plasma, care is necessary to ensure that the correct plasma is used to impart the required effect into the final product. Plasma treatment has been shown to be effective in assisting the desizing of cotton, imparting improved hydrophilic characteristics to hydrophobic synthetic fibres such as polypropylene, antibacterial treatments based on the deposition of silver nanoparticles, and as ambient temperature sterilisation of medical textiles. Some materials, such as untreated polypropylene, have poor adhesion to coatings and laminates but with plasma treatment prior to coating, the adhesion strength is greatly increased. Similar print adhesion improvements also can be achieved with materials that are subsequently ink-jet printed.

At the present time, a great deal of development work is underway to further explore the future possibilities in the use of plasma finishing, including the surface grafting of polymers that has proved to give enhanced and more permanent finishes to a wide range of textile fibres used in the technical textile sector [17].

7.5.3 Nanotechnologies

Nanotechnology is the use of fibres on the nanoscale. This generally refers to fibres with diameters in the range of 1–100 nm. The use of nanotechnology is still in its infancy but is the subject of a great deal of development work [18,19]. Present developments have shown significant improvements in textile breathability in comparison with standard woven textiles because air passes through the textiles more easily from the much larger surface area of the individual nanofibres.

The use of nanotechnology has been shown to provide improvements in several technical textile areas, such as antistatic properties, oil and water repellency, electrical conductivity, flame retardancy, and abrasion resistance. The application of chemicals in nanoparticulate form to technical textiles can result in marked improvements in performance compared with the same chemicals applied by more conventional techniques, but as yet, the full potential of this has yet to be realised.

At the present time, nanotechnology is still in its infancy on an industrial scale, but current research and process developments are increasingly achieving full advantages of this technology.

References

[1] Heywood D, editor. Textile Finishing. Bradford, UK: Society of Dyers and Colourists; 2003.
[2] Hall ME. Finishing of technical textiles. In: Horrocks AR, Anand SC, editors. Handbook of Technical Textiles. Cambridge: Woodhead; 2000. p. 152–72.
[3] Zeronian SH, Inglesby MK. Cellulose. Springer Publishing, New York, USA; 1995, p. 7.
[4] Miles LWC. Drying and setting. In: Heywood D, editor. Textile Finishing. Bradford, UK: Society of Dyers and Colourists; 2003. p. 57.
[5] Hersh SP, Montgomery DJ. Static electrification of filaments: experimental techniques and results. Text Res J 1955;5:279–95.
[6] Smith PA, East GC, Brown RC, Wake D. Generation of triboelectric charge in textile fibre mixtures and their use as air filters. J Electrostat 1988;21:81–98.
[7] Tsuji W, Okada N, cited by Sakurada I. Vinyl fibres. In Handbook of Fibre Science and Technology: Volume IV. Fibre Chemistry. New York, NY, USA: Marcel Dekker; 1985. p. 580.
[8] Lewin M. Flame retardance of fabrics. In: Lewin M, Sello SB, editors. Handbook of Fibre Science and Technology: Volume II. Chemical Processing of Fibers and Fabrics, Functional Finishes, Part B. New York: Marcel Dekker; 1983. p. 1–141.
[9] Horrocks AR. Flame retardant finishing of textiles. Rev Prog Color 1986;16:62–101.

[10] Horrocks AR. Flame retardant finishes and finishing. In: Heywood D, editor. Textile Finishing. Bradford: Society of Dyers and Colourists; 2003. p. 214–50.

[11] Horrocks AR. Overview of traditional flame-retardant solutions. In: Alongi J, Horrocks AR, Carosio F, Malucelli G, editors. Update on Flame Retardant Textiles: State of the Art, Environmental Issues and Innovative Solutions. Shawbury, UK: Smithers Rapra; 2013. p. 123–78.

[12] Dombrowski R. Flame retardant for textile coatings. J Coated Fabrics 1996;25:224.

[13] Horrocks AR. Flame retardant/resistant textile coatings and laminates. In: Horrocks AR, Price D, editors. Advances in Fire Retardant Materials. Cambridge: Woodhead Publishing; 2008. p. 159–87.

[14] Horrocks AR. Flame retardant and environmental issues. In: Alongi J, Horrocks AR, Carosio F, Malucelli G, editors. Update on Flame Retardant Textiles: State of the Art, Environmental Issues and Innovative Solutions. Shawbury, UK: Smithers Rapra; 2013. p. 207–38.

[15] Selen Kilinc F, editor. Handbook of Fire Resistant Textiles. Cambridge: Woodhead Publishing; 2013.

[16] Shishoo R, editor. Plasma Technologies for Textiles. Cambridge: Woodhead Publishing; 2007.

[17] Mather R. Surface modification of textiles by plasma treatments. In: Wei Q, editor. Surface Modification of Textiles. Cambridge: Woodhead Publishing; 2009. p. 296–317.

[18] Vigneshwaran N. Modification of textile surfaces with nanoparticles. In: Wei Q, editor. Surface Modification of Textiles. Cambridge: Woodhead Publishing; 2009. p. 164–84.

[19] Alongi J, Carioso F, Malucelli G. Smart (nano) coatings. In: Alongi J, Horrocks AR, Carosio F, Malucelli G, editors. Update on Flame Retardant Textiles: State of the Art, Environmental Issues and Innovative Solutions. Shawbury, UK: Smithers Rapra; 2013. p. 257–311.

Bibliography

Alongi J, Horrocks AR, Carosio F, Malucelli G, editors. Update on Flame Retardant Textiles: State of the Art, Environmental Issues and Innovative Solutions. Shawbury, UK: Smithers Rapra; 2013.

Gao Y, Cranston R. Recent advances in antimicrobial treatments of textiles. Text Res J 2008;78(1):68–72.

Holme, I. Textiles: A Global Vision. Vision 2020: Finishing for the Future. Proceedings of the Textile Institute Centenary Conference, Ian Holme.

Kaounides L, Yu H, Harper T. Nanotechnology innovation and applications in textile industry: current markets and future growth trends. Mat Technol 11/2007;22(4):209–37.

Pailthorpe M, David SK. Antistatic and soil-release finishes. In: Heywood D, editor. Textile Finishing. Bradford, UK: Society of Dyers and Colourists; 2003. p. 308–20.

Morent R, De Geyter N, Verschuren J. Non-thermal plasma treatment of textiles. Surface and Coatings: Elsevier; 2008.

Sawhney APS, Condon B, Singh KV, Pang SS, Li G, Hui D. Modern applications of nanotechnology in textiles. Text Res J 2008;78(8):731–9.

Virk K, Ramaswamy GN, Bourham M. Plasma and antimicrobial treatment of nonwoven fabrics for surgical gowns. Text Res J 2004;74(12):1073–9.

Wei Q, editor. Surface Modification of Textiles. Cambridge: Woodhead Publishing; 2009.

Coating of textiles

8

Roy Conway

Textiles Division, School of Materials, University of Manchester, Manchester, United Kingdom

Chapter Outline

8.1   Introduction

Coating covers a range of techniques that offer one of the most important means of conferring additional and/or novel properties and, hence, added value to technical textiles. This chapter considers the application of coatings to textiles, technical textiles being substrates processed for their properties rather than for their aesthetic appearance, and is based on the author's more than four decades of experience in the industry. A number of earlier reviews exist that the reader may find of use to complement this chapter [1,2] including that by Hall in the earlier edition of this chapter [3]. It is important to realise that the various chemicals and formulations used in industry are very varied and change rapidly over time, and the applicability to the requirements of the end products should be researched thoroughly. In addition, new chemicals are continually developed, and improvements to formulations are continually being made. The bibliography provides examples of useful sources of information in these respects.

Handbook of Technical Textiles. http://dx.doi.org/10.1016/B978-1-78242-458-1.00008-X

8.2   History of coating

The earliest coated textiles originated in Central and South America where natural rubber latex exuded from the trunks of certain trees. This milky substance is spread onto fabrics and draped in sunlight, which coagulated the latex to produce a flexible, elastic, and liquid-proof textile that was then used for waterproof covers, bags, and containers. The discovery of the Americas in the sixteenth century led to this knowledge becoming widespread in Europe and by the late eighteenth century, a rubber industry had been established.

Natural rubber could be dissolved in naphtha to produce a viscous solution that was used to make a waterproof cloth patented by Charles MacIntosh in 1823, even though the fabric was invented by surgeon James Syme. The original patent, granted in 1823, was number 4804.

MacIntosh made the waterproof fabric into coats in the family textile factory in Glasgow. The company merged with clothing company Thomas Hancock of Manchester that had been experimenting with rubber-coated fabrics since 1819. The early fabrics did have some problems with stiffness, smell, and a tendency to melt in hot weather, which were improved by Hancock who patented a method for vulcanizing rubber in 1843, solving many of these problems. The company continued in business until it was taken over by Dunlop Rubber in 1925.

Since then, many different compounds have been used to manufacture coated textiles, especially with the advent of polymer technology, which gives rise to an extremely wide range of materials that meet a diverse scope of requirements (see Table 8.1).

8.3   Coating materials

The range of materials used for textile coating is extremely diverse, and developments are continually being made to meet certain specifications for the final material. Charles MacIntosh's initial rubber coating in solvent naphtha has developed over the passage of time, leading to water-based coatings for soil resistance, fire retardance, and waterproofing; almost all solvent-based coating is now being eliminated. Continual

Table 8.1   **Brief list of polymer usages in coating**

Chemical	Use
Polyvinyl chloride (PVC)	PVC bags, tarpaulins, rubber bath mats
Polyethylene and other polyolefins	Low-cost waterproof sheeting, roofing membranes
Polyurethane	Artificial leather, lamination adhesive
Acrylics	Main base polymer for fire retardancy, flocking, textile stabilisation
Rubber (natural or synthetic)	Protective and waterproof clothing
Waxes	General waterproofing for outdoor wear and tents
Hot melt epoxy resins	Composite materials for aerospace, automotive

development in chemicals and formulations continues to add to the diversity of the uses of coated textiles. In this chapter, the main intention is to deal with coating techniques rather than the chemicals and polymers used in coating.

There are, however, some more commonly used materials that should be mentioned. A large number of textile coatings are prepared using acrylic-based polymers, especially in the field of soft furnishing for drapes and upholstery. This field requires a low-cost material to enhance the final product with the addition of a flock to provide a blackout requirement to a drape or to include flame retardance to an upholstery fabric. Additives to a basic acrylic formulation allow a wide variety of properties to be achieved.

For many years, a synthetic leather material for upholstery and automotive applications was required. One of the earliest materials was a nitrocellulose coating in solvent solution onto a woven fabric and sold under the trade name of Rexine, first registered as a trademark in 1899 [4]. Developments were made and virtually replaced the nitrocellulose with polyvinyl chloride because of the hazardous nature of the nitrocellulose material. Solvent-based polyurethanes were developed and used for upholstery to gave a soft leather-like handle to the fabrics but initially suffered from poor hydrolytic stability and poor abrasion resistance.

Current developments have given rise to hot melt epoxy coatings, which are used in composite materials along with carbon fibre technology for used in areas where high strength and low weights are required such as aerospace, motor racing, and ocean yacht racing [5,6].

Technology is developing in a great number of areas and new products are evolving all of the time. The following sections give some indications as to the types of processes used in the coating of textile materials. Similar processes can be used for other materials such as papers, films, and foils and in some cases for the coating of metals.

8.4   Coating techniques

The basic fact is that for a textile material, there are only two areas where the coatings may be applied. Either the surface of the material is to be coated or the coating is impregnated into the structure of the fabric. The dimensional stability of the textile must be taken into account to determine the technique to use. If a material is dimensionally stable, it can be coated using direct coating, but transfer coating is required for non-dimensionally stable fabrics.

8.5   Direct coating versus transfer coating

As the name implies, the *direct coating* process applies the coating directly onto the dimensionally stable fabric, using the appropriate coating technique for the product to be manufactured, depending the end use of the product and the rheological characteristics of the coating to be applied. A majority of the techniques described later can be used for direct coating.

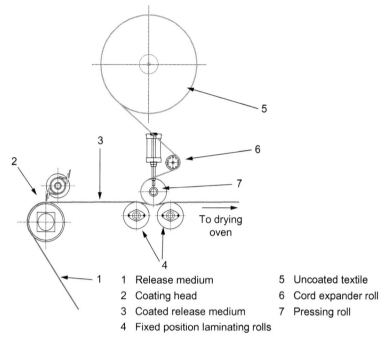

1 Release medium   5 Uncoated textile
2 Coating head   6 Cord expander roll
3 Coated release medium   7 Pressing roll
4 Fixed position laminating rolls

Figure 8.1 Transfer coating.

However, as already stated, transfer coating is accomplished for non-dimensionally stable fabrics by first applying the coating onto a release medium such as a wax-coated paper or a silicone rubber belt. Immediately following the coating, the textile is laid on the wet coating before being either dried or cured, whichever is appropriate. Once the "sandwich" is dried, the textile with its coating is separated from the release medium with each of the layers being rolled separately. The coated textile is then ready for any subsequent processing, and the release medium is ready for reuse for further coating. This process is illustrated schematically in Figure 8.1.

As with direct coating, the release medium may be coated using one of the techniques outlined in the following.

8.6   Blade coating

Blade coating is probably the most common technique used for the application of a chemical either onto or into the structure of a textile. In general terms, if the coating is required to go onto the surface of a textile, the technique to be used is blade over roll. This would normally be carried out using a "shoe" profile blade. Applying a coating into the structure of the textile would normally be carried out using the blade in air technique with a "knife" profile blade. Although the normal technique would use a particular blade profile, no hard-and-fast rule dictates this. The choice would depend on the coating being applied and the requirement for the end use of the product. Many

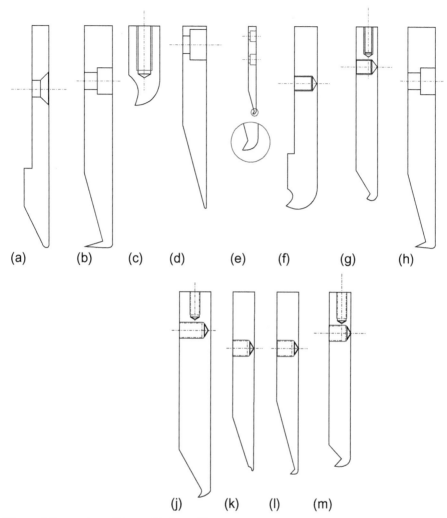

Figure 8.2 A variety of doctor blade profiles.

different coating companies would use their own tried-and-tested blade profile that might have been developed over many years of experience. Figure 8.2 shows a small selection of blade profiles with shoe profile blades B, C, E, F, G, H, J, L, and M. Knife profile blades are shown as A, D, and K.

8.6.1 Blade-over-roll coating

In coating, the nomenclature of each process is very straightforward and logical. In this case, *blade-over-roll coating* quite simple means that the coating is applied with the blade directly over a roll as shown in Figure 8.3. Using this process, the

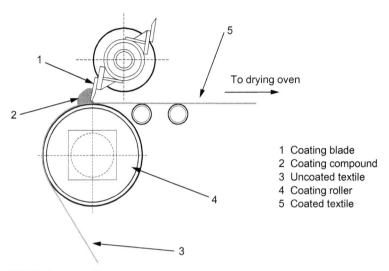

Figure 8.3 Blade-over-roll coating.

coating is more on the surface of the textile rather than into the textile structure. Usually the roll is driven at machine line speed with the addition of a speed trim facility to allow a variation of $\pm 5\%$ of line speed so that a small variation in the coating weight may be achieved. The blade is positioned either on top dead centre of the roll or slightly in front of the roll. The height of the blade above the textile to be coated determines the actual weight of coating to be achieved. Usually $1\,\mu m$ gives a coating weight of $1\,g/m^2$ when using a coating compound at 100% solids and a density of $1\,g/cm^3$. The required coating weight can be determined using this information. For example, a $1\,g/m^2$ coating weight can be achieved with a coating gap of $2\,\mu m$ when using a compound at 50% solid content. As with all coatings, care should be taken so that the amount of wet coating applied does not lead to any deformities such as blisters or bubbles when passed through any subsequent drying process. These deformities can occur when the oven temperatures are set too high at the entry section, causing the solvent to boil within the coating. If this does occur, the amount of wet coating should be reduced and/or the temperature gradients in the drying oven should be revised. Typically, for a five-zone oven, when using a water-based coating, the oven temperatures are set at 100, 120, 130, 140, and 150 °C to ensure that no surface imperfections are created and that all of the water is removed from the coating. The coating compound is applied directly in front of the doctor blade to form a bank of material and as the textile passes through, an even layer of compound is applied. See Figure 8.3 for a graphic representation of this process.

In order to obtain as smooth a coating as possible, is it normal to ensure that the tip of the shoe blade is the last part of the blade to be in contact with the coating compound. If the lower flat portion of the shoe is used, it is normal to have a smooth finish in the final dried coating. This is shown schematically in Figure 8.4.

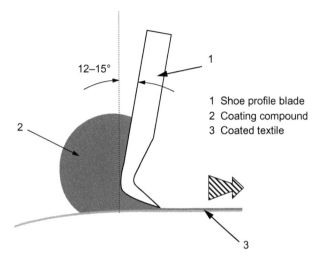

Figure 8.4 Typical shoe blade angle in blade-over-roll coating.

8.6.2 Blade-in-air coating

Blade-in-air coating is often called *floating knife coating* when the coating blade is depressed into the textile in an area where there is no direct fabric support. As with blade-over-roll coating, the coating chemical is added to the equipment to form a bank in front of the coating blade. Most often, the blade profile used in this technique is a knife blade (typically shown as (a) in Figure 8.2) rather than the shoe blade commonly used in blade-over-roll coating. Although a knife profile blade is the most common profile to be used, there are no hard-and-fast rules, and a shoe profile may be used. The choice depends on the rheology of the coating compound and the required amount of penetration of the coating into the substrate. The blade in air technique applies the coating into the textile structure rather than onto the surface of the substrate. The method to determine the coating weight is empirical, and several trial and error tests need to be carried out to determine the amount required in the final product. Adjustments to the coating weight could be made by a variation in the coating compound solids content, the tension in the textile substrate, and the angle of the coating blade in respect of the textile passing through the machine. Figure 8.5 shows blade-in-air coating in a schematic form.

8.6.3 Blade-over-blanket coating

The number of companies using the blade-over-blanket coating technique is decreasing throughout the world. The technique uses a seamless rubber blanket under tension that is driven at the machine line speed (see Figure 8.6). Usually there is no option to adjust this speed during processing. A shoe profile blade (typically shown as (b) in Figure 8.2) is commonly used with the coating compound placed in front of the blade to form a bank or rolling dam. Textiles coated using this technique are quite commonly open or less dense structures in which the blanket gives support to the fabric at the

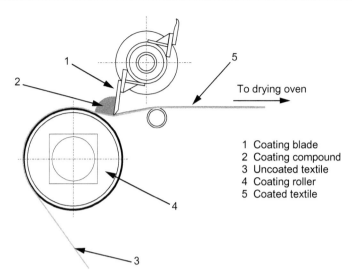

Figure 8.5 Blade-in-air coating.

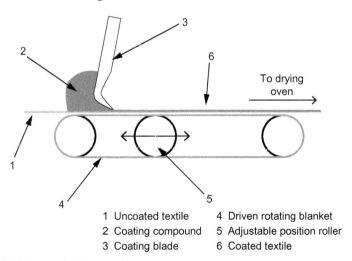

Figure 8.6 Blade-over-blanket coating.

coating head. Often, there is a horizontally adjustable roller placed in the middle of the blanket that can be used to give some adjustment to the blanket tension and consequently to the amount of support to the textile being processed.

8.7 Reverse roll coating

The equipment used in reverse roll coating in general is significantly more expensive than other coating techniques because it involves three precision ground rollers compared with only one in knife coating; each roller has its own independent drive.

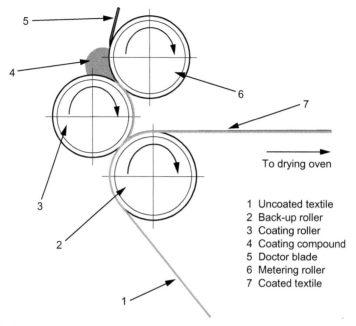

Figure 8.7 Reverse roll coating.

As Figure 8.7 shows, this requires three precision ground rollers, each with its own independent drive. The back-up roller is driven at the machine line speed. The coating roller is normally situated at 45° from top dead centre of the coating roller and is capable of running up to six times as fast as the back-up roller and line speed. There could be a fixed gap between the coating roller and the textile substrate, but quite often there is a zero gap that would wipe all coating from the roller on to the substrate. The amount of coating compound applied to this roller is governed by the metering roller. This roller operates at a very slow speed and is fitted with a doctor blade to clean the roller surface during operation. The weight of coating applied to the substrate is governed by a combination of the speed ratio of the coating roller to line speed, the gap between the coating roller and the substrate and the gap between the coating roller and the metering roller. This technique is more commonly used in paper coating, high-speed coating, and low-weight coating, often achieving as low as 1 g/m² of coating compared to between 5 and 10 g/m² for knife-in-air coating and 15 and 20 g/m² in knife-over-roll coating.

8.8   Roller coating

Roller coating (see Figure 8.8) is a technique that does not give as uniform a coating weight as other techniques because there is very little control over the amount of coating medium applied to the coating roll and, hence, fabric. It is a low-cost system used

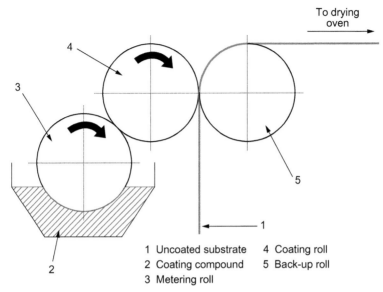

Figure 8.8 Roller coating.

in areas where an amount of coating weight variation is not very important in the final product. Quite often this is the preferred technique for the back coating of carpet products using a high filler content, resin system when the aim is to manufacture a product with an overall increased weight as well as to provide an anchorage for the carpet pile.

8.9   Rotary screen coating

Some products require that the coating is laid on the textile in a discontinuous or a pattern form. This is done by using the rotary screen coating technique shown in Figure 8.9. A *rotary screen* is made from a fine galvanised nickel mesh made into a seamless cylinder, commonly at a circumference of 640 mm. This cylinder is impregnated with a light-sensitive monomer when the required pattern is laid on the surface in the form of a black and white photographic type film. When subjected to ultraviolet light through the film, the areas exposed polymerise. Once exposure is completed, the mesh screen is washed, leaving the required pattern in the mesh.

The screen is attached to the coating machine using unique couplings and is rotated at a surface velocity equal to that of the fabric to be coated. A coating compound feed tube is placed in the centre of the screen, which has either narrow slots or a series of holes along its length to allow the compound to pass through. In operation, coating compound is then forced through the pattern in the screen onto the substrate. Coating compound rheology is important in rotary screen coating to prevent flooding or pattern distortion and is commonly controlled using a thixotropic additive into the coating compound formulation.

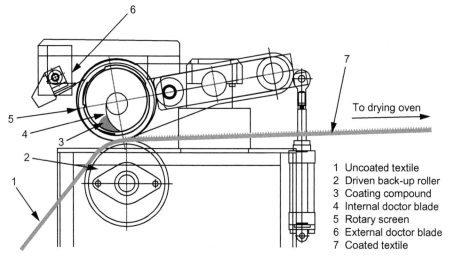

Figure 8.9 Rotary screen coating.

1 Uncoated textile
2 Driven back-up roller
3 Coating compound
4 Internal doctor blade
5 Rotary screen
6 External doctor blade
7 Coated textile

In some cases, coating by rotary screen is added using a fine mesh pattern screen that gives an overall small dot-like coating. A so-called whisper blade, or flexible gravure blade, is employed on the surface of the dots to spread them and produce a continuous coating. Although this has been carried out in the industry for some considerable time, only low-weight and relatively non-uniform coatings are produced.

8.10   Lick roll coating

Lick roll coating employs a roller running onto the surface of the substrate to deposit a coating. The lick roll, which may be a plain steel roll or an engraved roll, is driven and is capable of a variable speed in relation to the overall machine line speed (see Figure 8.10). It would also be capable of running in either a forward or reverse direction with respect to the substrate. This technique is normally used with low-viscosity coating compounds and as a low-cost technique for coating into the textile structure.

Normally, the lick roll is partially immersed into a bath of the coating compound, allowing the compound to be directly applied to the textile. In some installations, however, a doctor blade is fitted against the lick roll to measure the amount of coating compound.

8.11   Gravure roll coating

Gravure roll coating is a method in addition to rotary screen coating for the application of a coating in a discontinuous format. In this case, a steel roller is etched with the required pattern, which can vary from trihelical, truncated tetrahedron, or pyramidal formats, depending on the requirements of the final product. This simple process is schematically shown in Figure 8.11. It is commonly used for the application of

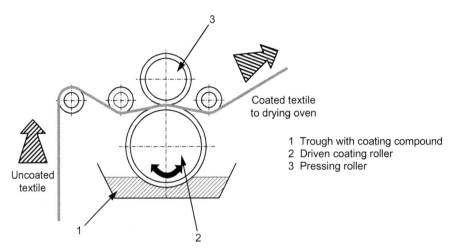

Figure 8.10 Lick roll coating.

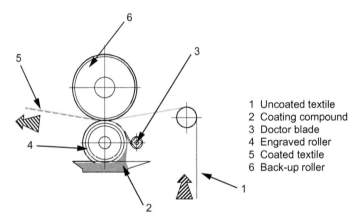

Figure 8.11 Gravure roll coating.

low-weight and low-viscosity compounds often as a sealing layer or protective layer to a previously processed textile.

In some cases, the engraved roller has a deposit of copper on the surface to allow for a simpler engraving process by using a mechanical engraving rather than an etch engraving. In this case, the copper surface once engraved has a chrome deposit to give a harder wearing roller as well as a surface much less prone to physical damage.

8.12   Extrusion coating

Extrusion coating is normally used in high-volume applications that have commercial advantages because the equipment requires a large capital outlay. Extrusion coating is not normally used for solution or dispersion coatings but for 100% solid content

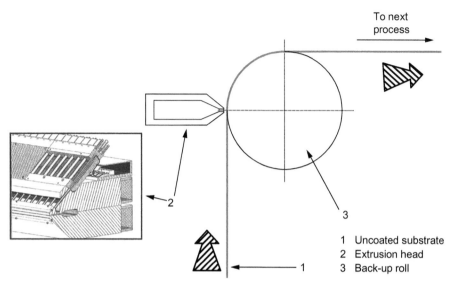

Figure 8.12 Extrusion coating.

compounds. The compounds normally are thermoplastic polymers, copolymers, or a mixture of different polymers.

Extrusion coating is capable of giving a wide range of coating weights from approximately 5 g/m² to 5 kg/m², depending on the requirements of the final coated product.

Quite often this technique is used for the manufacture of waterproof roofing membranes that coat a low-cost polymer onto a synthetic woven substrate.

The coating weight is governed by the extrusion head gap with the coating being extruded directly onto the surface of the substrate as shown in Figure 8.12. The production speed is restricted only by the capacity of the extrusion system.

8.13 Powder coating

Powder coating is normally achieved by applying a thermoplastic polymer in powder form to a textile substrate (see Figure 8.13). Although this is the normal process, some thermoset or cross-linking powders are available for it. As with all types of coating, the formulation is chosen to be the most appropriate for the final product. The polymers are normally subjected to a grinding process in order to achieve the powder form. Some polymer systems require that the grinding be done under cryogenic conditions using liquid nitrogen in order to prevent premature melting and coagulation. Powders would then be sieved to reduce the particle size distribution to achieve a uniform coat weight on the final product. It is common to see particle sizes in the range of 0–80, 80–200, 200–500, and 500–1000 μm although there are no limits to the ranges.

The powder is added to a hopper on the powder scatter coating head using a pneumatic powder elevator that was installed to keep a constant height of the powder in the

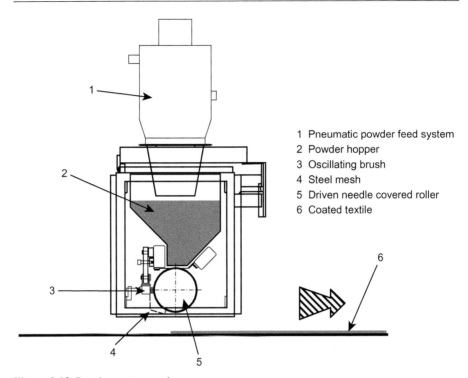

1 Pneumatic powder feed system
2 Powder hopper
3 Oscillating brush
4 Steel mesh
5 Driven needle covered roller
6 Coated textile

Figure 8.13 Powder scatter coating.

hopper. Under the hopper is a driven roller that is covered in card clothing. The card clothing is chosen to allow the powder particles to fill the interstices of the needles on the card clothing. As this driven roller rotates, a flexible doctor blade removes any excess powder and allows a uniform amount of polymer to pass.

An oscillating bar with a longer needle card clothing operates across the surface of the card clothing to remove all of the powder on the roller. Gravity then causes the powder to fall onto an oscillating mesh to further randomise the polymer before it is laid on the textile substrate underneath. This coated material is then passed into an infrared heated zone to melt the polymer before any further processing.

The deposition weight of the coating depends on the speed or rotation of the card-covered roller and the speed of the textile substrate passing through the machinery.

8.14 Spray coating

Spray coating is currently used for the application of low-weight coatings, mainly using a hot melt polymer. In the past, spray coating was used extensively to apply a binder chemical to wadding products and with other applications when uniformity of coat weight was not an important factor in the final product. In spray coating, a low-viscosity chemical compound is contained in a pressurised vessel and carried onto a substrate by a high-pressure stream of air through a nozzle. A wide variety of designs of the nozzle system allows varied pattern in the spray to be achieved, again depending on the requirements of the final product. The process is not currently considered to be most appropriate for coating technology when uniform coating weights are necessary.

The use of spray coating decreased over the past two decades with the introduction of other application techniques, principally blade coating and gravure coating that can achieve a significantly more uniform coat weight distribution. An increase in costs for the base chemicals used in the process has also led to a change in application techniques to give more uniform coatings and reduce waste.

8.15   Foam coating

The application of a coating to a substrate using foam technology allows previously uncoatable textiles to be coated. This is because of the fact that they would have a relatively open structure, and lower viscosity compounds generally pour through the open areas and produce an incomplete or very uneven coating. Many foam coatings are used to give a softer handle to the final product, which is essential on products such a drapes or curtains that require a soft handle but uniform coating. Depending on the actual coating chemical, foaming of the material is undertaken by either chemical or mechanical means, the latter being preferred when some formulations are unable to be foamed using chemical additives.

8.15.1   Mechanical foam techniques

Foamed coating compounds are produced by mixing the base chemical with air or, in some cases, an inert gas such as nitrogen. The base chemical contains a saponification agent to allow a foam to be produced. The air is introduced into the base chemical in a specific amount in order to allow the appropriate density of foam to be made. Typically, a reduction in density for an acrylic formulation for drapes is from 1200 to 200 g/l, thus giving the final product a much improved handle or feel. Once the coating has been applied, the product is passed through a drying oven to leave approximately 4% moisture when it is passed through a calendering system to remove the aeration and give a uniform solid coating with the required softness of handle.

There are currently two types of mechanical foam machine in use in the textile manufacturing process. The rotating disc system is the least expensive but does tend to give a foam that has a wide variety of bubbles. This can lead to some pinholes and blisters in the final product. The most uniform system for foam manufacture is the rotor and stator system in which a series of pins on the rotor are rotated at high speed with the pins on the stator remaining static (see Figure 8.14). Different manufacturers use either round or square pins, but there is little evidence of which is the more efficient. Generally, there is a positive pressure in the mixing head in the region of 3 bar, which would assist in the uniformity of the foam production. The vast majority of coating of foams is made using the mechanical foam technique.

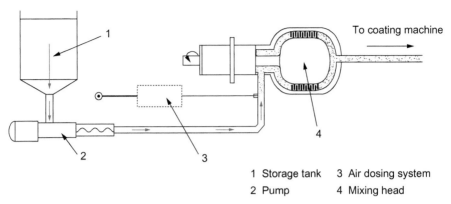

1 Storage tank   3 Air dosing system
2 Pump   4 Mixing head

Figure 8.14 Rotor/stator mechanical foam machine.

8.15.2 Chemical foam techniques

A chemical foam is made by including in the formulation a chemical that decomposes under heat and liberates a gas. Most common additives are azodicarbonamides that liberate nitrogen upon decomposition. The main problem with these chemicals is that their decomposition temperature is higher than any subsequent processing temperature of the coated textile. As a result of this, an addition to the formulation of a catalyst very significantly lowers the decomposition temperature and allows liberation of gases to aerate the coating compound. This system is most commonly used in coating polymers such as vinyls where a foam or cushion layer is required. For blown vinyl production, chemical foams have been found to be more appropriate than mechanical foams because a more stable and durable foam is made. Generally, chemical foams are not subjected to the crushing process used in mechanical foamed products because the foam presence is a crucial feature in the final material.

8.16 The future of textile coating

Textile coating will continue to be used in the foreseeable future, and present products will continue to be manufactured using the current techniques. Developments in coating formulations and a requirement for further products will ensure that textile coating continues into the future. Two developments that are becoming increasingly important are hot melting [7] and plasma processing [8] technologies; these are considered next.

8.16.1 Hot melt coating and laminating

The use of epoxy resins and developments in carbon fibre technology have led to an increase in the development of composite materials that have been used to replace metals in high-strength, low-weight applications. The use of these composite materials is increasing at a high rate, particularly in aerospace applications. The manufacturing process for these materials is a modified form of existing coating processes. Generally,

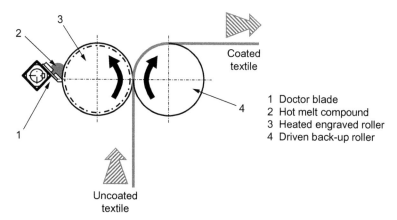

Figure 8.15 Typical hot melt coating for lamination.

1 Doctor blade
2 Hot melt compound
3 Heated engraved roller
4 Driven back-up roller

in the use of epoxy resin systems, the resin formulation containing a cross-linking agent is heated to give a viscous compound that is then coated, generally using blade or rotary screen technology. The coating blade must be insulated from the main machine structure to prevent heat loss and dissipation to other parts of the machine. Coating may be performed by either a direct coating technique or a transfer coating technique using an appropriate release paper. In the transfer coating process, it is common for the coating backup roller also to be heated. Once coated, the resin systems are cooled to form either a malleable coated fabric or a rigid structure, depending on the particular formulation being used. Hot melt coating using a reactive polyurethane and a heated gravure roller (see Figure 8.15) represents a major step forward in the manufacture of laminated products when good adhesion is paramount and resistance to washing and dry cleaning is required, particularly in garments. Because the gravure coating uses an engraved roller, the polymer layer is not continuous and allows the laminated product to retain flexibility. This is now used in automotive applications as an alternative to flame-laminated products because of the high temperature resistance when the polyurethane becomes fully cross-linked with atmospheric moisture, rendering it inert.

8.16.1.1 Plasma coating

The recent advent of atmospheric plasma [9] as opposed to vacuum processes has enabled the commercial exploitation of plasma treatment of textile surfaces to be realised [9]. Following such treatments, textiles can acquire entirely new material qualities by creating new functionalities, such as self-cleaning finishes, bioactive surfaces, and flame-retardant properties. With a plasma coating, it is possible to create coatings with long-term stability and that can allow less expensive textiles to be used as a replacement for more conventional materials by applying a conventional coating to the less expensive textile whereas major problems previously arose with adhesion. This is so because application of a plasma to the surface of the material can change the basic chemistry of the substrate, which creates new functional chemical groups that can interact with those present in the coating being deposited.

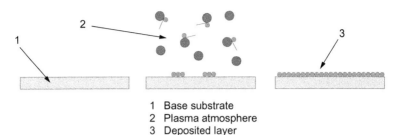

1 Base substrate
2 Plasma atmosphere
3 Deposited layer

Figure 8.16 Plasma coating.

The application of a plasma can put a layer onto the surface of the textile for protective barriers, a hydrophobic layer, a hydrophilic layer, or a friction-reducing layer (e.g. simulating a polytetrafluoroethylene layer), depending on the nature if the actual plasma being employed (Figure 8.16).

Plasma coating of yarns and fabrics is proving to be an important first step in the manufacture of smart textiles. Thus, it is evident that developments within this new area of plasma coating will make significant impact in the manufacture of high added-value materials in the future.

References

[1] Woodruff FA. Coating, laminating, bonding, flocking and prepregging. In: Heywood D, editor. Textile finishing. Bradford: Society of Dyers and Colourists; 2003. p. 447–525.
[2] Shim E. Coating and laminating processes and techniques for textiles. In: Smith WC, editor. Smart textile coatings and laminates. Cambridge: Woodhead Publishing; 2010. p. 10–41.
[3] Hall ME. Coating of technical textiles. In: Horrocks AR, Anand SC, editors. Handbook of technical textiles. Cambridge: Woodhead Publishing; 2000. p. 173–86.
[4] Rexine trade mark number UK00000227208 entered into the register on 17th November 1899.
[5] Klimke J, Rothmann D. Carbon composite materials in modern yacht building. Reinforced Plastics. London; 2010.
[6] Gay D. Composite materials, design and application. CRC Press, Taylor and Francis Group, London; 1997.
[7] Glawe A, Reuscher R, Koppe R, Kolbusch T. Hot-melt application for functional compounds on technical textiles. J Ind Text 2003;33:85–92.
[8] Shishoo R. Plasma technologies for textiles. Cambridge: Woodhead Publishing; 2007.
[9] Herbert T. Atmospheric pressure cold plasma processing technology. In: Shishoo R, editor. Plasma technologies for textiles. Cambridge: Woodhead Publishing; 2007. p. 79–128.

Bibliography

Fung, W., 2002. Coated and Laminated Textiles. Cambridge: Woodhead Publishing.
Geissmann, A., 2012. Coating Substrates and Textiles. New York, USA: Springer Publishing.
Heywood, D., 2003. Textile Finishing. Bradford: Society of Dyers and Colourists.

Sen, A.K., 2001. Coated Textiles: Principles and Applications. Technology and Engineering. London: CRC Press, Taylor and Francis Group.

Sen, A.K., 2007. Coated Textiles: Principles and Applications. Technology and Engineering, second ed. London: CRC Press, Taylor and Francis Group.

Smith, W.C., 2010. Smart Textile Coatings and Laminates. Cambridge: Woodhead Publishing.

Traubel, H., 1999. Coating of Textiles. New York, USA: Springer Publishing, pp. 11–20.

van Parys, M., 1994. Coating. Technology and Engineering. Guimaraes, Brussels, Belgium: Eurotex.

Coloration of technical textiles

I. Holme

International Dyer, World Textile Information Network, Leeds, United Kingdom

Chapter Outline

9.1   Introduction

Technical textiles are used in a very wide variety of end uses in which the functional performance requirements are paramount. Thus, technical textiles must possess the requisite physical and mechanical properties to maintain the structural integrity throughout all the manufacturing and fabrication processes and during the service life of the material. For some end uses, therefore, coloration is not strictly necessary because the aesthetic properties of the technical textile – for example, colour and pattern, lustre, texture, handle, and drape – may not always be required to appeal to the visual and tactile senses of the customer.

Coloration of technical textiles by dyeing or printing is primarily intended for aesthetic reasons but also provides a ready means of identifying different qualities or fineness of the materials. For example, the fineness of a surgical suture and its visibility at the implant site are easily identified by colour. High-visibility clothing and camouflage printing clearly provide the extreme ends of the coloration spectrum for technical textiles.

Coloration also introduces other functional properties distinct from the aesthetic appeal of colour. Colorants can hide fibre yellowing and aid fibre protection against weathering, both factors of importance when the physical properties of the technical textile must be maintained over a long service life under adverse end-use conditions.[1] Heat absorption is also increased when black materials are exposed to sunlight, an important factor for baling materials for agriculture.

Handbook of Technical Textiles. http://dx.doi.org/10.1016/B978-1-78242-458-1.00009-1

The dominant theme in textile coloration over the last decade has been and remains sustainability. Many strategies have been developed to minimise the environmental impact of the coloration step as well as over the life cycle of the coloured textile material. Accurate colour communication coupled with intelligent dyestuff, pigment, and optimised process selection can greatly minimise the consumption of resources (e.g. water, energy, and chemicals). In addition, decreasing pollution loads, minimising waste, enhancing product durability, and shortening the time to reach the market also continue to be important factors.[2]

Strenuous efforts have been made by colorant manufacturers to eliminate heavy metals in dyestuffs (when possible) and to adhere to strict control over organic impurity levels in organic colorants.[2-5] Auxiliary chemicals are engineered to be free of alkylphenol ethylene oxides (APEO) and dyestuffs to be free of banned amines and absorbable organohalogen (AOX). Thus, in textile dyeing, dyestuffs with a high colour yield on the fibres decrease the levels of colour in the effluent. Such dyestuffs are applied in ultralow liquor ratio dyeing machinery to minimise water, energy, and chemical consumption. Intelligent dyestuff selection results in compatible dyestuffs that can be applied together, are robust to the process application conditions, and can give rise to high levels of correct first-time dyeing, thereby avoiding the need for wasteful shading additions or stripping/redyeing to achieve the required colour.[6] An emerging trend is the increasing use of more sustainable chemical auxiliary products that are biobased (i.e. manufactured from natural renewable resourced materials).

9.2   Objectives of coloration

The objectives of coloration treatments are first to produce the desired colour in dyeing and the colours in the coloured design image in printing on the technical textile.[7] Second, coloration treatments are to ensure that the necessary colour fastness requirements for the end use are achieved. Third, the whole operation should be carried out at the lowest cost commensurate with obtaining the desired technical performance. After dyeing, the level and uniformity of colour across the width and along the length of a technical textile fabric must be within the defined colour tolerances to which the dyer and the customer agreed prior to coloration. Preservation of the original appearance and quality of the technical textile prior to coloration is also essential in order to ensure that the technical textile is of marketable quality.

9.3   Coloration of technical textiles

Coloration involves the application of colorants to the technical textile. It is a complex field because of the variety of fibres, filaments, yarns, fabrics, and other materials requiring coloration and the diverse nature of the end-use and performance requirements. Coloration may be carried out by dyeing the materials to a uniform colour or by printing to impart a design or motif to the technical textile. Fibres, yarns, and fabrics may also be multicoloured by specialised dyeing techniques, for example, space dyeing, or by weaving or knitting different coloured yarns.[8]

The colorants used may be either water-soluble (or sparingly water-soluble) dyes, or alternatively water-insoluble pigments.[9] Most dyes are applied to technical textiles in an aqueous medium in dyeing and in printing, although selected disperse dyes may also be dyed in supercritical fluid carbon dioxide (above the critical point under very high pressure) for a few specialised end uses, for example, polyester sewing threads.[10] Polyester fabrics can be beam dyed with disperse dyes in the DyeCOO supercritical carbon dioxide (sCO_2) dyeing system under a pressure of about 300 bar at temperatures up to 130 °C.[11,12] The advantages of sCO_2 dyeing are that at the end of the dyeing cycle, the dyed material does not have to be thermally dried and 95% of the carbon dioxide can be recycled for reuse in sCO_2 dyeing. This dyeing method also avoids the production of wastewater pollution. Developments in dyestuff chemistry may extend the sCO_2 dyeing method to dyeing fibres other than polyester, thereby providing a more sustainable dyeing process for technical textiles. By contrast with dyes, pigments are either physically entrapped within the filaments during synthetic fibre extrusion, for example, by mass pigmentation, to give a spun-dyed fibre[13] or caused to adhere to technical textiles in pigment printing through the use of an adhesive binder.[14]

Dyeing is normally carried out on textile materials from which surface impurities, for example, fibre lubricants, spin finishes, sizes, particulate dirt or natural colouring matters, and so on, have been removed by appropriate pretreatments (e.g. desizing and scouring) and to which a stable whiteness has been imparted by chemical bleaching.[15] However, many synthetic fibres do not normally require chemical bleaching prior to coloration because they may be whitened by incorporation of a fluorescent brightening agent during fibre manufacture.

Printing may be carried out mainly on technical fabrics that may be in their natural state, chemically bleached, or whitened with a fluorescent brightening agent after tinting or dyeing. Conventional dyeing and particularly printing are most conveniently and economically performed on fabrics that also allow greater flexibility through the selection of colours late in the technical textile production sequence to meet the varying market requirements. Printing may also be performed on high-performance garments using automatic flat screen or digital ink jet printing methods.[16]

9.3.1 Dyes

Over the last two decades, the manufacture of synthetic fibres has progressively moved from the United States and Western Europe into Asia, especially China, which now dominates synthetic fibre production. Manufacture of dyes and pigments has also followed the same path because of the increasingly tighter regulations on pollution in Western Europe and the introduction of the regulation of the Registration, Evaluation, Authorisation and Restriction of Chemicals (REACH) in the European Union.[2] At first, the very large dye and pigment manufacturing units established in China, India, Korea, and Taiwan with their lower labour costs and lower standards of regulation on discharges of chemicals into the environment resulted in economies of scale, lower production costs, and a progressive decline in the prices of commodity dyes and pigments worldwide. However, in 2014, the rising labour costs coupled with much stricter enforcement of the regulations on environmental pollution in China and India has caused a number of chemical intermediate manufacturers to be closed. This has led to a temporary shortage of certain intermediates, rising prices, and a consequent marked rise in dyestuff prices in Western Europe.[17,18]

The dyes used commercially on technical textiles may be selected from the very wide range of synthetic organic colorants based on aromatic compounds derived from petroleum.[9] Dyes are conjugated organic structures containing an alternating system of single and double bonds within the molecule that impart the ability to absorb certain wavelengths of visible light, so that the remaining light scattered by the dyed technical textile is perceived as coloured.[9,19]

The dye structure must contain a chromophore, a chemical group that confers on a substance the potential to become coloured, for example, nitro, nitroso, azo, and carbonyl groups. To become a useful dye, however, the molecule should contain other chemical groups, such as amino, substituted amino, hydroxyl, sulphonic, or carboxyl groups called auxochromes. These generally modify or intensify the colour, render the dye soluble in water, and assist in conferring substantivity of the dye for the fibre.[9,19] High substantivity aids in attaining a high degree of dye exhaustion onto the fibres during exhaust dyeing when dyes are progressively and preferentially adsorbed by the fibres from the dyebath to produce coloured technical textile materials. High dye-fibre substantivity also generally confers high colour fastness during end use, for example, high colour fastness when exposed to washing and light.

The manufacture of synthetic organic dyestuffs and pigments represents a multi-stage chemical manufacturing sequence that can take months to complete.[9,19,20] The raw materials from petroleum refining, such as ethylene, propylene, benzene, and toluene with naphthalene and anthracene, provide the organic feedstock for more than 90% of the synthetic dyestuffs industry. Primary chemicals derived from benzene, toluene, and naphthalene, such as phenol, nitrobenzene, aniline, and phthalic anhydride, and so on, are reacted to form chemical intermediates. Some 300 chemical intermediates constitute about 90% of the quantity of intermediates used. A series of unit processes are then conducted (e.g. nitration, sulphonation, oxidation, reduction, alkylation, and halogenation) in order to produce the chemical products from which the synthetic dyes and pigments can be manufactured.

Manufactured dyestuffs are formulated colorants typically containing only 30–40% active colorant. The rest of the dyestuff formulation consists of agents to dilute the colour strength, electrolytes to improve the dye exhaustion on the fibre, and dispersing agents to facilitate dispersing of the dye in the dyebath.[21] To avoid dusting dyestuffs into the air during weighing and dispensing, granular, grain, or pearl forms of dyestuffs are used. Antidusting agents are added to powdered dyestuffs for the same reason. Water-soluble dyes can be produced as true liquid dyestuffs (e.g. basic dyes), but liquid dispersions of disperse and vat dyestuffs are also produced. The latter may contain a viscosity modifier to minimise sedimentation/settling of the dye particles during storage and must be stirred prior to use. Liquid dyestuffs are preferred for continuous dyeing and for printing because of their convenience for use on automated weighing, metering, and dispensing systems.

Commercial dyestuffs are subjected to stringent quality control procedures for hue and colour strength, average particle size, and particle size distribution and are dried to a uniform moisture content and packaged in drums or plastic containers suitable for transportation, storage, and dispensing.[20,21] Resealable packaging may be used to prevent the ingress of moisture from the atmosphere, and in many colour kitchens, the ambient temperature and relative humidity are controlled in order to maintain reproducibility in weighing. Some dyestuffs are mixtures formulated for specific shades such as black or contain dyes from different application classes suitable for dyeing specific fibre blends, for example, 65/35 polyester/cellulosic fibre blends.[22]

The formulation of inks for digital ink jet printing is more complex, and the ink formulation is engineered to provide good runnability on a particular print head jetting system.[23,24]

9.3.2 Pigments

Pigments are synthetic organic or inorganic compounds that are insoluble in water, although some are soluble in organic solvents.[9,19] The pigment particles are ground down to a fine state of subdivision (0.5–2 μm) and stabilised for use by the addition of dispersing agents and stabilisers. Both powder and liquid (i.e. dispersions of) pigments are used for the coloration of technical textiles, and the pigment finish must be compatible with the application conditions used in mass pigmentation of manufactured fibres or in pigment printing of technical textile fabrics.[25,26]

Organic pigments are manufactured as solid particulate materials that are insoluble in the application medium (e.g. fibres, textile coatings, films, and laminates). Pigment properties such as the pigment colour are governed by the chemical constitution but are also largely controlled by their crystalline composition.[27] Pigments can occur as single or primary particles, aggregates, and agglomerates. Primary particles in the crude pigment may consist of single crystals with lattice disorders, or combinations of several lattice structures may be present. Aggregates consist of primary particles grown together at their surfaces and cannot be broken down by dispersion processes. However, agglomerates, which consist of groups of single crystals or aggregates joined at their corners and edges but not grown together, can be separated by a dispersion process.

Pigment synthesis may result in polymorphism; for example, copper phthalocyanine can exist in five modifications while quinacridone and azo pigments may be present in at least three different modifications.[27] Pigments therefore owe their colour to the combination of the way that the pigment particles, aggregates, and agglomerates scatter and absorb incident light. Light scattering/absorption depends on the size of the particles so that the colour and the application properties of a pigment (e.g. brightness, tinctorial strength, shade, purity, and colour depth) depend not only on the chemical constitution but also on the physical form (e.g. size, shape) of the pigment.

9.3.3 Fluorescent brightening agents

These are organic compounds that absorb some of the ultraviolet radiation in sunlight or other sources of illumination and re-emit in the longer wavelength blue-violet region of the visible spectrum.[28–30] Such compounds can be applied to technical textiles either by mass pigmentation (in manufactured fibres) or via machinery used for conventional dyeing of all types of materials and add brightness to the whiteness obtained from chemically bleaching the textile.

Many fluorescent brightening (whitening) agents can be coapplied in chemical bleaching treatments provided that the bleaching conditions do not affect the molecular structure or the properties of the fluorescent brightening agent.[29] There are many chemical classes of fluorescent brightening agents including derivatives of benzimidazole, benzoxazole, coumarin, distyrylarene, bis(triazinylamino)-stilbenedisulphonic acid, 1,3-diphenylpyrazoline, naphthalimide, and pyrene according to which fibre(s) is(are) to be whitened.[28–31]

Fluorescent brightening agents can be applied to technical textiles either by mass coloration (in manufactured fibres) or via the machinery used for conventional dyeing of all types of materials.[28–31] Fluorescent brightening agents add brightness to the whiteness obtained from the chemical bleaching of natural fibres as well as in their blends with manufactured fibres and are used to produce the highest standards of whiteness (high whites).

9.3.4 Range of colorants available

The Colour Index Online is published jointly by the Society of Dyers and Colourists and the American Association of Textile Chemists and Colorists and is the primary source of information on colorants. The Colour Index Online is available by subscription. It lists the majority of commercial dyestuffs and pigments (both past and present) although the situation continually changes as the manufacture of some colorants ceases, to be replaced by new products.[32] The wide variety of colorants available is based on many chemical types that make up the major application groups of colorants. The percentage distribution of each chemical class between major application ranges is illustrated in Table 9.1 based on all the dyes listed when the chemical class is known and including products that are no longer used commercially.[9] Azo colorants make up almost two-thirds of the organic colorants listed in the Colour Index Online. Anthraquinones (15%), triarylmethanes (3%), and phthalocyanines (2%) are also of major importance. Each colorant is assigned a specific Colour Index Constitution Number (CICN) that depends on the type of chemical structure of the colorant (e.g. monoazo), although for confidentiality reasons, the chemical structure of some colorants in the CICN in the Colour Index Online may not be disclosed. Approximately 70% of all commercial dyestuff consumption is applied to textile materials.

In the Colour Index Online dyes are classified by application class, colour, and number, for example, CI Disperse Blue 60, CICN 63285. The commercial name of the dyestuff varies, however, according to the dye maker. Dyestuffs being formulated

Table 9.1 Percentage distribution of each chemical class between major application ranges

Chemical class	Distribution between application ranges (%)								
	Acid	Basic	Direct	Disperse	Mordant	Pigment	Reactive	Solvent	Vat
Unmetallised azo	20	5	30	12	12	6	10	5	
Metal complex azo	65						12	13	
Thiazole		5	95						
Stilbene			98					2	
Anthraquinone	15	2		25	3	4	6	9	36
Indigoid	2					17			81
Quinophthalone	30	20		40				10	
Aminoketone	11			40	8		3	8	30
Phthalocyanine	14	4	8		8	9	43	15	3
Formazan	70						30		
Methine		71		23		1		5	
Nitro, nitroso	31	2		48	2	5		12	
Triarylmethane	35	22	1	1	24	5		12	
Xanthene	33	16			9	2	2	38	
Acridine		92		4				4	
Azine	39	39				3		19	
Oxazine		22	17	2	40	9	10		
Thiazine		55			10			10	25

Source: see Ref. [9].

products, with nominally the same CI number may vary in active colorant content and in the nature of the additives incorporated within the formulation.[33] Illuminant metamerism may also be observed on the dyed textile; this difference in colour is observed when two colours are viewed under different illumination conditions, even when using dyestuffs that nominally possess the same colour index number.

Ecotoxicological considerations are now very important, and strict regulations have been introduced within the European Union to ban certain arylamines formerly used in azo dye manufacture because they are carcinogenic to humans.[3] The German MAK classification has been widely adopted as a standard, and the major dye makers within the Ecological and Toxicological Association of Dyes and Organic Pigment Manufacturers (ETAD) also adhere to strict limits for organic impurities and heavy metal contents of their commercial dyestuffs. Such limits form part of many ecolabelling schemes (e.g. Oeko-Tex 100 Standard).[34]

The major classes of dyes and the fibres to which they are applied are illustrated in Table 9.2. The type of dyestuff used for dyeing technical textiles depends on the fibre(s) present in the material, the required colour and the depth of colour, the ease of dyeing by the intended application method, and the colour fastness performance required for the end use. Some dyestuff classes dye a number of different fibres, but in practice, one fibre predominates. For example, disperse dyes are mainly used to dye

Table **9.2**   **Major classes of dyes and the fibres to which they are applied**

Dye class	Major fibre type
Acid (including 1:1 and 1:2 metal complex dyes)	Wool, silk, polyamides (nylon 6, nylon 6.6)
Mordant (chrome)	Wool
Azoic	Cellulosic fibres (cotton)
Direct	Cellulosic fibres (cotton, viscose, polynosic, HWM, modal, cuprammonium, lyocell fibres), linen, ramie, jute, and other lignocellulosic fibres
Reactive	Cellulosic fibres (cotton, viscose, polynosic, HWM, modal, cuprammonium, lyocell, linen, ramie, jute), protein (wool, silk)
Sulphur	Cellulosic (cotton)
Vat	Cellulosic fibres (cotton, linen)
Basic	Acrylic, modacrylic, aramid
Disperse	Polyester, cellulose triacetate, secondary cellulose acetate, polyamide, acrylic, modacrylic, polypropylene, aramid

HWM = high wet modulus.

polyester fibres, although they can also be applied to nylon 6, nylon 6.6, acrylic, modacrylic, secondary cellulose acetate, cellulose triacetate, and polypropylene fibres but with limitations on the depth of colour and the colour fastness that can be attained.[35]

There are, however, many types of manufactured fibres, particularly synthetic fibres, that are currently virtually impossible to dye with conventional dyestuffs using normal dye application methods. This is because filaments that are designed for high-strength end uses are often composed of polymer repeat units that do not contain functional groups that could act as dyesites (e.g. polyethylene, polypropylene, polytetrafluoroethylene) and/or may be highly drawn to produce a highly oriented fibre with a high crystallinity.[36]

Other high-strength, high-performance fibres such as ultrastrong high-performance polyethylene and thermostable and fire-resistant fibres, such as polyamideimide fibres (e.g. Kermel), and fibres based on polyimide, polybenzimidazole, novoloid, polyacrylate, fluoropolymers, polyphenylene sulphide, poly(ether ether ketone) (PEEK) are most likely to be coloured via mass pigmentation (see Section 9.5). Glass fibres and oxidised polyacrylonitrile fibres (which are black in colour) can both be printed using a pigment-binder system.

Because dyes are considered to diffuse in monomolecular form into the fibre only through noncrystalline regions or regions of low order (i.e. high disorder), it follows that highly oriented fibres such as *meta*-aramid, *para*-aramid, and high-strength polyethylene fibres are extremely difficult to dye. In some cases, specialised dyeing techniques using fibre plasticising agents ("carriers") to lower the glass transition temperature (T_g) of the fibre (i.e. the temperature at which increased polymer segmental motion opens up the fibre structure) together with selected dyes may be employed to speed the diffusion of dyes into the fibres, but the colour depth may be restricted to pale-medium.[36–39] In addition, or alternatively, resort may be made to high-temperature dyeing in pressurised dyeing machinery at temperatures in the range 130–140 °C for the same purpose.

However, mass pigmentation for many high-strength fibres offers a more satisfactory production route to coloured fibres provided that the introduction of the pigments does not significantly impair the high-strength properties.[13] Some fibres, such as carbon and partially oxidised polyacrylonitrile, cannot be dyed simply because they are already black as a result of the fibre manufacturing process. However, contrasting colours may be printed on such fibres using pigment printing methods or by blending with other dyed fibres.

Microfibres are generally defined as fibres or filaments of linear density less than 1 dtex.[40] Such silklike fibres pose considerable difficulties for level dyeing of medium to heavy depths of colour. The high surface area per unit volume of microfibres increases the light scattering, necessitating the use of higher amounts of dye to achieve the same colour depth as on coarser fibres. The use of bright trilobal filaments or the incorporation of titanium dioxide delustering agents within the microfibres increases the light scattering, making the difference even more noticeable.[41,42]

Nanofibres are manufactured fibres with diameters from 1 to 10 nm (nano-sized fibres) and from 100 to 1000 nm (submicron-sized fibres).[43] Because of the handling difficulties in conventional dyeing and thermal drying of such fibres it is most likely that coloration will be carried out by mass coloration during fibre extrusion. Polyester, polypropylene, and polyamide nanofibres are used in technical textile materials such as nonwovens, protective clothing, and medical and filtration end uses.

9.4   Dye classes and pigments

9.4.1   Acid dyes

Acid dyes are anionic dyes characterised by possessing substantivity for protein fibres such as wool and silk and polyamide (e.g. nylon 6 and nylon 6.6) fibres or any other that contain basic groups. Acid dyes are normally applied from an acid or neutral dyebath.[44]

As the size of the acid dye molecule generally increases from level dyeing (or equalising) acid dyes to fast acid, half-milling, or perspiration-fast dyes to acid-milling dyes and thence to supermilling acid dyes, the colour fastness to washing increases because of the increasing strength of the nonpolar forces of attraction for the fibre.[44] Dyes of 1:1 and 1:2 metal complexes also behave like acid dyes from the viewpoint of application to the fibre and possess good colour fastness to washing and light. Level-dyeing acid dyes yield bright colours whereas the larger sized milling acid and supermilling acid dyes are progressively duller. The 1:1 and 1:2 metal complex dyes also lack brightness but can provide good colour fastness performance.[45]

Acid dyes contain acidic groups, usually sulphonate groups, either as $-SO_3Na$ or $-SO_3H$ groups, although carboxyl groups ($-COOH$) can sometimes be incorporated. Wool, silk, and polyamide fibres contain amino groups ($-NH_2$) that in an acid dyebath are protonated to yield basic dyesites ($-NH_3^+$). The acid dye anion $D \cdot SO_3^-$, where D is the dye molecule, is thus substantive to the fibre and is adsorbed, forming an ionic linkage, a salt link, with the fibre dyesite. Monosulphonated acid dyes can be adsorbed to a greater extent than di-, tri-, and tetrasulphonated acid dyes because the number of basic dyesites in the fibres is limited by the fibre structure. Thus, the colour buildup is generally greatest with monosulphonated dyes on wool. However, the dyestuff solubility in water increases with the degree of dye sulphonation.

Dyed wool fibre quality is improved by pretreatment before dyeing with Valsol LTA-N™, for example, an auxiliary that extracts lipids from the cell membrane complex of wool, thereby speeding up the intercellular diffusion of dyes.[46] In this Sirolan LTD™ (low temperature dyeing) process, wool may be dyed for the conventional time at a lower temperature or for a shorter time at the boil, decreasing fibre degradation in the dyebath.

Levelling agents may be used to promote level dyeing of acid dyes on wool and other fibres.[47] Anionic levelling agents enter the fibre first and interact with basic dyesites, restricting dye uptake. Amphoteric levelling agents, which contain both a positive and a negative charge, block the basic dyesites in the fibre but also complex with the acid dyes in the dyebath, slowing the rate at which the dyes exhaust onto the fibres. As the dyebath temperature is increased, the anionic (or amphoteric) levelling agent desorbs from the fibres, allowing the dye anions to diffuse and fix within the fibre. In addition, as the dyebath temperature is raised, the dye/amphoteric levelling agent complex breaks down, releasing dye anions for diffusion inside the fibre. The dyes are thus gradually adsorbed by the wool as the dyebath temperature is increased.

In the iso-ionic region (pH 4–5), the cystine (disulphide) cross-links in the wool are reinforced by salt links formed between charged carboxyl and amino groups in opposing amino acid residues in adjacent protein chain molecules within the wool fibre. This temporarily strengthens the wool, and the abrasion resistance of the dyed fabric is improved. Fibre yellowing is decreased, and brighter colours may be dyed.[48]

The buildup of acid dyes on silk and polyamide fibres is limited by the fewer number of fibre dyesites compared with that on wool. Acid dyes are attracted to the amine end groups (AEG) in polyamide fibres, such as nylon 6 and nylon 6.6. Nylon 6 has a more open physicochemical structure and a lower T_g compared with nylon 6.6. Thus, acid dyes diffuse more readily into nylon 6, but the colour fastness to washing of a similar dye on nylon 6.6 is generally superior because of the more compact fibre structure.[49,50] False twist texturing processes, which use contact heating of the yarns, open the fibre structure, and may modify some of the fibre cross-sections that were in contact with the heater, resulting in a slightly lower colour fastness to washing.

Chemical variations from changes in the AEG of nylon fibres can give rise to dyeability variations with acid dyes, a problem often referred to as *barré* or *barriness*.[49] Physical variations caused by temperature and/or tension differences in nylon fibres can similarly lead to differences in the uptake of disperse or 1:2 disperse premetallised dyes. Physical variations can be minimised by dyeing at higher temperatures (e.g. up to 120 °C) with nylon 6.6, or prolonging the dyeing time. High-temperature dyeing is used with the larger 1:2 metal complex acid dyes in order to achieve better fibre levelling and fibre penetration, which leads to improved colour fastness to washing. Acid dyes can also be printed on wool (which has been pretreated by chlorination), degummed silk, polyamide, and other fibres using conventional print thickeners in a print-dry-steam fixation and wash off and dry process.[51,52]

Colour fastness to washing with acid dyes on nylon fibres, especially nylon microfibres, is improved by after-treatments such as syntanning, that is, the adsorption of a sulphonated synthetic tanning agent that provides a physical barrier to desorption by blocking the fibre pore structure in the fibre surface regions and by providing electrostatic (ionic) repulsion to dye desorption.[49,50] Combining an appropriate syntan with a fluorochemical treatment can provide stain-blocking properties in addition.[53] However, syntan treatments degrade during high-temperature or steaming treatments and tend to yellow the dyed fabric, dulling the colour.

9.4.2 Mordant dyes

Chrome dyes are the only type of mordant dye of any commercial significance.[44] They are used in exhaust dyeing to dye wool or occasionally in polyamide fibres to deep dull colours of high colour fastness to wet treatments and light. The fibres are usually dyed by the afterchrome method in which the fibre is dyed by a chrome dye (similar to an acid dye), and then the dyed fibre is given a treatment at pH 3.5 in a second bath containing sodium or potassium dichromate.[54] The absorption of dichromate ions leads to the formation of a 1:1 and/or 1:2 chromium metal complex dye inside the fibre, which can lead to some problems in batch-to-batch reproducibility of shade, particularly if the pH control in chroming is variable. Low chrome dyeing procedures can promote exhaustion of the chromium onto the fibres, decreasing environmental pollution from the wastewater from dyeing.[44,54] A major disadvantage of the afterchrome method is that the final colour is not developed until the chroming stage, which can have difficulties in shade matching. Ammonia aftertreatment can improve the colour fastness to washing.

Statutory limits on the discharge of chromium to sewer from chrome dyeing and rinsing vary around the world, but typical values are[44]

- Hexavalent chromium: $0.0–0.5\ \mathrm{mg\,l^{-1}}$
- Trivalent chromium: $2.0–5.0\ \mathrm{mg\,l^{-1}}$
- Total chromium: $2.0–5.0\ \mathrm{mg\,l^{-1}}$

The handling of dichromates is a health issue. The residual labile chromium levels in wool and chromium discharges from dyeing and rinsing cannot meet the requirements of many ecostandards (e.g. the Oeko-Tex 100 standard). This is leading to a marked decline in the use of chrome dyestuffs except when machine-washable colour fastness is required on heavy black and navy shades on wool fabrics.

9.4.3   Basic dyes

Basic dyes are cationic dyes characterised by their substantivity for standard acrylic, modacrylic, basic dyeable polyester, and basic dyeable nylon fibres.[55] Basic dyes can be applied to protein and other fibres, for example, secondary cellulose acetate, but the light fastness of basic dyes on hydrophilic fibres is poor.[56] The major outlets for basic dyes are acrylic and modacrylic fibres on which basic dyes can impart bright colours with considerable brilliance and fluorescence. The ionic attraction between the basic dye and the sulphonic acid dyesites in acrylic fibres is strong, which yields high colour fastness to washing. The close-packed physicochemical nature of acrylic fibres and the strong dye-fibre bonding can result in poor migration and levelling properties during dye application but impart very high colour fastness to light.[36,56]

Acrylic fibres vary widely in their dyeability because of the different amounts of different comonomers used with poly(acrylonitrile) that modify the fibre glass transition temperature. This may range from 70 to 95 °C according to the source of the acrylic fibre manufacturer.[57] Cationic retarders are widely used on acrylic and modacrylic fibres to promote level dyeing, and the basic dyes used should preferably be selected with the same compatibility value. The compatibility value may range from 5 (slow diffusing) to 1 (rapid diffusing). A compatibility value of 3 has been recommended for package dyeing, but for printing, compatibility values are of much less importance. Print fixation is by the print-dry-steam wash off and dry process, although wet transfer printing techniques are also possible using selected dyes.[56]

The acidic dyesites in acrylic fibres become much more accessible over a narrow band of temperatures in the glass transition temperature range. Therefore, even when using cationic retarders, which initially adsorb on to the fibre blocking the dyesites, the dyebath temperature must be raised slowly through this temperature range to allow the cationic retarder to desorb slowly and allow the basic dye to be adsorbed uniformly.[56,57] Alternatively, constant temperature dyeing methods may be used in exhaust dyeing. Continuous pad-steam-wash off and dry methods can also be used for dyeing acrylic fabrics.

The basic dye uptake is limited by the number of acidic dyesites in the fibre, but approximately 95% of all colours on acrylic fibres are dyed using basic dyes. However, when very pale colours are to be dyed, it is common to use disperse dyes, which have superior migration and levelling properties in order to attain a level dyeing because the strong dye-fibre bonding renders this very difficult to achieve using basic dyes.[56] The buildup of disperse dyes on acrylic fibres is limited, and the colour fastness to washing and light are generally lower than for basic dyes. The colour fastness of basic dyes on acrylic fibres is superior to similar dyeings on modacrylic fibres. Some modacrylic fibres are prone to delustering during dyeing and may require relustering by boiling for 30 minutes in a high concentration of electrolyte (e.g. 50–200 g l^{-1} sodium chloride) or by dry-heat or treatment in saturated steam.[57]

9.4.4   Azoic colouring matters

Azoic colouring matters are formed inside the fibre (usually cellulosic fibres), generally by adsorption of an aromatic hydroxyl-containing compound, such as a naphthol or naphtholate (azoic coupling component) followed by coupling with a stabilised aromatic diazonium compound (the azoic diazo component also termed fast base or salt) to form a coloured insoluble azo compound.[7,58] The application method must be carried out with care, and the final colour is obtained only after soaping off to remove any traces of the azoic dye on the fibre surface that would give inferior colour fastness to washing and rubbing (crocking). Although azoic colouring matters are economical for the production of red and black, the colour range is now more limited because some diazo components based on certain aromatic amines have been withdrawn because of their possible carcinogenic nature. Precise colour matching with such a complex two-bath procedure can also give problems in practice. Thus, the use of azoic colouring matters has been in marked decline over the last decade.

9.4.5 Direct dyes

Direct dyes were the first class of synthetic dyes to dye cotton directly without the use of a mordant. Direct dyes are sulphonated bisazo, trisazo, or polyazo dyes and anionic dyes that are substantive to cellulosic fibres when applied from an aqueous bath containing an electrolyte.[7,59] Direct dyes based on stilbene, copper-complex azo, oxazine, thiazole, and phthalocyanine structures are also used. Less bright in colour than acid or basic dyes, the brightness diminishes with the molecular complexity of the dye. Phthalocyanine dyes are used for very bright blue and turquoise blue colours of good colour fastness to light, and copper-complex azo dyes also exhibit good fastness to light, although these dyes yield relatively dull colours.

Electrolyte in the form of sodium sulphate or sodium chloride is usually added into the dyebath to overcome the negative charge on the cellulosic fibre surface that otherwise would repel the approach of the direct dye anions. The sodium cations from the electrolyte neutralise the negative charge at the fibre surface, allowing the dye anions to be adsorbed and retained in the fibre.[59]

Coulombic attraction, hydrogen bonding, and nonpolar van der Waals forces may operate between the dye molecule and the fibre, depending on the specific direct dye and the fibre structure. Direct dyes are normally linear and planar molecules that allow multipoint attachment to cellulose chain molecules, but the forces of attraction between dye and fibre are relatively weak when the dyed fibre is immersed in water.[59,60] Thus, the colour fastness to washing is moderate to poor but to light may vary from excellent to poor, depending on the molecular structure.

The colour fastness to washing can be improved by aftertreatment of the dyed fibre, originally by the use of after-coppering (i.e. treatment with copper sulphate to form a metal-complex direct dye) or by diazotisation and development. Both approaches lead to a pronounced shade change, and copper in the wastewater from dyeing is environmentally undesirable. Modern aftertreatments make use of cationic fixatives that may complex with the anionic dye and/or form a metal complex and/or react with the cellulose fibre hydroxyl groups to form strong covalent bonds. Improvement in colour fastness to washing may be accompanied by some diminution in colour fastness to light.[56,61]

Direct dyes may be used on all cellulosic fibres, and selected dyes may be used to dye wool, silk, and nylon fibres in the manner of acid dyes. Addition of electrolyte into the dyebath increases dye aggregation in the dyebath. As the dyeing temperature is increased, the dye aggregation decreases, releasing individual direct dye anions for diffusion and adsorption inside the fibre. Direct dyes have been separated into three classes, Class A—salt-controllable, Class B—temperature-controllable, and Class C—both salt and temperature controllable. It is normal to select compatible dyes from within the same class, but selected dyes from Class A and B, and Class B and C may be dyed together in the same dyebath in exhaust dyeing. Direct dyes may also be applied to cellulosic fabrics by continuous pad-steam-wash off and dry methods.[7,59]

The depth of colour obtained using direct dyes is deeper on viscose, lyocell, and high wet modulus fibres and on mercerised cotton compared with bleached cotton, and careful dye selection is required to produce level, solid colours when dyeing blends of these fibres.

9.4.6 Reactive dyes

Reactive dyes, sometimes termed *fibre-reactive dyes*, are a very important class of dyes for dyeing cellulosic fibres[62] and are used to dye protein fibres such as wool[63] or silk.[64] Although relatively expensive, the reactive dyes provide a wide colour gamut of bright colours with very good colour fastness to washing. This is because during dye fixation, which is usually conducted under alkaline conditions, strong covalent bonds are formed between the dye and the fibre.[7,65] Reactive dyes may react by substitution (e.g. monochlorotriazinyl and dichlorotriazinyl dyes) or by addition (e.g. vinylsulphone dyes).

Increasing the number of fibre-reactive functional groups in the reactive dye molecule increases the cost of the dyestuff but also markedly increases the dye fixation efficiency on cellulosic fibres. Avitera SE dyestuffs (Huntsman Textile Effects) are trifunctional dyes that increase the dye fixation level to around 90% or higher.[66,67] This greatly decreases the necessity for prolonged washing off of unfixed dyestuff after dyeing. Cellulosic materials can be bleached and dyed using the enzyme-based Gentle Power Bleach system followed by Avitera SE dyeing and washing off all at low temperatures (e.g. typically at 60–65 °C).

Typical ranges of reactive dyes and their respective reactive groups are shown in Table 9.3. Recent developments have led to the introduction of homobifunctional reactive dyes (e.g. two monochlorotriazine groups) and heterobifunctional-reactive dyes (e.g. using monochlorotriazine plus vinylsulphone) in an attempt to increase the dye fixation on the fibre under alkaline conditions from 50% to 70% with one reactive group to 80% to 95% with two reactive groups[68]. Dye application methods on fabrics include exhaust dyeing, cold pad-batch-wash off, and continuous pad-steam-wash off and dry methods. Fixation by dry-heat, saturated steam, or superheated steam may also be used according to the type of reactive dye employed.[62]

Reactive dyes hydrolyse in contact with water and alkali, and during dyeing, some hydrolysed reactive dye is adsorbed by the fibre because it behaves like a substantive direct dye but with lower colour fastness to washing than the reactive dye that is covalently bonded. Thus, the dyeing stage must always be followed by an extended washing-off treatment to remove hydrolysed reactive dye.[62,69] Novel wash accelerators

Table 9.3 **Important reactive dye systems**

System	Typical brand name
Monofunctional	
Aminofluorotriazine	Novacron F (Huntsman Textile Effects)
Trichloropyrimidine	Drimaren X (Archroma)
Chlorodifluoropyrimidine	Drimaren K (Archroma)
Sulphatoethylsulphone	Remazol (DyStar)
Bifunctional	
Bis(aminochlorotriazine)	Procion H-E, Procion H-EXL (DyStar)
Bis(aminonicotinotriazine)	Kayacelon React (Nippon Kayaku)
Aminochlorotriazine-sulphatoethylsulphone	Sumifix Supra (Sumitomo)
Aminofluorotriazine-sulphatoethylsulphone	Novacron C (Huntsman Textile Effects)
Undisclosed	Levafix CA (DyStar)
Undisclosed	Remazol RGB (DyStar)
Undisclosed	Drimaren HF/CL (Archroma)
Trifunctional	
Undisclosed	Avitera SE (Huntsman Textile Effects)

based on innovative surfactants have been introduced to shorten the number of rinsing baths and decrease the rinsing time as well as economising on water and energy consumption. Provided that the residual hydrolysed reactive dye left within the cellulosic fibre after dyeing or printing is at a concentration $0.1\,g\,l^{-1}$, the best colour fastness to washing will be obtained.[70] If the concentration of hydrolysed reactive dye in the final hot wash bath is $0.003\,g\,l^{-1}$, this will prevent the staining of white grounds.[70]

Generally reactive dyes are produced in granular form, but vinylsulphone dyes are also available as liquids, which are more convenient for use in continuous dyeing and printing. Reactive dyes are sulphonated and highly water soluble and are exhausted onto the fibre using electrolyte (e.g. in the manner of direct dyes) and fixed using an appropriate alkali. The high water solubility creates an environmental problem in wastewater treatment plants because generally only 0–30% of the hydrolysed reactive dye is removed by such treatment.[71] Low-salt reactive dyes have also been introduced to decrease dyeing costs and to avoid corrosion problems in wastewater concrete pipework networks caused by high concentrations of sulphate anions.

9.4.7 Sulphur dyes

Sulphur dyes are chemically complex and are prepared by heating various aromatic diamines, nitrophenols, and so on with sulphur and sodium sulphide. Sulphur dyes are produced in pigment form without substantivity for cellulose.[7] Treatment with a reducing agent (e.g. sodium sulphide or sodium hydrosulphide) in an alkaline dyebath converts the sulphur dye into an alkali-soluble reduced (leuco) form with substantivity for cellulosic fibres. Once absorbed within the fibre, the dye is then oxidised, usually with hydrogen peroxide, back to the insoluble pigment form. Soaping off after dyeing is important to ensure the maximum colour fastness to washing and rubbing.[7,72]

Sulphur black and navy are the major dyestuffs used, and the colour gamut is limited to dull colours of moderate colour fastness to washing and light. The colour fastness to chlorine and to bleach fading is poor when using multiple wash cycles with detergents containing low temperature bleach activators such as tetraacetylethylenediamine (TAED, United Kingdom) or sodium nonanoyloxybenzenesulphonate (SNOBS, United States).[73] Sulphur dyes cost less than many other dyes, but the wastewater from dyeing may require specialised treatment before release to a conventional wastewater treatment plant.

Diresul RDT dyestuffs (Archroma) are prereduced sulphur-free sulphur dyes that are free of heavy metals and of AOX (absorbable organo-halides).[74,75] Diresul RDT dyestuffs can be applied to textile materials under nitrogen instead of air to avoid preoxidation and exhibit high exhaustion on cellulosic fibres, yarns, and fabrics. This decreases the pollution load in the effluent, and Diresul RDT dyestuffs have low fish toxicity and require relatively little salt, water, and energy during their application.

9.4.8   Vat dyes

Vat dyes are water insoluble but contain two or more keto groups ($>C=O$) separated by a conjugated system of double bonds that are converted into alkali-soluble enolate leuco compounds ($>C–O^-$) by alkaline reduction, a process called *vatting*.[7,76] The dye application method involves three stages, namely, alkaline reduction and dissolution of the vat dye (normally using sodium hydroxide and sodium dithionite (hydrosulphite)), absorption of the substantive leuco compound by the fibre, aided by electrolyte (e.g. sodium sulphate and wetting, dispersing, and levelling agent), followed by regeneration of the vat dye inside the fibre by oxidation in air or hydrogen peroxide. The dyed fibre is then thoroughly soaped off (washed with soap or special detergent at a high temperature) to remove any surface dye and to complete any dye aggregation inside the fibre in order to obtain the final colour. In enclosed batch dyeing, machinery-entrapped air within the machine can then be displaced by nitrogen gas. This prevents the preoxidation of the leuco vat dyestuff, which can lead to bronzing from vat dye particles on the fibre surface and to inferior levels of colour fastness to washing and rubbing.

Vat dyes are based on indanthrone, flavanthrone, pyranthrone, dibenzanthrone, acylaminoanthraquinone, carbazole, azoanthraquininone, indigoid, and thioindigoid structures.[7,76] Application methods include both batch exhaust dyeing and continuous pad-steam-wash off and dry methods, and sodium dithionite, formaldehydesulphoxylate, or hydroxyacetone may be used for the reduction stage.[76]

Vat dyes are relatively expensive but do offer outstanding colour fastness to light and washing on cellulosic fibres. In the unreduced form, some vat dyes behave like disperse dyes and can therefore be applied to polyester and polyester/cellulosic blends to achieve pale-medium colour depths. For polyester/cellulosic fabrics a pad-dry-bake-chemical pad-steam-oxidise-wash off and dry production sequence is used. For exhaust dyeing, the novel technique of electrochemical reduction has been claimed to provide a redox potential of up to ~960 mV suitable for the reduction of vat and sulphur dyes. Cathodic reduction of vat dyes with the addition of a mediator (a soluble reversible redox system (e.g. an iron [II/III]–amino complex) could potentially decrease the chemical costs for reduction and lower the chemical load in the wastewater by 50–75%.[77]

A few selected vat dyestuffs absorb near infra-red radiation (NIR) sufficiently well to enable the production of camouflage colours on cotton fabric for military concealment and deception (e.g. flexible net covers and dyed and printed military clothing).[78] Suitable vat dyes for near infra-red camouflage are generally based on anthraquinone–benzanthrone–acridine polycyclic ring systems and must exhibit high colour fastness standards to light, washing, and perspiration. Vat dyes suitable for camouflage colours on cotton such as pale and dark brown, pale and dark green, and grey have been produced. For synthetic fibres, the incorporation of a small amount of carbon black pigment in spin dyeing followed by conventional dyes appropriate to the polymer can sometimes also be used.

9.4.9   Disperse dyes

Disperse dyes are substantially water-insoluble dyes that have substantivity for one or more hydrophobic fibres (e.g. secondary cellulose acetate) and are usually applied as a fine aqueous dispersion.[7,39,79,80] The major chemical classes used are aminoazo-benzene-, anthraquinone-, nitrodiphenylamine, styryl- (methine), quinophthalone-, and benzodifuranone-based dyes. Disperse dyes are milled (ground) with a dispersing agent (e.g. polymeric forms of sodium dinaphthylmethane sulphonates) to a fine dispersion (0.5–2 μm) and may be supplied as grains, powders, or liquid dispersions.

Disperse dyes are essentially nonionic dyes that are attracted to hydrophobic fibres, such as conventional polyester, cellulose triacetate, secondary cellulose acetate, and nylon, through nonpolar forces of attraction.[39,53,79,80] In the dyebath, some of the disperse dye particles dissolve to provide individual dye molecules that are small enough to diffuse into the hydrophobic fibres. The aqueous solubility of disperse dyes is low, for example, 0.2–100 mg l^{-1} at 80 °C, – but increases with increase in dyebath temperature, in turn increasing the concentration of soluble dye available for diffusion so that the dyeing rate increases.

Dispersing agents are essential because they assist in the process of decreasing the dye particle size and enable the dye to be prepared in powder and liquid forms.[79] In addition, the dispersing agent facilitates the reverse change from powder to dispersion during dyebath preparation and maintains the dye particles in fine dispersion during dyeing. This prevents agglomeration of the dye particles in powder form and aggregation in the dyebath. The dye solubility in the dyebath can be increased by the use of levelling agents and carriers.

The general mechanism of dyeing with disperse dyes under exhaust dyeing conditions is illustrated in Figure 9.1. The diffusion and adsorption of dye molecules is accompanied by desorption of some of the dye molecules from the dyed fibre back into the dyebath to facilitate dye migration from fibre to fibre to achieve level dyeing.[56] The disperse dyes are considered to dye fibres via a solid solution mechanism that involves no chemical change.[38] Each disperse dye dissolves in the fibre more or less

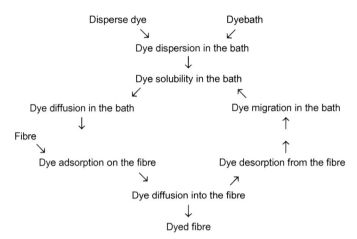

Figure 9.1 General dyeing mechanism for disperse dyes (source: see Ref. 56, p. 126).

independently of any other disperse dye present. Under constant dyebath conditions, the ratio of the amount of dye on the fibre to the amount of dye in the bath is a constant at equilibrium, which varies according to the particular dye, but ultimately a separate different saturation level (or solubility limit) is achieved for each dye above which no more of that dye can be taken up by the fibre. However, most commercial dyeings are never carried out to equilibrium.

Except for high-temperature (high-pressure) exhaust dyeing at 110–140 °C, disperse dyes may also be exhaust dyed using carriers (e.g. fibre plasticising agents that decrease the fibre T_g) at 100 °C.[37,79,80] This method is not very popular except when blend dyeing is practised and high-temperature dyeing, if used, would degrade one of the fibres. Odourless carriers are preferred, but there are problems with carrier spotting, and carrier retention by the fibres leads to fuming of the carrier because it is volatilised off during postheat setting treatments whereas some carriers may potentially pose environmental pollution problems.[37] Approximately 50% of disperse dye consumption is used for navy and black shades, and about 70% of disperse dyes are applied to polyester and polyester/cellulose blends by exhaust dyeing.[81]

Disperse dyes may also be applied by a pad-dry-thermofix-wash dry process in a continuous open width fabric treatment. This uses the sublimation properties of disperse dyes, which vapourise directly from the solid state without prior melting. Polyester can be dyed at 190–220 °C by such a process, originally the Thermosol® process by DuPont.[38] Sublimation of disperse dyes is also used in dry-heat transfer printing,[82] but conventional print application by printing and fixing in saturated or superheated steam is also practiced, followed by washing off and drying.[51]

To minimise potential problems linked to the presence of disperse dyes on the fibre surface after dyeing (e.g. poor colour fastness to washing, rubbing, and thermomigration), reduction clearing after dyeing with sodium hydroxide, sodium dithionite (hydrosulphite), and a surfactant followed by hot washing is normally practiced, particularly with medium-heavy depths of colour.[79] However, some classes of disperse dyes (e.g. diester and thiophene azo dyes) are clearable by alkali alone.[83]

9.4.10   Pigment formulations

The range of chemical compounds used as pigments varies widely and includes both inorganic pigments such as titanium dioxide and the oxides of antimony, iron, and zinc, as well as carbon.[84] Organic pigments may be based on a very wide variety of chemical structures, for example, azo, anthraquinone, dioxazine, indanthrone, isoindolinone, perylene, quinacridone, copper phthalocyanine, heterocyclic nickel complexes, and many others (see, e.g. Table 9.4).[85] The general considerations relating to the manufacture of pigments and their application to technical textiles are separately discussed in

Table 9.4 **High performance pigments for polypropylene fibres**

CI generic name	CI number	Chemical class
CI Pigment Yellow 93	20710	Azo condensation
Yellow 95	20034	Azo condensation
Yellow 109	56284	Isoindolinone
Yellow 110	56280	Isoindolinone
Yellow 155	Not known	Azo condensation
Yellow 181	11777	Monoazo
Yellow 182	Not known	Heterocyclic monoazo
Orange 61	11265	Isoindolinone
Red 122	73915	Quinacridone
Red 144	20735	Azo condensation
Red 149	71137	Perylene
Red 166	20730	Azo condensation
Red 177	65300	Anthraquinone
Red 202	Not known	Quinacridone
Red 214	Not known	Azo condensation
Red 242	20067	Azo condensation
Red 257	Not known	Heterocyclic nickel complex
Violet 19	73900	Quinacridone
Violet 23	51319	Dioxazine
Violet 37	51345	Dioxazine
Blue 15:3	74160	Copper phthalocyanine
Blue 60	69800	Indanthrone
Green 7	74260	Copper phthalocyanine

CI denotes Colour Index.
Source: see Ref. [85].

Section 9.3.2 (manufacture), Sections 9.5.1–9.5.3 (mass pigmentation), Section 9.5.5 (coating), and Section 9.9.1 (printing). The finishing of pigment formulations is technically complex, and strict quality control is required to ensure satisfactory application and performance of the pigment in practice.

9.5 Mass coloration of manufactured fibres

9.5.1 Dyes and pigments for mass coloration

In mass coloration, dyes or pigments are incorporated into the polymer melt (in the case of polyamide, polyester, and polypropylene) or into the polymer solution (in the case of viscose, secondary cellulose acetate, cellulose triacetate, and acrylic or modacrylic fibres) during fibre manufacture.[13,25] The dyes that are being used pass into a dissolved phase on incorporation into the polymer melt or solution, but pigments remain as finely dispersed particles. In the main, mass pigmentation is much more widely used for colouring manufactured fibres (both synthetic and regenerated fibres)[13,25] except on wet-spun acrylic fibres when gel dyeing is used (see Section 9.5.4).[55,56,60] However, mass pigmentation can be used for dry-spun acrylic polymers that are dissolved in a solvent (e.g. dimethyl formamide) and spun into hot air.

Technical criteria that are important for mass coloration include[85,86]:

- Pigment particle size and particle size distribution, both in the spinning mass and in the filament
- Pigment preparations
- Solubility of the colorant in the spinning mass
- Colour fastness properties
- Ability to remain stable under the mass-processing conditions of the polymer

Fine pigment particles ($\leq 1\,\mu m$) are required because coarse particles would interfere with the filterability of polymer solutions or melts and could impair the tensile strength of fibres, the diameter of which normally lies in the range $16–45\,\mu m$. The pigments used must have good resistance to organic solvents and good heat stability for use in fibre manufacture.

Products suitable for the bath dyeing and mass coloration of manufactured fibres are presented in Table 9.5.[25]

9.5.2 Mass coloration methods

Four main methods of incorporating colorants into manufactured fibres depend on the specific fibre production process.[13,25] These are:

- Batch process
- Injection process
- Chip blending
- Chip dyeing

Table 9.5 Products for the bath dyeing and mass coloration of manufactured fibres

Fibre	Bath dyes	Solvent dyes and pigments for mass coloration
Viscose	Direct Vat Reactive	Aqueous pigment pastes
Acetate	Disperse	Acetone-soluble dyes Pigment dispersions in acetate
Acrylic	Disperse Basic	Pigment dispersions in polyacrylonitrile
Polyamide	Disperse Acid Metal complex	Aqueous dyes Polymer-soluble dyes
Polyester	Disperse	Polymer-soluble dyes Pigment dispersions in polyester
Polypropylene	Acid chelatable	Pigment dispersions in various carriers

Source: see Ref. [25]

Pigment preparations or solvent dyes may be used in the batch method in which the whole of the spinning mass is coloured, the colorant concentration being equal to that in the coloured filament after extrusion. In the injection process, a coloured concentrate (masterbatch) is continuously metered into the spinning mass that is then extruded.[85–87] A masterbatch is a concentrated mixture of pigments and/or additives encapsulated during a heat process into a carrier polymer that is then cooled and cut into a granular shape.[87] Homogeneous mixing within the molten bulk of the synthetic polymer followed by filament extrusion can thus be used to produce spuncoloured filaments and fibres. Except for pigments to impart colour, the masterbatch could also contain other additives according to the fibre performance requirements. These could include one or more of the following additives:

• Ultraviolet stabilisers
• Flame retardants
• Antistatic agents
• Antioxidants
• Antimicrobials
• Fibre lubricants

Such additives are important for synthetic fibres used in technical textile end uses.[86] In rugged applications such as upholstery, sportswear, and apparel, for example, the durability of colour, stability to ultraviolet radiation, flame retardancy, and antibacterial properties are important. In polypropylene artificial turf, coloured pigments and UV stabilisers are used, and the material must achieve the required standards of fibre resilience, colour fastness to light and weathering, and antimicrobial performance. Automotive textiles require the highest levels of colour consistency and product performance. The colour fastness of automotive seating, carpet, and nonwoven panels must withstand the high-temperature conditions generated within the enclosed vehicle interior when it is parked in strong sunlight and/or in a high ambient temperature.[87]

Chip blending is suitable only for melt coloration, for example, for polyamide, polyester, and polypropylene. Polymer chips are homogeneously mixed with the colour concentrate prior to extrusion. Chip dyeing is a more specialised technique used with nylon 6. The polymer chips are precoloured with polymer-soluble dyes and then melt spun.[25]

Mass pigmentation is widely used as the major coloration route for polypropylene fibres.[1,13,25,85,86] The flexibility of the smaller-scale nature of polypropylene fibre manufacture compared with the large continuous polymerisation/extrusion processes for polyamide and polyester fibres, the lower polymer melt temperature of polypropylene and the difficulties in dyeing this hydrophobic, low linear density fibre in conventional dyeing equipment, ensure that mass pigmentation is the major coloration route. Mass coloration of many other melt-spun fibres is carried out on large-scale reactor vessels that require extensive cleaning to avoid subsequent colour contamination problems with the next colour to be spun. This is not economic unless large fibre weights per colour may be spun for a specific end use. However, attempts continue to find a satisfactory conventional dyeing method for modified polypropylene fibres using exhaust dyeing or continuous dyeing techniques.[1,89]

9.5.3   Mass coloration and colour fastness properties

The colour fastness properties of manufactured fibres produced by mass dyeing, particularly by mass pigmentation, are generally superior to those obtained on technical textiles by conventional dyeing and printing. Mass pigmentation is used where high colour fastness to light and weathering is required, for example, in tenting fabrics, awnings, sun blinds, carpets, synthetic sports surfaces, and so on and when the colour can be economically produced.

The organic pigments and a few inorganic pigments used for mass pigmentation of polyester and polypropylene fibres must exhibit[85,86]:

- Sufficient thermal stability
- Colour fastness to light adequate for the intended end use
- No migration (no blooming or contact bleed)
- Compatibility with other additives (e.g. the UV stabilisers widely used in polyolefins) with no photodegradation effects on the polymer
- No adverse effect on the fibres' mechanical properties

High-performance organic pigments are used for brilliant colours and good overall colour fastness performance on polypropylene fibres (see Table 9.4).

Inorganic pigments such as titanium dioxide (rutile), zinc oxide, and antimony oxide are used for white, and carbon (small particle size channel black) is used for black whereas some iron oxide browns are used for cost reasons. The tensile strength of mass-pigmented polypropylene fibres after 300 h of exposure in an Atlas 600 WRC accelerated light fastness testing equipment is markedly higher than the uncoloured fibre, demonstrating the protective effect of the presence of the pigments against the deleterious effects of ultraviolet radiation on the fibres.[85]

Colour retention in awning fabrics for outdoor applications is critical, and two types of climatic conditions are most critical where pigments are incorporated in acrylic and modacrylic technical textiles[90]:

- Hot and dry (typically 38–49 °C, relative humidity ≤20%) with high average sunshine
- Warm and humid (32–38 °C, relative humidity ≥70%) with average sunshine

In hot and dry conditions, both the polymer and the pigment may degrade whereas in warm and humid conditions, pigment degradation is more likely. However, pigments that are degraded and fade under warm and humid conditions may often be quite stable under hot dry conditions.

9.5.4   Gel dyeing of acrylic fibres

For wet-spun acrylic fibres, the manufacture of producer-dyed fibres involves the passage of acrylic tow in the gel state (i.e. never-dried state) through a bath containing basic dyes.[55,56,87] This was used in the Courtelle Neochrome process (formerly Courtaulds, now AKZO Nobel) to produce dyed acrylic fibres from a continuous fibre production line typically at a speed of 50 m min^{-1} with economic batch weights per colour of 250–500 kg.

The liquid basic dyes are metered in at a rate appropriate to the acrylic tow mass and speed with the recipe based on a computerised colour match prediction system that allows the selection of a very wide range of colours using a choice of the technically best dyes (i.e. easiest to apply and highest colour fastness), lowest cost dyes, or least metameric dyes.[91] As the freshly coagulated acrylic tow passes through the dyebath, the basic dyes diffuse inside the gel-state acrylic tow in a matter of seconds. The dyed tow is then drawn and steamed, crimped, and cut to the appropriate staple length for use in technical textiles or may be used alternatively in filament form.[88]

9.5.5   Pigments for technical coated fabrics

Many technical textiles are coated in order to provide the high performance specification demanded by the end use. The many considerations governing the use of coloured organic pigments in coatings and the range of pigments available have been reviewed by Lewis.[92]

The pigments incorporated in silicone rubber mixes for textile colour coating applications are usually inorganic pigments such as red iron oxide and titanium dioxide for white articles.[93] Organic pigments are not normally suitable for high-temperature coating applications. Carbon black pigments are not generally used in silicone rubber mixes because they can affect the activity of peroxides added as curing agents.

9.6   Conventional dyeing and printing of technical textiles

9.6.1   Pretreatments prior to conventional coloration

Before fibres, yarns or fabrics are to be dyed via batch exhaust or continuous pad-fixation methods or printed by print-dry-steam-wash off-dry or pigment print-dry-cure methods, it is important to remove any natural, added, or acquired impurities from the fibres so that these impurities do not interfere with coloration. Natural fibres such as cotton, wool, silk, and flax (linen) fibres contain appreciable quantities of impurities naturally associated with the fibres, which are removed by scouring and other treatments, for example, degumming silk to remove sericin (gum) and carbonising wool to remove vegetable matter (e.g. burrs, seeds).[94] Particularly important for cotton and flax (linen) fibres is the efficient removal of the natural waxes present mainly on the fibre surface in order to impart hydrophilicity to the fibres, thereby ensuring satisfactory wetting out with high levels of uniformity and reproducibility in subsequent dyeing or printing treatments.[95]

Manufactured fibres and filaments, that is, artificial fibres from synthetic or re-generated polymers, are produced under carefully controlled conditions so that water-soluble/emulsifiable spin finishes or fibre lubricants are the main impurities and are simply removed by scouring.[94]

9.6.2   Singeing

For many woven or knitted technical fabrics, it may be necessary to singe the fabric surface by passage through a gas flame or an infra-red zone at open width to remove protruding surface fibres. This gives a clear fabric surface and more uniform color-ation because a hairy fabric surface can impart a lighter coloured surface appearance (termed *frostiness*) after dyeing or printing. Singeing may be integrated with subse-quent wet processes such as desizing.[96]

An alternative to singeing is to treat cellulosic fibre fabrics, such as cotton with cellulase enzymes under acidic or neutral pH conditions in a biopolishing treatment. Biopolishing attacks and weakens the protruding surface fibres that are then removed by mechanical action to produce a smoother fabric surface. The Cellusoft Combi pro-cess (Novozymes) combines biopolishing and dyeing in one process, saving water, energy, and processing time.[97]

9.6.3   Desizing

The warp yarns of most woven fabrics are generally sized with a film-forming poly-mer yielding a more cohesive structure.[98] Sizes are used to increase yarn strength, decrease yarn hairiness, and impart lubrication to staple fibre warp yarns in order to minimise the number of warp breaks during weaving. In the weaving of synthetic continuous filament yarns, the size provides good interfilament binding to prevent filament snagging as well as providing yarn lubrication and antistatic performance to decrease the buildup of electrostatic charges on the yarns during high-speed weaving. Yarns for knitting are not sized, and a desizing treatment is not therefore required.

All sizes, as well as any other sizing components – for example, waxes, softeners, and lubricants that may be hydrophobic – must be removed by appropriate desizing treatments.[99] Synthetic water-soluble sizes (e.g. acrylates, polyvinyl alcohol) may be simply removed by washing, whereas natural sizes (e.g. starch, modified starch) may require chemical degradation treatments, such as oxidation, hydrolysis, and so on. Starch-based sizes may be degraded by hydrolysis and removed using enzymatic treatments specific for starch (e.g. malt diastases or bacterial α-amylases) under controlled conditions of pH and temperature. Enzymatic desizing cannot be combined with alkaline scouring and hydrogen peroxide bleaching because of enzyme degradation during such pretreatments. If considered appropriate, oxidative desizing may be integrated with scouring or with scouring and bleaching to provide a shorter integrated treatment whereas in large, vertically integrated plants, undegraded synthetic sizes that are removed in desizing may be recycled using ultrafiltration and reused.

After every desizing treatment, all traces of size or degraded size are removed by thorough washing followed by efficient mechanical removal of liquid water (e.g. hot mangling or vacuum extraction). This is followed by thermal drying on steam-heated cylinders (called *cylinder* or *can drying*) or by hot air drying on a stenter (also known as a tenter, or frame (USA)).

9.6.4   Scouring

Scouring is a critical treatment for natural cellulosic fibres – for example, cotton and flax (linen) and their blends with other fibres – because all traces of hydrophobic waxes (whether naturally occurring or applied during the manufacturing sequence) must be effectively removed in order to achieve satisfactory wetting in all subsequent dyeing, printing, finishing, coating, lamination, and bonding operations.[95] Scouring is normally accomplished by hot alkaline treatment (e.g. sodium hydroxide) followed by a thorough hot wash off. It is also important for the removal of fatty materials from wool and any manufactured fibres using hot detergent solutions.

The use of bioscouring treatments with pectinase enzymes is growing in importance for both woven and knitted cotton fabrics because the removal of pectins and cotton waxes from the fibres are facilitated.[100] The three types of pectins in cotton are homogalacturonans and rhamnogalacturonans I and II, all of which may be degraded by pectinases, which can be based on pectin esterases, polygalacturonases, and polygalacturonate lyases. Bioscouring at 50–60 °C is followed by raising the temperature to 90 °C to aid in melting and removing cotton waxes and oils. A lower fabric weight loss and a softer handle coupled with water and energy savings are obtained by bioscouring compared with traditional alkaline scouring with caustic soda (NaOH).

9.6.5   Bleaching

Fibres must be uniformly white if pale colours or bright colours are required; hence, natural fibres such as cotton, silk, wool, and linen must be chemically bleached in order to achieve a satisfactory stable whiteness.[15,94] Hydrogen peroxide under controlled alkaline conditions is normally used in pad-steam-wash off, pad-batch-wash off, or immersion bleaching treatments, although other oxidising agents such as peracetic acid, cold alkaline sodium hypochlorite, or sodium chlorite under acidic conditions may also be used. Residual peroxide residues in the fibres after hydrogen peroxide bleaching are destroyed by aftertreatment with a catalase enzyme.

However, chlorine-based bleaching agents give a poor bleaching environment, and chlorine is retained by the fibre, necessitating an antichlor aftertreatment with a reducing agent (e.g. sodium sulphite) followed by washing to remove any residual odour in the fabric. Retained chlorine from chlorine-based bleaches or from chlorinated water used for dyeing can give rise to shade changes when dyeing cotton with reactive dyes. Regenerated cellulosic fibres such as viscose, cuprammonium, polynosic, modal, high wet modulus, and lyocell fibres are marketed by fibre producers with a satisfactory whiteness for dyeing and printing.

Bleaching cotton, regenerated cellulosic fibres (e.g. viscose, modal, and lyocell), and their blends can also be done by biobleaching using an enzyme bleaching system.[100] The Gentle Power Bleach system of Huntsman Textile Effects uses an enzyme (Invazyme-LTE) from Genencor that enables biobleaching to be conducted at neutral pH and 65 °C. The bleaching liquor contains an ester (e.g. propylene glycol diacetate) that under the action of a perhydrolase enzyme generates peracetic acid (the bleaching agent). In addition, a surfactant and/or emulsifier, a peroxide stabiliser, and a sequestering agent are used together with a pH buffer system to maintain the pH in the pH 6–8 region. This more sustainable biobleaching system results in a reduction in water usage by at least 30% and a decrease in energy consumption by 40%. The cotton weight loss is cut almost in half when compared with traditional alkaline hydrogen peroxide bleaching, and the natural softness of the cotton is retained. A stabilised liquid catalase enzyme is used to destroy any residual peroxide residues on the fibres that might otherwise create problems in subsequent dyeing.

9.6.6   Fluorescent brightening

If high whites are required, both natural and regenerated cellulose fibres may be chemically bleached and treated with a fluorescent brightening agent.[28,29] Fluorescent brightening agents are based on diaminostilbene derivatives, triazoles, aminocoumarins, and many other organic compounds and are absorbed by the fibre. Fluorescent brightening agents are organic compounds that absorb the ultraviolet radiation present in daylight and re-emit light in the blue-violet region of the visible spectrum. During the absorption and re-emission processes, some energy is lost; hence, the light emitted is shifted to a longer wavelength. The effect is to add brightness to the whiteness produced by chemical bleaching. Alternatively, blue tints may be used, for example, ultramarine.

9.6.7 Mercerisation

Mercerisation of cotton is a fibre-swelling/structural relaxation treatment that may be carried out on yarns but more usually on fabrics.[101,102] Hank or warp mercerisation of yarns often creates dyeability differences because of yarn tension variations that occur during mercerisation. During mercerisation in a 22–27% caustic soda solution, both mature and immature cotton fibres swell so that the secondary wall thickness is increased. The fibre surface appearance and the internal structure of the fibre are modified. This improves the uniformity of fabric appearance after dyeing, and there is an apparent increase in colour depth after mercerisation that has been claimed to give cost savings of up to 30% on pale colours (e.g. 1–2% owf (on weight of fibre)) and even 50–70% on heavy depths when using some reactive dyes.[103,104] Dead cotton fibres (i.e. those with little or no secondary wall) are, however, not improved after mercerisation.

Woven fabric mercerisation is normally carried out under tension on chain or chainless fabric mercerising ranges[102] whereas tubular fabric mercerising ranges are widely used for weft knitted cotton fabrics.[105] Mercerisation leads to a number of changes in fibre and fabric properties[101–105]:

- A more circular fibre cross-section
- Increased lustre
- Increased tensile strength, a major factor for technical textile fabrics
- Increased apparent colour depth after dyeing
- Improved dyeability of immature cotton (greater uniformity of appearance)
- Increased fibre moisture regain
- Increased water sorption
- Improved dimensional stability.

After mercerisation, the structure of native cotton fibres, cellulose I, is converted into cellulose II, which is the stable fibre form after drying.[104] The sorptive capacity of mercerised cotton is greater when the fabric is mercerised without tension (slack mercerising) to give stretch properties to the fabric. An increase in drying temperature can also decrease the sorptive capacity, especially at temperatures above 80 °C.[106]

9.6.8 Anhydrous liquid ammonia treatments

This form of cotton fabric pretreatment is much less common than mercerising and has been most widely used in Japan.[104,107] Impregnation in anhydrous liquid ammonia at −38 °C in an enclosed machine followed by a swelling/relaxation stage and removal of the ammonia by thermal drying and steaming converts the cellulose I crystalline form back to either cellulose I or cellulose III, depending on the structural collapse of the fibre, and the final traces of ammonia are removed in the steamer. This ammonia-dry-steam process can be used to give better improvements in cotton fabric properties than mercerisation, although the increase in colour depth after dyeing in this process is usually somewhat lower than that achieved after mercerisation. The high capital cost of the machinery for anhydrous liquid ammonia treatment and ammonia recovery as well as environmental considerations have limited the wider exploitation of this technique.

9.6.9   Heat setting

Synthetic thermoplastic fibres, yarns, and fabrics may be heat set, steam set, or hydro-set in order to obtain satisfactory dimensional stability during subsequent hot wet treatments.[108] Fabrics may be set prior to coloration or set after coloration.[49] Hydrosetting in hot water is rarely carried out, but false twist textured yarn can be steam set in an autoclave using a double vacuum double steam cycle to attain satisfactory removal of air and, hence, uniformity of temperature in the treatment. Steam setting avoids the slight fibre yellowing that can occur in fabric form through fibre surface oxidation during hot air setting on a stenter, and the handle is softer.

Heat setting on modern stenters is often done by first drying and then heat setting in one passage through the stenter. The temperature and time of heat setting must be carefully monitored and controlled to ensure that consistent fabric properties are achieved. During heat setting, the segmental motion of the chain molecules of the amorphous regions of the fibre are generally increased, leading to structural relaxation within the fibre structure. During cooling, the temperature is decreased below the fibre T_g, and the new fibre structure is stabilised. Because the polymer chain molecules have vibrated and moved into new equilibrium positions at a high temperature in heat setting, subsequent heat treatments at lower temperatures do not cause the heat set fibres to relax and shrink so that the fabric dimensional stability is high.

Presetting fabric prior to dyeing alters the polymer chain molecular arrangement within the fibres and, hence, can alter the rate of dye uptake during dyeing.[38,49,50] Process variations (e.g. temperature, time, or tension differences) during heat setting may thus give rise to dyeability variations that become apparent after dyeing. Fabric setting after coloration can lead to the diffusion of dyes such as disperse dyes to the fibre surface and to sublimation, thermomigration, and blooming problems, all of which can alter the colour and markedly decrease the colour fastness to washing and rubbing of technical textiles containing polyester fibres.[83]

9.6.10   Quality control in pretreatment

In all the pretreatments given to fabrics, it is necessary to control the process carefully in order to minimise fibre degradation and yellowing.[15] Mechanical damage (i.e. holes, poor dimensional stability, and fibre damage from overdrying) must also be avoided. Fibre yellowing makes it very difficult to dye pale bright colours whereas any fibre degradation may subsequently increase during dyeing and printing treatments and result in inferior physical properties (e.g. low tensile strength, tear strength, and abrasion resistance) or to poor colour fastness performance. Inadequate pretreatment can lead to poor wettability, uneven coloration, and inferior adhesion of coatings in technical textiles.

9.7   Total colour management systems

9.7.1   Specification of colours and colour communication

The time-honoured method of specifying colours involves sending physical standards (e.g. dyed or coloured patterns) to the dyehouse followed by dyeing samples to match the colour in the laboratory.[109] The samples are submitted to the colour specifier for acceptance or rejection. If they are not accepted, this process must be repeated in order to obtain satisfactory colour matching to the standard. The approved laboratory sample is then used as the basis for the initial dye recipe for bulk dyeings. A sample from bulk dyeing may then be submitted to the specifier for final approval prior to delivery. Similar principles apply to the production of laboratory strike offs of each colour in a print for approval by the colour specifier prior to machine printing.

This process is a time-consuming procedure that has now been shortened considerably in order to avoid colour changes through storage and handling of physical samples and to achieve a quick response and provide just-in-time delivery to the customer. The use of a colour specifier program can dramatically decrease the cycle time for approval. This involves the use of instrumental colour measurement for formulation and quality control, accreditation procedures for enabling self-approval of colours, and colour specification using reflectance data.[109–111] In the most advanced colour management systems, it is now possible to visualise a colour on a colour monitor so that colour communication between the colour specifier and the dyer is vastly simplified. A number of standardised colour specification systems, for example, Pantone and Munsell Chip systems, can also be used.[111]

Modern colour communications systems may use a digital camera that exactly illustrates the viewer's perception of the coloured sample under a standard reference light source (e.g. D65 daylight) on a calibrated colour monitor.[112] This emulates how the image looks under reflected light using transmitted radiation. The colour data are then transmitted to other locations without losses or corruption onto a second or other colour monitor identically calibrated to the first colour monitor. Thus, the receiver (e.g. textile dyer/printer) sees the same colour as the sender. Colour printers can then be used to reproduce an exact image of the original on paper. Viewing cabinets (e.g. VeriVide) should have dimmable light sources that can be adjusted to match monitor screen brightness. Viewing cabinets for visual colour-matching purposes are now available using light-emitting diode (LED) technology that can represent different illumination conditions without the need to change lamps/tubes.

The whole area of total colour management in textiles from colour perception, colour description/specifying systems, instrumental colour measurement, colour quality evaluation, colour management, and effective colour communication throughout the textile supply chain has been discussed in detail elsewhere.[113]

9.7.2   Colour measurement

The principles of colour measurement and numeric systems for expressing colour have removed the subjectivity involved in visual assessment on colour of textile materials.[113–116] Objective colour matching using a spectrophotometer for colour measurement is superior to visual colour-matching assessment because instrumental colour measurement allows the colour to be specified in terms of reflectance data, and the precision of colour matching to the standard reflectance data can then be calculated numerically in colour difference units (ΔE).[114]

Colour tolerancing systems can be used to provide better agreement between visual assessment and the instrumentally measured colour difference, and colour acceptability limits can be agreed in advance in order to facilitate decisions on colour.[114–117] Colour acceptability limits are numerical values at which the perceived colour differences become unacceptable to the specifier, that is, single-number shade-passing systems. Instrumental shade-sorting systems are widely used to separate fabric lengths on fabric rolls into similar colours.[118]

The spectrophotometers currently in use may give different measured reflectance values according to the instrument design. The geometry, wavelength scale, bandwidth, or light source may differ between different makes of spectrophotometer.[109,114,115] The two most common geometries used for direction of illumination/direction of view are diffuse/8° and 45°/0°. Readings given by these two geometries are not compatible unless the ideal perfect diffuser is measured. Illumination is provided either by a tungsten filament halogen cycle incandescent lamp or a pulsed xenon discharge lamp.

Tungsten filament lamps do not have significant ultraviolet emission compared with pulsed xenon discharge lamps. The measurement of samples containing fluorescent brightening agents or some fluorescent materials is thus affected by the inclusion or exclusion of the ultraviolet component of the illumination.[109]

Spectrophotometers use either single-beam, double-beam, or dual-beam optics. The latter are now more widely used to compensate for the variability of the spectrum of pulsed xenon discharge lamps.[109,119] The collection optics may gather light across the sample image, focusing it onto the spectrophotometer aperture, or it is imaged at infinity, gathering light along a narrow range of angles. This latter method is claimed to provide a greater depth of field and insensitivity to sample surface imperfections. Considerable progress has been made by the manufacturers of spectrophotometers to ensure that the instruments give reliable, repeatable measurements, and the inter-instrument agreement (i.e. the colour difference values for a set of colour standards measured with two or more instruments of the same model) is now very low (typically 0.04 ΔE). Inter-instrument agreement and precision of colour measurement are now very important for the colour measurement of some technical fabrics when the colour matching must be very close to the standard as for automotive fabrics where reproducibility to within 0.3 ΔE is typically demanded.

The spectrophotometers in use may have 16 data points at 20 nm intervals over the 400–700 nm range, 31 data points from 400 to 700 at 10 nm intervals, or 40 data points from 360 to 750 nm at 10 nm intervals.[109] In addition, the bandwidth may be 20 nm, 10 nm, or less. As a result, it is difficult to convert measurements from one format to the other. Although conversion is possible, the colour difference measured at two different bandwidths on the same sample can be as high as 2.0 CIELAB units. However, the largest aperture should be used for colour measurement because it should provide more repeatable measurements.

CIELAB is the approximately uniform colour space with three mutually perpendicular opponent-colour axes, namely L*[lightness(white-black)], a* (red-green), and b* (yellow-blue).[115]

If L^*_B, a^*_B and b^*_B are the CIELAB rectangular coordinates of the colour of a batch dyeing and L^*_S, a^*_S and b^*_S those of the colour standard (i.e. the target colour to be achieved) then $\Delta L^* = L^*_B - L^*_S$, $\Delta a^* = a^*_B - a^*_S$ and $\Delta b^* = b^*_B - b^*_S$. The total colour difference between the standard and the batch is then calculated from the difference in lightness, chroma and hue as

$$\Delta E_{ab}^* = \left[(\Delta L^*)^2 + (\Delta a^*)^2 + (\Delta b^*)^2 \right]^{1/2}$$

where ΔE^*_{ab} is the CIELAB colour difference between the standard and the batch. The total colour difference is normally referred to as ΔE.

Colour measurement of dyed textiles should be performed on conditioned samples because the colour may change with temperature and moisture content, and the colour change may exceed the pass/fail tolerance.[109] The sample presentation should also specify the number of fabric layers and the backing material to be used because these factors also affect the perceived colour. Colour measurement of loose fibres generally causes greater variations in measurement than on woven fabrics, so it is advisable to use multiple measurements to improve the repeatability. There are considerable differences in colour between cut and uncut pile yarn when measurements are made on the fibre tips of a cut hank of yarn or on the side of the yarn.[114]

Portable colour measurement systems use handheld spectrophotometers with data storage and extensive memory capacity for monitoring the colour from batch to batch and within batch (e.g. side-centre-side and end-to-end colour variations within a fabric roll).[120] These systems are very convenient for quality control purposes both within the laboratory and in the production plant, but sophisticated online colour measurement systems are also employed on continuous fabric dyeing ranges.[121] A typical total colour control system is illustrated in Figure 9.2.[114]

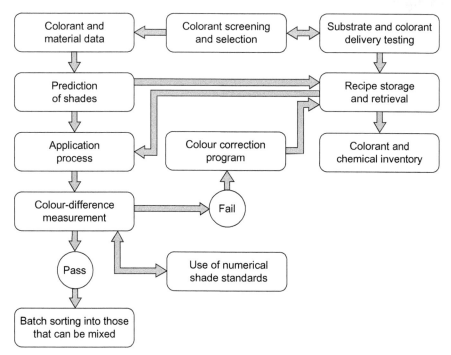

Figure 9.2 Total colour control (source: see Ref. 114, p. 85).

9.7.3   Laboratory matching

In the modern dyehouse, dyestuffs and chemicals are purchased at an agreed upon quality standard.[33,114] Dyestuffs for use in laboratory matching in many dyehouses are dissolved or dispersed in stock solutions above an automated laboratory dispensing unit. The colour specifier may supply a physical sample, the reflective curve of which is measured on the spectrophotometer in the laboratory, or reflectance data are supplied directly to the laboratory.

The formulation of colorant recipes normally involve the use of a trichromat of red, yellow, and blue dyestuffs to match the required target colour unless it can be matched using fewer dyestuffs. Quality control is essential to ensure the accuracy of the recipe preparation, the repeatability of the dyeing process, and the colour measurement process. Recipe formulation in colour matching is performed on computer match prediction systems that use the Kubelka–Munk theory.[81,122] In this theory, part of the light passing through a pigmented layer is absorbed by the pigments, and the medium in which the pigments are dispersed and part of the light is scattered by the pigments. Using various simplifying assumptions the Kubelka–Munk theory leads to the equation:

$$K / S = \left(1 - R_\infty\right)^2 / 2R_\infty$$

where K = absorption coefficient; S = scattering coefficient; and R_∞ = the reflectance of a colour sample of optically infinite thickness.

The absorption and scattering coefficients of a coloured textile can thus be represented using the absorption and scattering coefficients of the individual dyes or pigments.

$$\begin{aligned}\left(K / S\right)_{mixture} &= \left(K / S\right)_1 + \cdots + \left(K / S\right)_n \\ &= K_1 / S_{sub} + \cdots + K_n / S_{sub} + K_{sub} / S_{sub}\end{aligned}$$

where K_1 to K_n = the absorption coefficients of the dyes; K_{sub} = the absorption coefficient of the substrate (i.e. fibre); and S_{sub} = the scattering coefficient of the substrate (i.e. fibre).

For a dyed fibre system, the reflectance of the coloured sample can be predicted from the ratio of the absorption coefficient of each dye in the mixture and the scattering coefficient of the substrate (i.e. K_n/S_{sub}). The absorption coefficients of the dyes K/S_{sub} are usually related to the dye concentration in a nonlinear relationship with the dye exhaustion decreasing with an increase in dye concentration.[122] A calibration database is therefore prepared using a range of dye concentrations on the appropriate fibre from which the absorption coefficient of the dye is correlated to the corresponding dye concentration in the dyebath.

Using this approach, computer match prediction systems can compute and formulate dye recipes that will match the target colour under D65 daylight.[81,122] The reflectance curves of the target colour and the recipe formulation under D65 daylight, cool white fluorescent (CWF), or incandescent sources (tungsten lighting) can be visually displayed simultaneously on a monitor, and the effects of different illuminants on metamerism can be visually displayed on screen.

Computer match prediction is then carried out using the appropriate database of the dyes used in the dyehouse on the substrate to be dyed.[114] If the substrate to be dyed in bulk has a different dyeability to that originally used to construct the database, adjustment of the recipe must be made. The colourist in the laboratory must then select from the computer match prediction the most appropriate recipe, taking into account such factors as recipe cost, the anticipated technical performance in bulk dyeing, colour fastness, metamerism, and the closeness of the predicted colour match to the specified colour.

The appropriate dye recipe is then dispensed from freshly made stock solutions of dyestuffs using manual or electronic pipettes or via a preprogrammed automatic laboratory dye-dispensing unit. A sample of the material is then dyed in an automated laboratory dyeing machine under conditions that simulate those to be used in bulk dyeing, for example, the same pH, chemicals and auxiliaries, liquor ratio, and temperature–time relationship. The laboratory matching is given the same rinsing or after treatment that will be used in bulk dyeing and is dried, conditioned, and the colour measured on the spectrophotometer. If the colour is a commercial match to the specified colour, then bulk dyeing is initiated. If the colour is outside the commercial colour tolerance because of differences in substrate dyeability and so on, a corrected recipe is predicted, and the process repeated to obtain a commercial colour match followed by bulk dyeing.[114,115,123]

Where there is close control over the colour strength of the dyestuffs and consistent substrate dyeability, it is often possible to operate so-called blind dyeing in which the computed dye recipe in the laboratory is used immediately for bulk dyeing.[123] This shortens the time required, decreases dyehouse costs, and offers quick response and rapid delivery to customers. When repeat dyeings of the same colour are required, it is usually possible to input the reflectance data gained from bulk dyeings in order to refine the database and thereby achieve a higher level of right-first-time dyeings. Correct first-time, right-on-time, and right-every-time dyeing is the goal of the dyer because this is the lowest cost dyeing system that provides quick response for customers.[123]

If a bulk dyeing proves to be off shade and the original recipe requires a correction (i.e. an addition) that results in extending the dyeing time, the dyer suffers a considerable financial penalty. If the colour is too dark and the bulk dyeing must be stripped and redyed, the dyer incurs additional cost penalties and often impairs the quality, physical properties, and surface appearance of the dyed material. In general, the shorter the processing time is under hot wet conditions in the dyebath, the lower is the fibre degradation and, hence, the quality of the dyed material is superior, all-important considerations for many technical textiles.

9.8   Dyeing machinery

Except for mass coloration discussed in Section 9.5, conventional dyeing machinery is used to dye technical textile materials in the form of all types of fibres, tows, yarns (e.g. warp, hank, package), fabric, garments, carpets, and so on.[25,124–127] Fabric may be dyed in rope form (i.e. as a strand) or at open width (i.e. flat), or circular weft knitted fabrics may be dyed in tubular form. Modern trends are to dye in yarn or fabric form because this allows the decision on colour to be made later in the manufacturing chain and lowers the cost of dyeing. Many technical fabrics are dyed at open width to avoid inserting creases into the fabric. Such creases can be difficult to remove in synthetic fibre fabrics because of hydrosetting occurring during dyeing.

9.8.1 Batch-dyeing machinery for exhaust dyeing

In conventional exhaust dyeing, dye molecules are transported to the fibre surface where the substantivity of the dye for the fibre ensures that adsorption takes place to produce a higher surface concentration of dye.[128] This higher surface concentration of dye promotes dye diffusion into the fibre, the rate of dye diffusion depending on the dye concentration. Dye molecules diffuse within the disordered regions of the fibres to provide adequate fibre penetration.[36] Level dyeing is attained through the migration of dye by desorption from inside the fibre back into the dye liquor and thence adsorption and diffusion in another part of the fibre, this levelling process being aided by the relative motion of the dye liquor and the fibre. After an appropriate time, a level, well-penetrated dyeing is obtained. This process is facilitated if compatible dyes, that is, dyes that diffuse at the same rate, are initially selected for use with a selected levelling agent. Dye fixation is ensured either by (a) electrostatic attraction between the dye and dyesites inside the fibre, (b) covalent bond formation between the dye and the fibre, (c) increasing the size of the dye or by dye aggregation leading to mechanical entrapment, or (d) conversion of the water-soluble dye into a water-insoluble form.

The three types of dyeing machinery for batch or exhaust dyeing are based on machines in which[124,125,127]:

1. The material moves, but the liquor is stationary (e.g. jig and winch [beck] machines for fabric dyeing)
2. The liquor moves, but the material is stationary (e.g. hank or package dyeing of yarns and beam dyeing of fabrics)
3. Both the liquor and the material move (e.g. jet, soft flow, and overflow jet dyeing machines for fabrics)

In general, a lower liquor ratio decreases the water and energy consumption and decreases the volume of wastewater, lowering effluent treatment costs. A lower liquor ratio can also facilitate a more rapid dyeing cycle and increase the dye exhaustion on the fibre because the total dye liquor is circulated more rapidly.

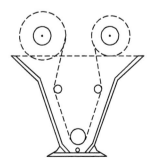

Figure 9.3 Jig dyeing machine (source: see Ref. 125, p. 29).

In the jig (or jigger), fabric is dyed at open width by traversing fabric from one roll through a dyebath and onto a second roll (see Figure 9.3).[125] When the second roll is full, the rolls are stopped, the fabric motion is reversed, and the procedure repeated as required. The dye liquor is added in portions to ensure satisfactory levelness from end to end of the fabric. A hood prevents the release of steam and helps to maintain the fabric temperature on the rolls where the majority of the dyeing takes place. Most jigs operate at 98–100 °C (i.e. under atmospheric conditions) but pressurised jigs operating at up to 140 °C are used for some technical fabrics, for example, polyester sailcloth. In the pressurised jig, the rolls and the dyebath are enclosed in an autoclave (i.e. pressure vessel) that can be closed with a ring seal. Jig dyeing operates at a liquor ratio of 3–5:1, and rinsing/washing off is more rapid on modern twin bath jig designs in which a combination of water spray and vacuum is used for removal of loose dyestuff.[129]

Jig dyeing is particularly suitable for technical fabrics that may be subjected to creasing and hence are preferably dyed at open width. In addition, this process is preferred for many technical fabrics that have a dense structure and are difficult to pump dye liquor through multiple fabric layers, as in beam dyeing (e.g. sailcloth, filter fabrics).

The winch or beck dyeing machine may be used for dyeing in rope form, normally at temperatures of up to 98–100 °C (see Figure 9.4).[125] High-temperature winches have also been used in the past but have now been replaced by high-temperature (i.e. pressurised) jet dyeing machines. The latter are less likely to crease the fabric and have improved liquor interchange that aids the more rapid attainment of level dyeing by promoting dye migration.

In beam dyeing, the fabric is wound at open width onto a perforated stainless steel beam and is kept in place by the use of end plates (see Figure 9.5).[125] Dye liquor is pumped through the multiple layers of fabric, usually from out to in, but in to out is also used. The fabric winding tension must be low to avoid stretching the fabric and to avoid high pressure being applied to the innermost fabric layers; otherwise, the fault known as *watermarking* or *moiré* may be observed. This is usually seen as a shimmering surface pattern on dark colours that changes with the orientation of the fabric relative to the observer. Too low a winding tension can, however, lead to telescoping

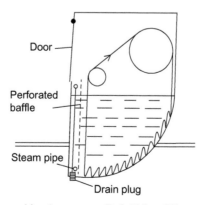

Figure 9.4 Winch dyeing machine (source: see Ref. 125, p. 38).

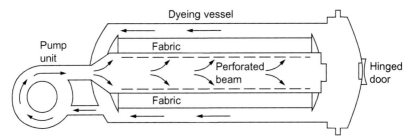

Figure 9.5 Sectional diagram of a high-temperature beam dyeing machine (source: see Ref. 45, p. 117).

and movement of the fabric on the beam. It is normal to wind a fent (or short sacrificial fabric length) onto the beam first to prevent perforation marks on the fabric layers next to the beam. The perforation marks show as a pattern of dark spots corresponding to the perforated holes in the beam and are often caused by poor dispersing or aggregation of disperse dyes. The liquor ratio in beam dyeing is around 10:1 although it can be decreased by the use of spacers, and most beam dyeing machines are pressurised and capable of dyeing at temperatures of 130–140 °C. Beam dyeing is particularly effective for dyeing thin permeable fabrics composed of synthetic filaments (e.g. nylon and polyester fabrics).

Pressurised machines are generally used for dyeing technical textiles composed of synthetic fibre materials. Soft flow jets provide a more gentle mechanical action on the fabric rope; the main fabric transportation is carried out over a winch reel followed by a soft flow jet system (see Figure 9.6).[125,130] In some jet dyeing machinery, the liquor flow is split between two jets that exert a lower pressure on the fabric. This is claimed to decrease linting (loss of fibre) and surface distortion in staple fibre fabrics. Garments are generally dyed in atmospheric or pressurised rotary drum machines or in atmospheric overhead paddle dyeing machines.[125,131]

Aerodynamic fabric transport systems use a blower to aid the fabric circulation through the jet dyeing machine. These systems enable liquor ratios as low as 1.3:1 to be used on hydrophobic filament fabrics such as polyester, although higher liquor ratios in the range 5–10:1 are used in conventional jets on hydrophilic cellulosic fabrics.[130]

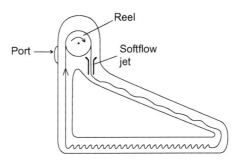

Figure 9.6 Softflow jet dyeing machine.

9.8.2 Pad-batch dyeing

Both woven and knitted fabrics can be dyed using this method with the padding (impregnation) stage conducted under ambient or hot conditions.[62,132] With reactive dyes, fabric prepared for dyeing is usually padded under ambient conditions and then wound on a roll rotating on an A-frame. The dye concentration and the pickup of dye liquor by the fabric must be controlled to ensure uniform fabric coloration. Pneumatic pressure control applied to the pad rollers in the padding mangle ensures that the mean nip pressure is constant across the width of the rollers to impart a consistent, uniform pickup of the dye liquor on the fabric. Automatic control of the dye concentration in the pad liquor is also ensured.

The length of padded fabric on the roll may be covered with polyethylene film to prevent evaporation and is then rotated slowly (to avoid migration of the dye liquor to the outer layers of the fabric roll) for an appropriate time to achieve dye fixation. The time required depends on the depth of colour, the reactivity of the reactive dye, and the temperature and the alkalinity but is generally in the range 4–24 h (often conveniently overnight). The dye fixation stage is then followed by a thorough hot wash off to remove the unfixed dyestuff and the alkali, and the fabric is then dried.

9.8.3 Continuous dyeing machinery

Continuous dyeing can be performed by dye impregnation, fixation, washing off, and drying on ranges designed for handling woven fabrics at open width and for carpet and warp dyeing ranges.[126] Continuous dyeing was originally designed for efficient and economical dyeing of long runs of fabric, and many modifications have been made to facilitate more rapid turnaround and less downtime between a higher number of shorter production runs per colour. Automatic metering of liquid dyes to control the dye concentration in the dyebath and intelligent padding systems are used to control the pickup of dye liquor by the fabric. Thermofixation (dry-heat), superheated steam, or saturated steam fixation units in which the temperature is monitored and controlled may be used to obtain dye fixation. Reproducible colour is obtained after washing off loose dyestuffs and drying. Colour change during a production run is accomplished online within minutes using automated wash down systems. Continuous colour monitoring on the dyed fabric output on some ranges can be used to adjust the colour being dyed on the input fabric. Either the whole width of the fabric may be continuously monitored, or a traversing spectrophotometer measuring side-centre-side variations may be employed.

For the continuous dyeing of woven polyester/cotton fabric, the polyester fibres are first dyed with disperse dyes using a pad-dry-thermofix treatment.[126] The cotton is then dyed using direct, reactive, or vat dyes using a pad-steam process. The dyed fabric is cleared of loose surface dyestuff by treatment on an open width washing range and then dried over steam-heated cylinders and batched. A scray is used at the fabric entry to the range to accumulate fabric so that the range continues to run at 50–150 m min^{-1} while the next roll of fabric is sewn onto the fabric being dyed.

9.9   Printing

9.9.1   Printing styles

Printing leads to the production of a design or motif on a substrate by application of a colorant or other reagent, usually in a paste or ink, in a predetermined pattern. In printing, each colour is normally applied separately and a wide variety of printing techniques is used. In direct printing, dyes are applied in a print thickener containing auxiliaries and are diffused into the fibres and fixed in a steamer or high-temperature steamer.[84] Surface dye and thickener are then thoroughly washed off in a continuous open width washer, and the materials are dried.

In pigment printing, water-insoluble pigments are applied with a heat-curable binder system followed by drying and curing, and the physical properties of the pigment print depend greatly on the adhesive properties of the print binder system. Pigment printing is a simple, popular method particularly for printing blended fibre fabrics.[14,84] In the drive to introduce more eco-friendly printing systems, Archroma has introduced its Printofix Ecological Printing System for pigment printing that comprises a complete package of Global Organic Textile Standards (GOTS)-approved auxiliaries and colorants that do not contain formaldehyde.[133]

Two other major styles of printing are employed, namely, discharge and resist printing.[134] In discharge printing, a dyed fabric is printed with a discharging agent (a reducing agent) that decolourises the dye, leaving a white motif against a dyed ground. Alternatively, a dye resistant to discharging can simultaneously be applied to give a second colour in the discharged area of the motif. In resist printing, the resisting agent (which may act mechanically or chemically) is printed onto the undyed fabric. This prevents the fixation of the ground colour that is then developed by dyeing, padding, or overprinting. A white resist or a coloured resist can be achieved if a selected dye or pigment is added to the resist paste and is then subsequently fixed.

Whichever form of printing is to be carried out, the textile substrate must be prepared for printing to make the substrate clean and receptive to aqueous-based print pastes and print binder systems.[135] A typical preparation sequence for a woven cotton fabric, for example, would include singeing, desizing, scouring, bleaching, mercerising, drying, weft straightening, and batching. Mercerising improves the dye yield and lustre on cotton, while a milder alkaline causticisation treatment is used on viscose and lyocell fabrics. However, for pigment printing, mercerising may be omitted. Adequate fabric wettability and absorbency are critical factors for high-quality printing either by screen or digital ink jet printing technologies.

9.9.2   The print image

The design image may be painted by hand onto transparent film for photographic development of diapositive images, or sophisticated CAD (computer-aided design) systems are used in conjunction with a design input scanner. The design image is then manipulated on screen, the colours are electronically separated and the digitised design images for each colour transferred to an engraving system for transfer onto a rotary or flat screen or for direct use in digital ink jet printing.[136]

Laser engraving of lacquer screens has drastically decreased the time required to produce a screen for fabric printing. The electroforming of nickel growth in galvano screen production can now be controlled to occur mainly in the top direction.[137] This enables the ratio between hole size, mesh count, and screen thickness to be moved beyond the conventional boundaries of rotary screen printing. Thus, the dam shape stays thin and becomes more streamlined. As a result, the open area and hole size of Stork Prints Nova 195® screens (thickness 115 μm) can be increased from 16% to 19% open area, and the hole size increased from 52 to 57 μm. This provides a better flow of the print paste, resulting in a higher printing resolution, more printing definition, and improved surface printing.

9.9.3 Fabric screen printing machinery

The global fabric printing market was estimated to be around 29 billion square metres of fabric in 2013 and is still dominated by traditional analogue textile printing using rotary screen and flat screen technologies.[138] The major output of traditional screen printed fabrics in 2011 came from China (30%); India (17.5%); other Asian countries (18.8%); the Americas (12.9%); Europe, Russia, and the CIS (11.1%); and the Middle East/Africa (9.7%). The major types of fabric printed are cotton (52%), polyester (18%), viscose (14%), polyester/cellulose blends (12%), nylon (3%), and wool/silk fabrics (<1%).[139]

The market segmentation of analogue textile printing by colorant type consists of pigments (50%), reactive (20%), disperse (15%), vat (8%), and acid/other dye-stuffs (7%) for various end-use applications.[139] Pigments are used in home furnishing, decoration, fashion, outdoor, and technical textiles (including camouflage fabrics). Reactive dyestuffs are used in fashion, bed shade, and camouflage fabrics whereas disperse dyestuffs are used in fashion (including polyester swimsuits), automotive seats, technical textiles, and other applications. Vat dyestuffs are used in both fashion and technical textiles, and acid/other dyestuffs are used in fashion (including nylon swimsuits) and flags and banners.

The older hand block printing, stencil, and hand screen printing processes are slow and are now largely confined to highly specialised or craft applications.[140] Engraved roller printing, although still used in some Asian countries such as India, is restricted in scope because of the high roller engraving costs, extended setup times, and its primary application to narrower width fabrics compared with screen printing. The major analogue printing systems used for technical fabrics are continuous rotary screen printing for longer production runs and automatic or semiautomatic flat screen (bed) printing for more specialised and intricate designs.[141,142] Dry-heat transfer (or sublimation) printing is used to a much smaller extent, mainly for garment applications and flags and banners. The most rapidly growing sector of textile printing is digital ink jet printing technology that is particularly well established in the polyester soft signage, flag and banner, and sportswear applications as well as in direct to garment (DTG) printing.

Rotary screen printing is the dominant printing method now employed for 60–70% of printed fabric production.[141–143] The design motif for each colour is developed as open mesh on the rotary screen by the use of film, laser, or black wax engraving systems. Both lacquer screen and galvano screens are used. Each rotary screen secured in end rings is fixed in bearings and rotated in precise register and in a predetermined sequence. A separate coloured paste is supplied to the inside of each cylindrical rotary screen in continuous fabric printing; the paste is forced through the open mesh of the design areas with the aid of a stationary squeegee that may be a conventional, airflow, or magnetic squeegee. In this way, the print paste is transferred through the rotating cylindrical screen onto the moving fabric that is temporarily gummed to an endless driven moving blanket (apron). A water-based adhesive spread evenly with a rubber squeegee may be applied to the blanket at entry. The fabric to be printed is pressed against the tacky blanket via a pressure roller. Hot-air drying of the adhesive prior to printing may be necessary. More commonly a tacky semipermanent or "permanent" adhesive based on acrylic copolymers may be applied to the blanket. Permanent thermoplastic adhesives are rendered tacky when heated and can remain serviceable for the printing of long rotary screen print runs.

Machines capable of continuously printing up to 24–36 colours are available, although most designs involve fewer than 8 colours, and rotary screen printing on textile materials up to 5 m in width can be done at speeds up to 80 m min^{-1} but typically at 40 m min^{-1}. Most fashion garment designs involve fewer than 8 colours, but furnishing fabrics can have up to 16 colours.

Automatic or semiautomatic flat screen (flat bed) printing is used for many designs when very precise images are required using polyester, polypropylene, or polyamide monofilament fabric as the screen material.[142] In automated flat screen printing, the fabric is gummed to the print table and progresses forward intermittently, one design repeat at a time, after the lowering, printing, and lifting of the screens to print all the colours sequentially, using a squeegee system to force the print paste into the fabric. The printing speed is lower than that normally obtained in rotary-screen printing.

Modern rotary and flat screen printing machines may be supplied with enough colour for each screen, with all the operations of weighing, metering, dispersing, and mixing of dyes and auxiliaries with stock thickener (suitably diluted), or of pigments with binders and softeners controlled by a robotised colour kitchen based on a computed print recipe. The colours are supplied in drums and supply pipes are used to furnish the colour for each screen in rotary screen printing.

At the end of the print run, excess print paste is removed and stored and may then be reused in a subsequent print run or disposed of in a landfill. Print monitoring systems are available for monitoring and comparing the fabric print with the electronically stored digitised design image. In the most sophisticated machines, it is possible to use such monitoring systems for correction of faults online, and many machines are designed for quick changeover of colours, for example, by screen and blanket washing and drying facilities online, and for rapid changeover of screens at a design change.[143] Automated setting up, monitoring, controlling, and correcting systems and washing and drying online have all dramatically improved the productivity and repeatability in rotary and flat screen printing.

Transfer printing may be performed by sublimation transfer, melt transfer, or film release methods.[82] In sublimation (or dry-heat) transfer printing, volatile dyes (typically disperse dyes) are preprinted onto a paper substrate and are heated in contact with the textile material, typically polyester fabric. The dyes sublime and are transferred from the vapour phase into the fabric in this dry-heat transfer printing method.

This process may be assisted by the application of a vacuum. Melt transfer is principally used on garments; a waxy ink is printed on paper, and a hot iron is applied to its reverse face to melt the wax onto the fabric surface. In the film release method, the print design is held in an ink layer that is transferred completely to the textile from a release paper using heat and pressure. The design is held onto the textile by the strong forces of adhesion between the film and the textile.

9.9.4 Digital ink jet printing

Digital ink jet printing is a rapidly expanding sector of the global textile print market with an estimated cumulative annual growth rate (CAGR) of 25–30% being predicted compared with 2.5% for analogue (screen) printing.[144] This could lead to 1 billion square metres of digital ink jet printed fabric by 2017 from a base of 600 to 700 million square metres in 2012.

The principles of fabric pretreatment for digital ink jet printing, ink formulation, and many other aspects of ink jet printing technology have now been comprehensively documented in a textbook,[145] and more recent developments have been reviewed in *Digital Textile* magazine.[146]

The initial fabric pretreatment and subsequent pretreatment for digital textile printing (PFDP) are critically important to ensure that the print definition, colour, and brightness as well as the colour fastness performance are at a high level.[135,147,148] Fabric pretreatment with thickeners and other chemicals appropriate to the class of dyestuffs being ink jet printed is essential to ensure good print definition and high dye fixation. Such thickeners and chemicals cannot be incorporated into the ink formulation because of problems affecting dye solubility, ink stability, corrosion of the jet nozzles, and undesirable rheological properties that would adversely affect the jetting performance and print runnability.

Normally, ink jet printers are supplied with up to eight different inks from individual ink reservoirs.[149] The eight colours are based on cyan, magenta, yellow, and black (CMYK) supplemented by other colours. The merging of the coloured ink droplets occurs on the pretreated textile surface forming the process colours in the design image. The use of process colours in ink jet printing imposes some limitations on the colour gamut obtainable compared with the spot colours used in conventional screen printing of textiles. The spot colours are prepared by premixing the colours in advance of screen printing.

Aqueous dye-based ink jet ink formulations typically contain dye (0.3–10 wt %), water-soluble organic solvents (5–15 wt %), and surfactant (0.1–0.2 wt %) with acid/alkali for pH adjustment, water, and additives.[23,150,151] The latter are added for specific purposes (e.g. perfumes, antiseptics, UV absorber, hygroscopic agent [humectant], fastness enhancer, fixing accelerator). The ink properties are optimised in order to generate high operating performance in specific print heads. The water-soluble organic solvents and surfactant are present to control the ink viscosity and surface tension as well as the absorption speed of the ink on the pretreated textile substrate. The main considerations covering the formulation of reactive, disperse, acid, and direct dye inks have been discussed elsewhere.[113]

Over the last decade, ink viscosities have increased from 3–5 to 7–15 mPas by adding glycols, thereby decreasing the relative amount of colorants (dyes or pigments) in the ink.[152] Print quality is negatively impacted with a decrease in image colour vibrancy and poor drying behaviour. The water-based Rheological Modified Inks® (RMI) by Sawgrass Technologies exhibit shear-thinning behaviour with the ink viscosity decreasing as the shear rate during high-speed jetting increases. With the RMI inks, the jetted fluid forms coaxial cylindrical layers under high shear, and this leads to the ability to greatly control dot formation. The rheology modifiers in RMI ink formulations enable up to 80% more colorants and 20–40% less glycols to be used in these inks.

Pigmented ink formulations depend on the print head type and the ink viscosity required.[153] Pigment inks have thus been prepared either without a textile binder or with a conventional or an unconventional textile binder. A typical pigment ink formulation contains finely dispersed pigment particles, a polymeric binder, water, cosolvent, surfactants, humectants, an antifoam agent, a viscosity control agent, a biocide to prevent spoilage, and a penetrant to speed drying on textile materials. The particle size and size distribution of the pigment affects the image quality, particularly colour density, and influences the settling of the pigment in the ink, the colloidal stability, and clogging of the jet nozzles and thus can impact on the jetting reliability. The detailed considerations surrounding pigmented ink formulations have been discussed in detail elsewhere.[113]

The introduction of DTG printing in 2003 led to an increasing demand for a white ink to print on to dark coloured garments (e.g. T-shirts).[154] In addition, the white ink can be totally or partially overprinted with coloured pigment ink jet inks to give coloured or "masked" pastel-type effects in the design. Titanium dioxide-based white pigment ink dispersions can be used for such masking effects to avoid the problems associated with the classic discharge printing approach.

In 2009, digital transfer printing of polyester by sublimation printing accounted for approximately 52% of all digital textile printing.[133] This included the polyester soft signage market and apparel printing such as short-run polyester fashion printing and sportswear printing (e.g. team shirts). Digital ink jet printing of disperse dye inks on to transfer paper uses disperse dyes that sublime at temperatures of 180–220 °C and have no affinity for the transfer paper(cellulose). In this temperature range, the disperse dyes are readily absorbed by the polyester fabric which is now widely used in the soft signage market as a replacement for vinyl signage materials.

In addition, the dye sublimation transfer printing (dye-sub) method that uses a heat-transfer calender, pretreated polyester soft signage can also be ink jet printed by direct-to-fabric printing using high light fastness disperse dye inks and in-line infra-red heaters or a heat chamber method to effect the dye sublimation process.[156] Alternatively, conventional analogue (screen) printing may be used on high-quality fashion apparel, automotive seating, and other end uses that require high colour fastness.

Ink jet printing is a non-impact printing method that uses print heads containing a bank of nozzles supplied with ink that is squirted through the nozzles at a constant speed by applying a constant pressure.[157] The ink jet is unstable and breaks up into small droplets (typically in picolitres) shortly after emergence from the nozzle. In the continuous ink jet (CIJ) technology, the droplets either impact on the textile surface or are deflected to a gutter for recirculation according to the image being printed.

Electrical charging of the droplets coupled with the use of an electric field controls the trajectory of the droplets. CIJ technology continuously ejects the droplets.

The other main form of ink jet printing uses drop-on-demand (DOD) technology.[157] This differs from CIJ technology in that ink drops are ejected solely when required to form the design image on the textile. Two main drop ejector mechanisms are used to generate the ink droplets. These are the piezoelectric ink jet (PIJ) and the thermal (or bubble) ink jet (TIJ) systems. In the PIJ system, a piezoelectric actuator decreases the ink volume in a chamber, causing an ink drop to be ejected from the nozzle. In the TIJ system, an electrical heater inside each nozzle raises the temperature of the ink to the point of bubble nucleation, and the explosive expansion of the vapour bubble causes the ink to be propelled out of the nozzle.

There are now many manufacturers of digital ink jet printing machines, many incorporating the same type of print head to provide low, medium, and high production printing systems that use either scanning or single pass modes.[158] DTG ink jet printers are much smaller than the fabric ink jet printers because they are generally used to print a small print area on a garment.

Significant developments have taken place in print head technology.[159–161] In 2003, most print heads had a maximum of 512 nozzles. This was increased to 1024 per print head in 2011, and the Kyocera KJ4B print head now has 2656 nozzles per print head. This has led to fabric ink jet printers with fewer print heads in a more compact print bar and to increased reliability because of fewer possible failure points in terms of components and connections. There are now digital ink jet printers capable of printing up to 1000 square metres per hour and because of its versatility in terms of rapid changes in the designs printed and the speed of production, this technology will ultimately compete with rotary screen printing technology.[162]

The digital ink jet printers used for printing carpets and carpet tiles use DOD electromagnetic solenoid-valve technology print heads that are considerably larger than the piezo print heads used in most digital ink jet printing of fabrics and garments.[163] The drop volumes ejected by valve technologies are measured in nanolitres ($1\,nL = 10^{-9}\,L$) rather than in picolitres ($1\,pL = 10^{-12}\,L$) in piezo technologies and give much lower print resolutions, around 16–76 dots per inch (dpi) compared with 720 dpi or higher for fabric printing. The lower print resolutions (16 dpi) are used on broadloom carpets and carpet tiles, 25 dpi on carpet mats and blankets, and 76 dpi on terry towelling, upholstery, and automotive fabrics. The low resolution valves (<20 dpi) use premixed spot colours, but higher print resolution valves can also use process colours (CMYK) as well as premixed spot colours.

Wide-format digital ink jet printers that can print fabric widths of up to 2 m and even up to 5 m using acid, reactive, disperse dye, and sublimation inks or pigmented inks are now available.[162] The fabric is normally pretreated and placed in the machine in roll form, printed, and then the dyes are fixed, usually by steaming in a separate machine, washed off, and dried. Both piezoelectric and bubble jet printing systems may be used; any unused colour is diverted into the ink reservoir and recycled. Generally, four, six, eight, and up to nine colours may be printed; the more colours used, the larger is the colour gamut that can be printed. Pigment inks are normally cured by infra-red heating, but ultraviolet-curable inks are now available, and this more sustainable method of curing will no doubt increase in importance in the future because of the savings in energy consumption.

In general, digital ink jet printing systems are designed principally for use with natural fibres (e.g. cellulosic, wool, and silk fabrics) and polyester fabrics.[145] On some machines, the ink jets are periodically cleaned automatically with solvent to avoid jet blockage, particularly in disperse dye systems. Such printing machines may be run overnight without operator supervision with preloaded design images and essentially instantaneous design changeover. Other systems are already being used for printing flags and banners and clearly have great potential for printing short production runs of advanced technical textile fabrics.

9.10   Colour fastness of technical textiles

The performance of a dyed or printed technical textile when exposed to various agencies during end use is normally assessed by appropriate colour fastness testing.[164,165] High standards of quality and performance in such tests are often related to the higher cost of the dyes and pigments used that possess superior colour fastness properties. There are national standards, for example, British standards (BS), European standards (EN), and international standards (ISO), but most countries are now experiencing a move towards harmonisation of test methods and performance standards, so that EN and ISO standards will be used in the future. In North America, the American Society for Testing and Materials (ASTM) and the American Association of Textile Chemists and Colorists (AATCC) have test methods. In addition, there are also test methods that have been devised by industry for use for specific applications, for example, automotive textiles.

Many colour fastness tests have been devised to simulate the likely end-use conditions, and the Society of Dyers and Colourists in the United Kingdom and the AATCC in the United States have been at the forefront of colour fastness testing developments. The colour fastness test usually defines a standard test method to be followed and gives the method of assessment to be used, but the performance level that is satisfactory for the end use must be agreed upon by the dyer or printer and the colour specifier. Companies often set their own in-house performance criteria for their ranges of dyed or printed technical textiles based on their general working knowledge and experience in the field, technical liaison with dye makers, field testing under actual end-use conditions, and assessment of materials that are the subject of complaints.

Colour fastness tests are designed to reproduce the conditions experienced by the textile from exposure to specific agencies during the manufacturing sequence (e.g. post-heat setting) and during the lifetime of the technical end use (e.g. leaching, domestic washing, weathering). The International Standards Organization's tests for colour fastness are listed in ISO 105 with the test methods grouped alphabetically according to a particular property as follows[166]:

A – General Principles
B – Light and Weathering
C – Washing and Laundering

D – Dry Cleaning
E – Aqueous Agencies
F – Adjacent Fabrics
G – Atmospheric Contaminants
H – Textile Floor Coverings
J – Colour Measurement
N – Bleaching Agencies
P – Heat Treatment
S – Vulcanising
X – Miscellaneous Agencies
Z – Colorant Characteristics

In general for technical textile materials manufactured in or imported into Europe, the ISO test methods are used. For textiles destined for the US market, the AATCC test methods described in the *AATCC Technical Manual* (on searchable CD) are normally used.[167]

It is not possible to discuss the many colour fastness tests that are now in use worldwide, but complete details are available in many other publications, which give details of a wide range of standard tests.[164,165] The major types of colour fastness test relate to the colour fastness to wet treatments (e.g. to washing), to light and weathering, to rubbing (crocking), to atmospheric contaminants, and to organic solvents (e.g. dry cleaning).

The colour fastness performance in standardised wash tests is rated by visual assessment of the change in colour of the coloured material and the degree of staining onto adjacent materials in the wash liquor (e.g. a multifibre fabric test strip[168] or specific adjacent fabrics), using ISO grey scales under standardised lighting conditions against a neutral grey background in a viewing cabinet. The grey scale ratings range from 5 (no change, i.e. excellent performance) through half-point ratings, for example, 4–5, down to 1 (large change, i.e. poor performance). The colour change may also be measured objectively using a spectrophotometer and the colour difference converted into a grey scale rating.[169] It is also possible to determine the grey scale rating for change in colour of a test specimen and assessment of staining onto an adjacent fabric by using digital imaging techniques such as the DigiEye[3] (VeriVide Ltd, United Kingdom).[170–172] The digital camera should have an effective resolution of not less than 3.0 megapixels (Mpixels). This digital imaging method of grading colour fastness will be incorporated into other ISO test methods in due course.

Colour fastness to light is normally assessed using a high-intensity filtered xenon arc lamp to simulate natural daylight; sample strips are mounted and partly covered on cards that can be individually rotated in a bank of holders around the accelerated fading lamp, depending on the type of machine, in an atmosphere of controlled temperature and relative humidity. A set of blue wool standard fabric strips that fade at known rates must also be used in each test. The setup conditions for specific test methods used for many types of technical textiles are preprogrammed on modern light fastness testing machines.[167]

The degree of fade in British and European standards is based on visual assessment of the degree of colour fade on the test sample compared with the equivalent degree of fade on the blue wool samples. The light fastness rating changes from 8 (highest) to 1 (lowest). Weathering may be conducted in field trials with samples being exposed to sunlight either covered or uncovered by glass or by exposure for a standard number of hours in an accelerated fading machine.

Colour fastness to rubbing (crocking) is assessed by rubbing a standard white fabric against the dyed sample under a constant pressure for an agreed upon number of strokes. The test may be conducted under wet or dry conditions, and the machine may be operated by hand or performed automatically in the latest machines.[167] Assessment of the degree of staining on the white fabric is assessed using the ISO grey scale for staining. For pigment printed materials, the rub fastness depends on the properties of the adhesive binder used.

9.11 Dyes and pigments for special effects in technical textiles

Certain dyes and pigments can be used in chromic materials that can exhibit a distinct colour change when exposed to an external stimulus.[173,174] The colour change may be reversible and controllable, enabling such chromic materials to be used in select technical textile end uses (e.g. intelligent systems and functional smart textiles). The most widely investigated chromic systems for textile applications are:

- Thermochromism: colour change caused by a temperature change
- Photochromism: colour change induced by ultraviolet light
- Ionochromism: colour change caused by interaction with an ionic species
- Electrochromism: colour change caused by electric current flow.

The most common form of ionochromism relates to halochromism in which the colour change is caused by variation in pH (hydrogen ion concentration).

Thermochromic systems applied to textiles have been based on leuco dye or liquid crystal systems, usually protected from external mechanical forces by microencapsulation and applied by screen printing.[173,174] Such microencapsulated systems may be regarded as pigments because they are applied to the textile as discrete solid particles. Leuco dye thermochromic systems change from coloured to colourless above specified mean transition temperatures, typically 30 °C, but systems are available for <10 to >45 °C with a range of colours. Liquid crystal systems involve changes in the liquid crystal structure (e.g. from a smectic phase to a chiral nematic phase) in which a helix forms where the pitch length changes rapidly with increasing temperature leading to the colour change. Leuco dye thermochromic systems have been used in T-shirts; smart textiles could use the colour change to indicate that the wearer of a garment is too hot (e.g. baby clothes). Medical textiles also offer potential for exploitation of thermochromic textiles.

T-type photochromic dyes under UV irradiation change from colourless to coloured because of the formation of an intensely coloured species through isomerism of the colourless molecule.[173,174] Fishing lines can thus be coloured in sunlight above the water line but colourless below the water line. Potential applications could include responsive camouflage and UV-sensing. P-type photochromic dyes may have uses in the biomedical field.

Ionochromic dyes use colorants that are pH-sensitive and could be used as analytical pH indicators.[174] Thus, the colour change on a wound dressing potentially could be used to monitor the healing process because the pH of the skin of burn patients is known to change during the healing process.

Electrochromic materials exhibit a colour change when subjected to an electric current and offer possible applications in smart technical textiles. An appropriately constructed controllable electrochromic textile could be used for adaptive camouflage, wearable displays, or biomimicry or chameleonic products, for example.[174]

Another area of interest in technical textiles is the use of optical effect pigments applied primarily by printing techniques to produce striking visual effects such as lustre, sparkle, iridescence, phosphorescence, or multicoloured effects depending on the viewing angle of the observer.[175] Aluminium and bronze pigments are used for metallic effects whereas mica-based nacreous or pearlescent pigments are used to produce iridescence through interference colours. Useful pearlescent systems that are coated with a highly refractive metal oxide (e.g. titanium oxide or iron oxide) can be created from thin films of materials with a low refractive index. This results in pigment particles with four interfaces, the layer thickness of which is controlled to provide the requisite interference colours. Phosphorescent pigments interact with light to give a delayed fluorescence that can provide glow-in-the-dark effects when printed on technical textiles.

References

[1]   Shore J. Coloration of polypropylene. Rev Prog Coloration 1975;6:7–12.
[2]   Easton JR. Key sustainability issues in textile dyeing. In: Blackburn RS, editor. Sustainable textiles: life cycle and environmental impact. Abington, Cambridge, UK: Woodhead Publishing Ltd; 2009. p. 139–54.
[3]   Wakankar DM. Regulations relating to the use of textile chemicals and dyes. In: Gulrajani ML, editor. Advances in the dyeing and finishing of technical textiles. Abington, Cambridge, UK: Woodhead Publishing Ltd; 2013. p. 105–32.
[4]   Cattoor T. European legislation relating to textile dyeing. In: Christie RM, editor. Environmental aspects of textile dyeing. Abington, Cambridge, UK: Woodhead Publishing Ltd; 2007. p. 1–29.
[5]   Boyter HA. Environmental legislation USA. In: Christie RM, editor. Environmental aspects of textile dyeing. Abington, Cambridge, UK: Woodhead Publishing Ltd; 2007. p. 30–43.
[6]   Bide M. Environmentally responsible dye application. In: Christie RM, editor. Environmental aspects of textile dyeing. Abington, Cambridge, UK: Woodhead Publishing Ltd; 2007. p. 74–92.
[7]   Stevens CB. I) Dye classes: General structure and properties in relation to use (II) colour fastness. In: Nunn DM, editor. The dyeing of synthetic-polymer and acetate fibres. Bradford, UK: The Dyers Company Publications Trust; 1979. p. 1–75.
[8]   Fluss K-H. Space dyeing—survey of methods. Bayer Farben Revue, E 1976; 26: 19–33.
[9]   Shore J. Colorants and auxiliaries, organic chemistry and application properties, Volume 1—colorants. Bradford, UK: The Society of Dyers and Colourists; 1990.
[10] Bach E, Cleve E, Schollmeyer E. Past, present and future of supercritical fluid dyeing technology. Rev Prog Color 2002;32:88–102.
[11] Scrimshaw J. CO_2 dyeing gets commercial roll-out. Int Dyer 2010;195(7), August 6–7.

[12] Holme I. Waterless dyeing: how near, how far? Int Dyer 2012;198(4) May 12–15.

[13] Ackroyd P. The mass coloration of man-made fibres. Rev Prog Color 1974;5:86–96.

[14] Giesen B, Eisenlohr R. Pigment printing. Rev Prog Color 1994;24:26–30.

[15] Hickman WS. Preparation. In: Shore J, editor. Cellulosics dyeing. Bradford, UK: The Society of Dyers and Colourists; 1995. p. 81–151.

[16] Maguire King K. Ink jet printing of technical textiles. In: Gulrajani ML, editor. Advances in the dyeing and finishing of technical textiles. Sawston, Cambridge, UK: Woodhead Publishing Ltd; 2013. p. 236–57.

[17] Anon. EU dyestuff prices up 80%. Int Dyer 2014;199(5):9.

[18] Holme I. Dyeing and finishing protein fibres. Int Dyer 2014;199(3):12–15.

[19] Zollinger H. Colour chemistry, synthesis, properties and applications of organic dyes and pigments. New York: VCH; 1987.

[20] Booth G. The manufacture of organic colorants and intermediates. Bradford, UK: The Society of Dyers and Colourists; 1998.

[21] Holme I. The provision, storage and handling of dyes and chemicals in dyeing and finishing plants. J Soc Dyers Col 1978;94(9):375–94.

[22] Shore J. Blends dyeing. Bradford, UK: The Society of Dyers and Colourists; 1998.

[23] Noguchi H, Shirota K. Formulation of aqueous inkjet ink. In: Ujiie H, editor. Digital printing of textiles. Abington, Cambridge, UK: Woodhead Publications Ltd; 2006. p. 233–51.

[24] Fu Z. Pigmented ink formulation. In: Ujiie H, editor. Digital printing of textiles. Abington, Cambridge, UK: Woodhead Publications Ltd; 2006. p. 218–32.

[25] Clarke G. A practical introduction to fibre and tow coloration. Bradford, UK: The Society of Dyers and Colourists; 1983.

[26] McLaren K. The colour science of dyes and pigments. 2nd ed. Bristol: Adam Hilger; 1986.

[27] Hunger K. The effect of crystal structure on colour application properties of organic pigments. Rev Prog Color 1999;29:71–84.

[28] Sarkar AK. Fluorescent whitening agents. Watford: Merrow; 1971.

[29] Williamson R. Fluorescent brightening agents. Amsterdam: Elsevier; 1980.

[30] Anliker R, Müller G. Fluorescent whitening agents. In: Coulston F, Korte F, editors. Environmental quality and safety, Suppl Vol IV. Stuttgart: Thieme; 1975.

[31] Fluorescent brightening agents. In: J Shore, Colorants and auxiliaries: organic chemistry and application properties, Volume 2 auxiliaries. 2nd ed. 2002; Society of Dyers and Colourists: Bradford, UK; 760–812.

[32] Colour Index (www.colour-index.com), The colourist, 2014, (4), 6.

[33] Holme I. The provision, storage and handling of dyes and chemicals for textile dyeing, printing and finishing. Vienna: UNIDO Textile Monograph UF/GLO/78/115; 1980.

[34] Anon. ETAD updates organic impurities thresholds. Int Dyer 2013;198(7):10.

[35] Nunn DM, editor. The dyeing of synthetic-polymer and acetate fibres. Bradford, UK: The Dyers Company Publications Trust; 1979.

[36] Holme I. Fibre physics and chemistry in relation to coloration. Rev Prog Color 1970;1:31–43.

[37] Murray A, Mortimer K. Carrier dyeing. Rev Prog Color 1971;2:67–72.

[38] Burkinshaw SM. Chemical principles of synthetic fibre dyeing. London: Blackie; 1995.

[39] Mock G. Dyeing of polyester fibres. In: Hawkyard C, editor. Synthetic fibre dyeing. Bradford, UK: Society of Dyers and Colourists; 2004. p. 45–81.

[40] Denton MJ, Daniels PN, editors. Textile terms and definitions. 11th ed. Manchester, UK: The Textile Institute; 2002.

[41] Hilden J. The effect of fibre properties on the dyeing of microfibres. Int Text Bull, Dyeing/Printing/Finishing 1991;(3)19, 22, 24, 26.

[42] Bide M. Dyeing of microfibres. In: Hawkyard C, editor. Synthetic fibre dyeing. Bradford, UK: Society of Dyers and Colourists; 2004. p. 235–65.

[43] Developments in Nanofibers for the New Millennium. In: Hongu T, Phillips GO, Takigami M, editors. New millennium fibers. Abington, Cambridge, UK: Woodhead Publishing Ltd; 2005. p. 269–88.

[44] Duffield PA. Dyeing wool with acid and mordant dyes. In: Lewis DM, Rippon JA, editors. The coloration of wool and other keratin fibres. Bradford, UK: John Wiley & Sons, in association with The Society of Dyers and Colourists; 2013. p. 205–28.

[45] Ingamells W. Colour for textiles—a user's handbook. Bradford, UK: The Society of Dyers and Colourists; 1993.

[46] Anon. A revival of interest in low-temperature dyeing. Wool Record 1996; 155(3618):35.

[47] Welham AC. The role of auxiliaries in the dyeing of wool and other keratin fibres. In: Lewis DM, Rippon JA, editors. The coloration of wool and other keratin fibres. Bradford, UK: John Wiley & Sons, in association with The Society of Dyers and Colourists; 2013. p. 75–98.

[48] Lewis DM. Damage in wool dyeing. Rev Prog Color 1989;19:49–56.

[49] Ginns P, Silkstone K. Dyeing of nylon and polyurethane fibres. In: Nunn DM, editor. The dyeing of synthetic-polymer and acetate fibres. Bradford, UK: The Dyers Company Publications Trust; 1979. p. 241–356.

[50] Lewis DM, Marfell DJ. Nylon dyeing. In: Hawkyard C, editor. Synthetic fibre dyeing. Bradford, UK: The Society of Dyers and Colourists; 2004. p. 82–121.

[51] Miles LWC, editor. Textile printing. 2nd ed. Bradford, UK: The Society of Dyers and Colourists; 1994.

[52] Broadbent PJ, Rigout MLA. Wool printing. In: Lewis DM, Rippon JA, editors. The coloration of wool and other keratin fibres. Bradford, UK: John Wiley & Sons in association with The Society of Dyers and Colourists; 2013. p. 393–430.

[53] Cooke TF, Weigmann H-D. Stain blockers for nylon fibres. Rev Prog Color 1990;20:10–18.

[54] Welham AC. Advances in the afterchrome dyeing of wool. J Soc Dyers Col 1986;102(4):126–31.

[55] Cox R. Acrylic and modacrylic fibres. In: Hawkyard C, editor. Synthetic fibre dyeing. Bradford, UK: The Society of Dyers and Colourists; 2004. p. 122–63.

[56] Holme I. Dye-fibre interrelations in acrylic fibres. Chimia 1980;34:110–30.

[57] Beckmann W. Dyeing of acrylic and modacrylic fibres. In: Nunn DM, editor. The dyeing of synthetic-polymer and acetate fibres. Bradford, UK: The Dyers Company Publications Trust; 1979. p. 357–91.

[58] Shore J. Dyeing with azoic components. In: Shore J, editor. Cellulosics dyeing. Bradford, UK: The Society of Dyers and Colourists; 1995. p. 321–51.

[59] Shore J. Dyeing with direct dyes. In: Shore J, editor. Cellulosics dyeing. Bradford, UK: The Society of Dyers and Colourists; 1995. p. 152–88.

[60] Vickerstaff T. The physical chemistry of dyeing. 2nd ed. London, UK: Oliver and Boyd; 1954.

[61] Hook JA, Welham AC. The use of reactant-fixable dyes in the dyeing of cellulosic blends. J Soc Dyers Col 1988;104(9):329–37.

[62] Shore J. Dyeing with reactive dyes. In: Shore J, editor. Cellulosics dyeing. Bradford, UK: The Society of Dyers and Colourists; 1995. p. 189–245.

[63] Lewis DM. Dyeing wool with reactive dyes. In: Lewis DM, Rippon JA, editors. The coloration of wool and other keratin fibres. Bradford, UK: John Wiley & Sons in association with The Society of Dyers and Colourists; 2013. p. 251–90.

[64] Gulrajani ML. Dyeing of silk with reactive dyes. Rev Prog Color 1993;23:51–6.

[65] Rys P, Zollinger H. Reactive dye-fibre systems. In: Johnson A, editor. The theory of coloration of textiles. 2nd ed. Bradford, UK: The Society of Dyers and Colourists; 1989. p. 428–76.

[66] http://www.fibre2fashion.com/sustainability/huntsman/product.asp.

[67] Holme I. Reactive dyes: innovations and challenges. Int Dyer 2012;197(1):13–15.

[68] Lewis DM. Developments in the chemistry of reactive dyes and their application processes. Color Technol 2014;130(6):382–412.

[69] Bradbury MJ, Collishaw PS, Moorhouse S. Exploiting technology to gain competitive advantage. Int Dyer 1996;181(4):13, 14, 17, 20, 22–23.

[70] Schneider R. Minimization of water consumption in washing-off processes. In: 18th IFATCC (International Federation of Textile Chemists and Colorists) Congress, Copenhagen, Denmark, IFATCC; 1999. p. 10–5.

[71] Waters BD. The regulator's view. In: Cooper P, editor. Colour in dyehouse effluent. Bradford, UK: The Society of Dyers and Colourists; 1995. p. 22–30.

[72] Senior C. Dyeing with sulphur dyes. In: Shore J, editor. Cellulosics dyeing. Bradford, UK: The Society of Dyers and Colourists; 1995. p. 280–320.

[73] Phillips DAS, Duncan M, Graydon A, Bevan G, Lloyd J, Harbon C, et al. Testing colour fading of cotton fabrics by activated oxygen bleach-containing detergents: an inter-laboratory trial. J Soc Dyers Col 1997;113(10):281–6.

[74] Annen Farben mit Schwefelfarbstoffen, Melliand Textilber, 1998, 79(10), 752, E199.

[75] Holme I. Recent developments in colorants for textile applications. Surf Coat Int, Part B Coating Transactions 2002;85(B4):243–64.

[76] Latham FR. In: Shore J, editor. Cellulosics dyeing. Bradford, UK: The Society of Dyers and Colourists; 1995. p. 246–79.

[77] Bechtold T. Electrochemistry in vat dyeing and sulphur dyeing—concepts and results, 18th IFATCC (International Federation of Textile Chemists and Colorists) Congress, Copenhagen, Denmark, IFATCC, 1999, 42–6.

[78] Burkinshaw SM, Hallas G, Towns AD. Infra-red camouflage. Rev Prog Color 1996;26:47–53.

[79] Blackburn D. Dyeing of secondary acetate and triacetate fibres. In: Nunn DM, editor. The dyeing of synthetic-polymer and acetate fibres. Bradford, UK: The Dyers Company Publications Trust; 1979. p. 76–130.

[80] Broadhurst R. Dyeing of polyester fibres. In: Nunn DM, editor. The dyeing of synthetic-polymer and acetate fibres. Bradford, UK: The Dyers Company Publications Trust; 1979. p. 129–240.

[81] Leaver AT. Novel approaches in disperse dye design to meet changing customer needs. In: American Association of Textile Chemists and Colorists 1999 International Conference, Charlotte, USA, AATCC; 1999. p. 367–74.

[82] Rattee ID. Transfer printing. In: Miles LWC, editor. Textile printing. 2nd ed. Bradford, UK: The Society of Dyers and Colourists; 1994. p. 58–98.

[83] Leadbetter PW, Leaver AT. Recent advances in disperse dye development. In: Introducing a new generation of high fastness disperse dyes, 15th IFATCC (International Federation of Textile Chemists and Colorists) Congress, Lucerne, Switzerland, IFATCC; 1990.

[84] Gutjahr H, Koch RR. Direct print coloration. Miles LWC, editor. Textile printing. 2nd ed. Bradford, UK: The Society of Dyers and Colourists; 1994. p. 139–95.

[85] Kaul B, Ripke C, Sandri M. Technical aspects of the mass-dyeing of polyolefin fibres with organic pigments. Chem Fibres Int 1996;46(4):126–9.

[86] Kaul BL. Mass pigmentation and solution dyeing of synthetic fibres. In: Hawkyard C, editor. Synthetic fibre dyeing. Bradford, UK: The Society of Dyers and Colourists; 2004. p. 218–34.

[87] http://www.masterbatches.com/bu/mb/internet.nsf/directname/textile.

[88] Emsermann H, Foppe R. Dyeing processes. In: Masson JC, editor. Acrylic fiber technology and applications. New York: Marcel Dekker; 1995. p. 285–312.

[89] Ruys L, Vandekerckhove F. Breakthrough in dyeable polypropylene. Int Dyer 1998;183(8):32–6.

[90] Lulay A. Apparel end uses. In: Masson JC, editor. Acrylic fiber technology and applications. New York: Marcel Dekker; 1995. p. 313–39.

[91] Holme I. Dispensing system enables continuous quick response. Int Dyer 1991;176(1):10–1.

[92] Lewis PA. Coloured organic pigments for coating fabrics. J Coated Fabr 1994;23(3):166–201.

[93] Banerjee B. Rubbers and elastomers used in textile coatings. In: Akovali G, editor. Advances in polymer coated textiles. Shawbury, Shrewsbury, Shropshire, UK: Smithers Rapra Technology Ltd; 2012. p. 25–78.

[94] Trotman ER. Textile scouring and bleaching. London, UK: Griffin; 1968.

[95] Holme I, Panti IA, Patel BD, Xin H. Chemical pretreatment of cotton fabrics for higher quality and performance. Belgium: West-European Textiles Tomorrow, International Symposium, University of Ghent; 1990, p. 35–59.

[96] Driver H. Fabric singeing—the vital first step. Text Technol Int 1993;178–80.

[97] Nagaraju RK, Khera JG, Nielsen PH. The combined bioblasting concept. Int Dyer 2013;199(4):36–7, 39.

[98] Holme I. Sizing for high speed weaving. Text Horizons 1985;5(6):42, 44.

[99] Holme I. Chemical pretreatment—current technology and innovations. Text Horizons Int 1993;13(4):27–9.

[100] Holme I. Enzymes power textile pretreatment. Int Dyer 2014;199(2):14–6.

[101] Marsh JT. Mercerising. London, UK: Chapman and Hall; 1951.

[102] Freytag R, Donzé J-J. Alkali treatment of cellulose fibers. In: Lewin M, Sello SB, editors. Handbook of fiber science and technology: volume 1. Chemical processing of fibers and fabrics, fundamentals and preparation part A. New York: Marcel Dekker; 1983. p. 93–165.

[103] Greenwood PF. Piece mercerizing: a modern process for knitted cottons. Text Inst Ind 1976;14(12):373–5.

[104] Holme I. The effects of chemical and physical properties on dyeing and finishing. In: Preston C, editor. The dyeing of cellulosic fibres. Bradford, UK: The Dyers Company Publications Trust; 1986. p. 106–41.

[105] Euscher G. Medium knit mercerizing. Text Asia 1982;13(9):57–60, 62.

[106] Gailey I. Causes of unlevel dyeing of cotton cellulose. The influence of mercerizing and bleaching processes on the fine structure of cellulose. J Soc Dyers Col 1951;67:357–61.

[107] Stevens CV, Roldan(-Gonzalez) LG. Liquid ammonia treatment of textiles. In: Lewin M, Sello SB, editors. Handbook of fiber science and technology: volume 1. Chemical processing of fibers and fabrics, fundamentals and preparation Part A. New York: Marcel Dekker; 1983. p. 167–203.

[108] Hearle JWS, Miles LWC. The setting of fibres and fabrics. Watford, UK: Merrow; 1971.

[109] Lau KC. Dynamic response to colour specifications. J Soc Dyers Col 1995;111(5):142–5.

[110] Sargeant C. Colour range management. J Soc Dyers Col 1995;111(9):272–4.

[111] Sargeant C. Colour visualisation and communication—a personal view. Rev Prog Color 1999;29:65–70.

[112] Sargeant C. Colour display systems. Rev Prog Color 2002;32:58.

[113] Xin JH, editor. Total colour management in textiles. Abington, Cambridge, UK: Woodhead Publishing Ltd; 2006.

[114] Park J. Instrumental colour formulation—a practical guide. Bradford, UK: The Society of Dyers and Colourists; 1993.

[115] McDonald R, editor. Colour physics for industry. 2nd ed. Bradford, UK: The Society of Dyers and Colourists; 1997.

[116] Gulrajani ML, editor. Colour measurement: principles, advances and industrial applications. Abington, Cambridge, UK: Woodhead Publishing Ltd; 2010.

[117] X-Rite. A guide to understanding color tolerancing. Grandville, Michigan, USA: X-Rite, Inc.; 1994.

[118] Li YSW, Yuen CWM, Yeung KW, Sin KM. Instrumental shade sorting in the last three decades. J Soc Dyers Col 1998;114:203–9.

[119] Battle D. The measurement of colour. In: McDonald R, editor. Colour physics for industry. 2nd ed. Bradford, UK: The Society of Dyers and Colourists; 1997. p. 57–80.

[120] Reininger DS. Textile applications for hand-held colour measuring instruments. Text Chem Color 1997;29(2):13–7.

[121] van Wursch K. On-line colorimetry in continuous dyeing. J Soc Dyers Col 1995;111(5):139–41.

[122] Xin JH. Controlling colourant formulation. In: Xin JH, editor. Total colour management in textiles. Abington, Cambridge, UK: Woodhead Publishing Ltd; 2006. p. 136–59.

[123] Glover B, Collishaw PSP, Hyde RF. Creating wealth from textile coloration. In: 15th IFATCC (International Federation of Textile Chemists and Colorists) Congress, Lucerne, Switzerland, IFATCC; 1990.

[124] Park J. A practical introduction to yarn dyeing. Bradford, UK: The Society of Dyers and Colourists; 1981.

[125] Wyles DH. Functional design of coloration machines. In: Duckworth C, editor. Engineering in textile coloration. Bradford, UK: The Dyers Company Publications Trust; 1983. p. 1–137.

[126] Park J, Smith SS. A Practical introduction to the continuous dyeing of woven fabrics. Upperhulme, Leek, UK: Roaches (Engineering); 1990.

[127] Madaras GW, Parish GJ, Shore J. Batchwise dyeing of woven cellulosic fabrics. Bradford, UK: The Society of Dyers and Colourists; 1993.

[128] McGregor R, Peters RH. Effect of rate of flow on dyeing, I—diffusional boundary layer in dyeing. J Soc Dyers Col 1965;81(9):393–400.

[129] Henningsen E. Serious alternative to continuous dyeing, Textile Month, 1998 February 19–20

[130] White M. Developments in jet dyeing. Rev Prog Color 1998;28:80–94.

[131] Bone JA, Collishaw PS, Kelly TD. Garment dyeing. Rev Prog Color 1998;18:37–46.

[132] Fox MR, Sumner HH. Dyeing with reactive dyes. In: Preston C, editor. The dyeing of cellulosic fibres. Bradford, UK: The Dyers Company Publications Trust; 1986. p. 142–95.

[133] Holme I. Working towards sustainable textiles production. Africa & Middle East Text 2011;(1):8.

[134] Berry C, Ferguson JG. Discharge, resist and special styles. In: Miles LWC, editor. Textile printing. 2nd ed. Bradford, UK: The Society of Dyers and Colourists; 1994. p. 196–239.

[135] Hawkyard C. Substrate preparation for ink-jet printing. In: Ujiie H, editor. Digital printing of textiles. Abington, Cambridge, UK: Woodhead Publishing Ltd; 2006. p. 201–17.

[136] Holme I. Quick response printing, African textiles, 1997/98 Dec/Jan 20, 30

[137] Gebhard A. Qualities combined in new printing screens. Int Dyer 2012;198(3):23–4.

[138] Provost J. Printing on wool. Digital Text 2013;4:10–4.

[139] Scrimshaw J. Digital textile has a long way to grow. Digital Text 2011;3:30–2.
[140] Miles LWC. Traditional methods. In: Miles LWC, editor. Textile printing. 2nd ed. Bradford, UK: The Society of Dyers and Colourists; 1994. p. 1–17.
[141] Ellis HA. Printing techniques—the choice. Text Horizons 1985;5(4):37–8, 40.
[142] Hawkyard CJ. Screen printing. In: Miles LWC, editor. Textile printing. 2nd ed. Bradford, UK: The Society of Dyers and Colourists; 1994. p. 18–57.
[143] Holme I. Right first time, African textiles, 1998 August/September 39–40.
[144] Provost J. Giant strides. Digital Text 2013;(1)12–6.
[145] Ujiie H, editor. Digital printing of textiles. Abington, Cambridge, UK: Woodhead Publishing Ltd; 2006.
[146] Digital Textile. Leeds, UK: World Textile Information Network.
[147] Kim YK. Effect of pretreatment on print quality and its measurement. In: Ujiie H, editor. Digital printing of textiles. Abington, Cambridge, UK: Woodhead Publishing Ltd; 2006. p. 252–75.
[148] Provost J. Be prepared. Digital Text 2011;(2)9–12.
[149] Raymond M. DuPont Artistri 2020 Textile printing system. Digital printing of textiles. Abington, Cambridge, UK: Woodhead Publishing Ltd; 2006, p. 69–83.
[150] Provost J. Fluid engineering. Digital Text 2011;(4)8–10.
[151] Provost J. The challenges of reactive ink jet printing. Digital Text 2009;(2):6–9.
[152] Mheidle M. New digital inks for extreme printing speed. Digital Text 2001;(5):18–9.
[153] Fu Z. Pigmented ink formulation. In: Ujiie H, editor. Digital printing of textiles. Abington, Cambridge, UK: Woodhead Publishing Ltd; 2006. p. 218–32.
[154] Provost J. Pigment may be the future of digital. Digital Text 2009;(1):9–12.
[155] Provost J. Quick release mechanism. Digital Text 2009;(5):8–11.
[156] Provost J. Simply sublime? Digital Text 2013;(3):10–4.
[157] Mariano Freire E. Ink jet printing technology (CIJ/DOD). In: Ujiie H, editor. Digital printing of textiles. Abington, Cambridge, UK: Woodhead Publishing Company Ltd; 2006. p. 29–52.
[158] Provost J. Drop on demand-print heads for digital textile printing. Digital Text 2009;(6):8–10.
[159] Provost J. Print head developments-turbo-charged performance. Digital Text 2011;(1):8–12.
[160] Provost J. New print heads increase choice for textile machinery manufacturers. Digital Text 2013;(2):38–9.
[161] Provost J. Firing up and industry, print head development—the key component driving growth in digital textile printing. Digital Text 2014;(5)12–9.
[162] Directory of digital textile equipment. Digital Text, 2014; (6), 60–83.
[163] Provost J. Pile drivers. Digital Text 2013;2:10–7.
[164] Methods of test for colour fastness of textiles and leather (BS 1006:1990. ISO 105).
[165] AATCC Technical Manual (on searchable CD).
[166] ISO 2009, http://www.iso.org/iso/iso_catalogue/catalogue_tc/catalogue_tc_browse.htm?commid=48172.
[167] Bide M. Colour measurement and fastness assessment. In: Gulrajani ML, editor. Colour measurement: principles, advances and industrial applications. Abington, Cambridge, UK: Woodhead Publishing Ltd; 2010. p. 196–217.
[168] Beal W. New multifibre from the Society. J Soc Dyers Col 1987;103(5/6):177.
[169] Smith KJ. Colour-order systems, colour spaces, colour differences and colour scales. In: McDonald R, editor. Colour physics for industry. 2nd ed. Bradford: The Society of Dyers and Colourists; 1997. p. 121–208.

[170]   BS EN ISO 105-A11 Textiles-tests for colour fastness Part A11: determination of colour fastness grades by digital imaging techniques.

[171]   Cui G, Luo MR, Rigg B, Butterworth M, Dakin J. Grading textile fastness part 3: development of a new fastness formula for assessing change in colour. Color Technol 2004;120(5):226–30.

[172]   Cui G, Luo MR, Rigg B, Butterworth M, Maplesden M, Dakin J. Grading textile fastness part 4: an interlaboratory trial using DigiEye systems. Color Technol 2004;120(5): 231–5.

[173]   Aitken D, Burkinshaw SM, Griffiths J, Towns AD. Textile applications of thermochromic systems. Rev Prog Color 1996;26:1–8.

[174]   Christie RM. Chromic materials for technical textile applications. In: Gulrajani ML, editor. Advances in the dyeing and finishing of technical textiles. Sawston, Cambridge, UK: Woodhead Publishing Ltd; 2013. p. 3–36.

[175]   Sekar N. Optical effect pigments for technical textile applications. In: Gulrajani ML, editor. Advances in the dyeing and finishing of technical textiles. Sawston, Cambridge, UK: Woodhead Publishing Ltd; 2013. p. 37–46.

Three-dimensional fabric structures. Part 1 – An overview on fabrication of three-dimensional woven textile preforms for composites

X. Chen[1], L.W. Taylor[1], L.-J. Tsai[2]
[1]University of Manchester, Manchester, UK; [2]Department of Textiles and Clothing, Fu Jen University, New Taipei City, Taiwan

Chapter Contents

10.1   Introduction

Weaving is an ancient technology that is unique in interlacing different linear materials from perpendicular directions in many different ways to form an integrated structure. For a long time, it was necessary to incorporate materials from only the warp and weft directions to form a sheet of material (i.e. fabric). Fabrics in two-dimensional (2-D) sheet form have many properties such as being drapable, flexible, warm, and strong, and all of these make them suitable to be used as materials for clothing and other domestic end uses. When high-performance fibres (such as glass, carbon, and aramid) are used for constructing such 2-D fabrics, the woven fabrics have many technical applications, including textile composite reinforcements for the aerospace industry (Long, 2005) and body armours for personal protection for the military and police (Chen and Sun, 2009).

The weaving process is also capable of making structures that have substantial dimension in the thickness direction formed by layers of fabrics or yarns, generally termed *three-dimensional (3-D) fabrics*. Many of these structures can be made on

Handbook of Technical Textiles. http://dx.doi.org/10.1016/B978-1-78242-458-1.00013-3

conventional weaving machines with little or no modification (Chen and Tayyar, 2003). The immediate advantages of fabrics with a notable thickness dimension include the structural integrity of the woven structure, the satisfaction of geometric shape, and volume required for many end-use applications (Chen and Hearle, 2008). In addition, the weaving technology is capable of leaving both crimped and straight yarns in fabrics to suit the application (Ko, 1999). Currently, both conventional and specially made weaving machines or devices are used to make various 3-D fabrics, mainly for the composite industry. There have been successful attempts in developing new weaving devices, particularly for making 3-D woven fabrics (Mohamed, 2008; Khokar, 2008). These new technologies arrange warp yarns in a 3-D shape and allow weft yarns to be inserted at different levels in one or two directions. The advantages of fabrics made by using such technologies are many. It is fair to say that making net-shaped preforms without the need of huge trimming after weaving is easier than the conventional weaving method, which reduces the waste of materials. Another obvious advantage is that the weaving devices can make 3-D fabrics that are much thicker than the conventional technology. By comparison, the conventional weaving technology can also make various types of 3-D fabrics but it confines the thickness, especially when making solid panels. The conventional weaving technology is also popular for making 3-D fabrics, perhaps for two reasons. The first is that making some 3-D fabrics needed is not always beyond the capability of conventional weaving machines. The other is that the broad base of conventional weaving machines is readily available for 3-D fabric production.

In any case, the 3-D woven fabrics used as composite preforms have an important role to play in the development of advanced textile composites (Miravete, 1999). This chapter reviews the 3-D woven structures and the manufacturing route for such structures to provide useful information for preform and composite manufactures to create precisely engineered composites.

10.2 Classifications of textile assemblies

The comparison and classification of textile preforms is an arduous task because of the many forms of 3-D fabrics available. In addition, the textile preforms can be classified according to different criteria, such as the yarn orientation, manufacturing process, and the geometric features of the textile preforms. Scardino (1989) listed the hierarchy

Table 10.1 Fibre architecture for composites (Scardino, 1989)

Level	Reinforcement system	Textile construction	Fibre length	Fibre orientation	Fibre entanglement
I	Discrete	Chopped fibre	Discontinuous	Uncontrolled	None
II	Linear	Filament yarn	Continuous	Linear	None
III	Laminar	Simple fabric	Continuous	Planar	Planar
IV	Integrated	Advanced fabric	Continuous	3-D	3-D

of the fibre assemblies into four levels (i.e. the fibre, yarns, 2-D fabrics, and 3-D fabrics) as shown in Table 10.1.

The engineering requirements, fibre orientation, structural integrity, and fibre volume fraction of textile preforms are of paramount importance in constructing textile assemblies. Fukuta and Aoki (1986) considered the methods of manufacture, including the dimension of the structure, fabric architecture, yarn dimensions, and directions within the preform. These can be seen in Table 10.2.

Khokar (2001) reported on 3-D woven fabrics that require shedding in yarns in the fabric length direction and in the through-the-thickness direction, which permit the weft yarns to be inserted in two directions into the fabric. Based on that, he classified the woven fabrics as shown in Table 10.3. The key criterion for 2-D and 3-D woven fabrics according to this classification is shedding. Soden and Hill (1998) classified the woven fabrics according to the weaving process that comprises the warp binder, warp interlinked, in-plane interlacing yarns, and stuffing warp yarns to produce an integrated structure.

It is agreed that 3-D woven fabrics can be manufactured from the conventional weaving machines and from specially made weaving machines/devices. Regardless of the type of machines used, it is fair to say that the weaving technology is capable of constructing 3-D fabrics with many different geometrical shapes. Chen (2007) studied the configurations and geometries of the 3-D woven fabrics and classified them into four different categories (i.e. solid, hollow, shell, and nodal), which are listed in Table 10.4.

Table 10.2 Fukuta and Aoki's classification system (Fukuta and Aoki, 1986)

Dimension	Axis				
	Non-axial	Mono-axial	Riaxial	Triaxial	Multi-axial
1-D		Roving yarn			
2-D	Chopped strand mat	Pre-impregnation sheet	Plant weave	Triaxial weave	Multi-axial weave, knit
3-D Linear element		3-D braid	Multi-ply	Triaxial 3-D weave	Multi-axial 3-D weave
Plane element		Laminate type	H or I beam	Honey	

Table 10.3 **Classification system (Khokar, 2001)**

	Loom	**Fabric dimension**	**Architectural structure**	**Yarn interlacement**
1	Conventional 2-D loom	2-D fabrication	Orthogonal	Warp and weft interlacing in one plane, horizontal
2	Conventional 2-D loom	2.5-D fabrication	Conventionally woven	Interlacing of warp, warp pile, and weft pile in 2 mutually perpendicular planes
3	Conventional 2-D loom	3-D fabrication	Orthogonal with through-the-thickness, also known as *multi-layer*	Interlacing orthogonal warp and weft with through-the-thickness
4	Conventional 2-D loom	3-D fabrication	Non-woven	Non-interlacing warp, weft, and through-the-thickness
5	Specialised 3-D loom	3-D fabrication	Orthogonal	Three sets of orthogonal yarns, vertical, and horizontal shed obtaining a structure of columns and rows

Table 10.4 **3-D textile structures and weave architecture (Chen, 2007)**

Structure	**Architecture**	**Shape**
Solid	Multi-layer Orthogonal Angle interlock	Compound structure with regular or tapered geometry
Hollow	Multi-layer	Uneven surfaces, even surfaces, and tunnels on different levels in multi-directions
Shell	Single layer Multi-layer	Spherical shells, and open box shells
Nodal	Multi-layer Orthogonal Angle interlock	Tubular nodes and solid nodes

10.3   3-D solid woven preforms

Manufacturing 3-D solid woven fabrics incorporates and manipulates yarns in the length, width, and through-the-thickness directions. The employment of the through-the-thickness yarn within the architecture differs greatly, depending on the end application of the preform. The through-the-thickness yarn is incorporated at varying levels and angles within orthogonal, angle-interlock, and multi-layer woven architectures to obtain the desired mechanical properties, such as resistance against delamination and impact damage. For instance, an orthogonal structure is a 3-D structure containing straight yarns in the three principal directions, providing a stiffer and stronger preform against tensile loading. Within the multi-layer structure, the through-the-thickness yarns are able to be stitched with any other layers, leaving much room for structural manipulation. Multi-layer fabrics have higher yarn crimp than other 3-D fabric counterparts.

10.3.1 Orthogonal woven architecture

The thickness of the orthogonal structure is indicated by the number of layers of the warp or weft yarn; there is one more layer of weft yarn than of the warp yarn. Yarns are basically running straight in the warp, weft, and the through-the-thickness directions. The through-the-thickness yarns travel vertically between any two weft yarn layers. The most common situation is when they travel between the top and bottom weft yarn layers as shown in Figure 10.1a. When the through-the-thickness yarns interlink with weft yarn layers at any other levels, tapered and figured solid preforms can be created based on the orthogonal principle. Some varieties of the orthogonal structures are indicated in Figure 10.1b and c.

There are two ways an orthogonal structure can be bound together according to Chen and Potiyaraj (1999). The orthogonal structures that use one set of through-the-thickness warp to bind the structures together are called the *ordinary orthogonal structures*, and those that use two sets of opposite-travelling binding warp yarns are known as the *enhanced orthogonal structures*. Obviously, the enhanced orthogonal structures introduce higher density of the through-the-thickness yarns in the thickness direction. Figure 10.2 shows the simulated images of an ordinary orthogonal structure and an enhanced orthogonal structure with three warp yarn layers, and the binding weave for both is $\dfrac{3}{2}\dfrac{1}{1}Z$ twill weave. It is expected that more biding yarns in the thickness direction would lead to higher resistance to composite delamination.

It is also possible that the orthogonal structures are made using some weft yarns as the binder (Banister and Herszberg, 1995).

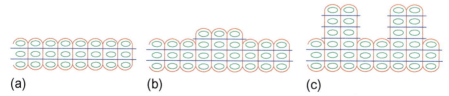

(a)   (b)   (c)

Figure 10.1 Orthogonal woven structures with variations. (a) Regular shape. (b) Tapered shape. (c) Figured shape.

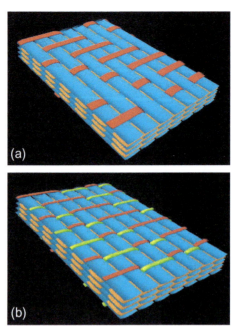

Figure 10.2 Ordinary and enhanced orthogonal structures with three-warp yarn layers.
(a) Image of a three warp-layer ordinary orthogonal fabric. (b) Image of a three warp-layer
enhanced orthogonal fabric.
Courtesy of TexEng Software Ltd.

10.3.2   Angle-interlock woven architecture

Angle-interlock structures contain a set of straight weft yarns and a set of warp yarns
that weave with the weft in a diagonal direction in the thickness. In most cases, the
warp yarns bind diagonally from top to bottom, and it is possible not to bind the
layers of weft yarns for the full depth (Chen and Potiyaraj, 1999). Straight warp
ends could be added to the angle-interlock structure. Figure 10.3 shows the cross-
sections of angle-interlock fabrics of seven layers of weft yarns with the warp binding

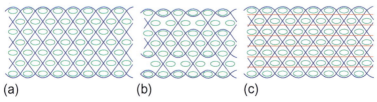

Figure 10.3 Cross-sectional views of angle-interlock fabrics (along the warp direction) with
seven layers of weft yarns. (a) Top-to-bottom binding. (b) Binding to 5th layer. (c) Warp
wadded fabric.
Courtesy of TexEng Software Ltd.

the structure in different ways. It has been reported that when wadding warp yarns are incorporated into the angle-interlock fabrics, the tensile modulus and tensile strength in the warp direction are sharply increased (Han, 1999).

It has been reported that the angle interlock fabrics have low shear rigidity (Chen et al., 1992). The reason is as follows. An angle-interlock fabric, such as the one shown in Figure 10.3a, has only one set of warp yarn weaving with the seven layers of weft yarns. The warp yarns contact the weft yarns only at the top and bottom layers; hence, the frictional resistance offered to shearing the angle-interlock fabric is lower compared to other forms of 3-D woven fabrics that use the same amount of materials. Low shear rigidity has led to good mouldability of the angle-interlock fabrics, and applications of such fabrics have been found in making helmet shells and in female body armour (Roedel and Chen, 2007; Chen and Yang, 2010a,b). When wadding warp yarns are incorporated structure, the tensile modulus and strength in the warp direction are sharply increased (Han, 1999).

Changes in the interlocking depth in this type of fabric lead to production of tapered preforms.

10.3.3 Multi-layer woven architecture

The distinctive feature of the multi-layer fabrics is that they have clearly defined fabric layers in the thickness of the fabric. Each layer is composed of a set of warp ends and a set of weft yarns. The layers are connected by weaving either the existing yarns (self-stitching) or external sets of yarns (central stitching) (Grosicki, 1977; Newton et al., 1996). A stitch can be introduced from any layer to any other layer, but that introduction conventionally needs to satisfy two conditions: (a) the stitching points should be properly hidden from the fabric layer to which it is stitched and (b) the introduction of a stitch should not alter the weave pattern of the fabric layer into which the stitch is introduced. These conditions are basically to ensure fabric appearance and may or may not be followed for fabrics whose appearance is not important. For a more stable fabric, the stitching yarns should be stitched only into the adjacent layer above or below.

Because of the structural characteristics, all warp and weft yarns in a multi-layer fabric are crimped, which leads to low initial modulus under tensile loading along the warp and weft directions. Straight wadding yarns can be added between every two fabric layers in either warp or weft or both directions; the additions of wadding yarns leads to higher modulus of the fabric (Goerner, 1989). Figure 10.4 shows a stitched four-layer fabric.

Figure 10.4 Cross-sectional view of a stitched four-layer fabric (along the warp direction).

10.4   3-D hollow woven preforms

3-D hollow woven preforms can be generally divided into two types, one with flat top and bottom surfaces and the other with uneven surfaces (Chen et al., 2004; Chen and Wang, 2006). Both can be manufactured based on the conventional weaving technology.

10.4.1   Hollow fabrics with flat surfaces

This type of hollow fabric incorporates three or more layers of fabrics. When three layers of fabrics are used, the layer connecting the top and bottom layers is woven with longer length than are the top and bottom layers. The length of the middle layer is determined by the thickness of the fabric and the configuration of the fabric cross-section. The cross-sectional shape can be trapezoidal, triangular, or rectangular (Koppelman and Edward, 1963) as shown in Figure 10.5. When this type of fabrics is made from yarns with suitably rigid yarns, such as glass, the structure is self-opening when resin is wicked into the fabric.

The hollow fabrics can be made in a single level or multiple levels. Diesselback and Stahl (1983) produced a triple-level flat surface structure with a triangular inner core. It was reported that because of their "expanding" nature, these fabrics are air inflated once they are taken off the loom.

Rheaume (1970, 1976) invented a 3-D hollow structure with flat surfaces comprising of two independent levels of plain woven fabric connected by vertical sections of fabrics. This is a special case of hollow fabric whose whole structure is foldable. The structure is woven flat and then opened when it is taken off the loom. The hollow fabric and the double I-shaped structure made from it are illustrated in Figure 10.6. During weaving, each fabric section was woven using a designated shuttle, enabling the structure to be woven in a flattened 2-D configuration.

Chen and Wang (2006) mathematically modelled the hollow fabrics with flat surfaces and created an algorithm for such fabrics to be designed and manufactured. Based on the specification of the cross-sectional geometry and the weave type for each layer, the algorithm generates the overall weave for the hollow structure that can be used directly for manufacture.

A similar type of hollow fabric is the woven version of the "spacer fabric", which is composed of two parallel layers of fabrics connected by vertical yarns/fibres. The spacer fabrics are mainly made using the knitting technology, and their air permeability

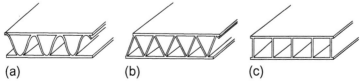

(a)   (b)   (c)

Figure 10.5 3-D hollow fabrics with flat surfaces. (a) Trapezoidal. (b) Triangular. (c) Rectangular.
Koppelman and Edward (1963).

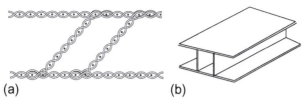

Figure 10.6 Flattening of a hollow fabric with flat surfaces and product. (a) Flattening of a double I fabric (cross-section). (b) Double I-shaped structure.
Rheaume (1970).

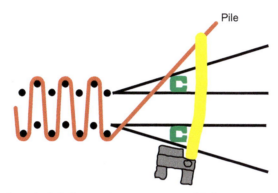

Figure 10.7 Weaving principle for creating woven spacer fabrics.
Van de Wiele website (2010).

and thermal conductivity depend much on the fabric parameters and fibres used (Yip and Ng, 2008). When such fabrics are made by weaving, the face-to-face velvet weaving technology is used, such as the Van de Wiele looms, the working principle of which is illustrated in Figure 10.7 (Van de Wiele website, 2010). According to this principle, the two substrate layers of fabrics are made simultaneously while the pile warp yarn travels between the layers to serve as the spacer yarn.

10.4.2 Hollow fabrics with uneven surfaces

Takenaka and Eiji (1988) fabricated a 3-D hollow fabric based on the multi-layer principle according to which the adjacent layers of fabrics are combined and separated at arranged intervals as shown in Figure 10.8. The combining and separating of adjacent layers allow the entire fabric to be opened after weaving to become a 3-D structure with cellular-shaped cells in the cross-section. Because of the opening of the structures, the top and bottom surfaces of this type of hollow structure are not flat. Chen et al. (2004) studied the mathematical modelling of this structure and created an algorithm for the computerised design and manufacture of this type of hollow fabric.

In a more systematic study (Chen et al., 2004, 2008; Tan and Chen, 2005), the 3-D hollow fabrics with uneven surfaces were defined based on the number of fabric layers involved and the lengths of the two types of cell walls. A *free cell wall* of a

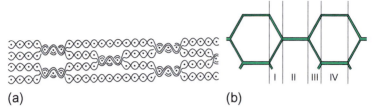

Figure 10.8 Illustration of cross-sections of hollow fabrics with uneven surfaces. (a) Flattened form. (b) Opened form (partial).
Panel a: Takenaka and Eiji (1988). Panel b: Chen et al. (2004).

hexagonal cell is one that is formed by one layer of fabric, whereas a *bonded cell wall* is a wall that is created by combining two adjacent fabric layers. Any hollow fabric of this type can be denoted by the general expression $xL(y+z)P$, where x is the number of layers of fabrics involved in making the hollow fabric, y is the length of the bonded walls, and z is the length of the free walls, both measured in number of picks. In this notation, L stands for layers of fabrics and P for picks, respectively. When the bonded walls have the same length as the free walls, the notation reduces to $xLyP$. As an example, the $8L3P$ hollow fabric involves eight fabric layers and has regular-shaped hexagonal cells in the cross-section and whose free wall and bonded wall both involve three picks.

10.5   3-D shell-woven fabrics

3-D shell-woven fabrics form doubly curved shell structures that maintain fibre continuity. Such fabrics may assume spherical shapes and cubic shapes. Such shell-shaped fabrics can be made using different techniques, such as weaving with discrete takeup, weave combination, moulding, and the origami method.

10.5.1   Weaving with discrete takeup

Busgen (1999) invented a method for the creation of 3-D shell fabrics by direct weaving. Modifications were made to the conventional loom on the takeup and let-off mechanisms. The conventional one-piece takeup roller was replaced with one made of many discs that are electronically controlled to perform individual takeup movement. The discrete takeup of the fabric results in double curvatures in it. Weft density of the fabric can be reduced at the top of the curvature, and some extra weft yarn can be added to ensure uniform material density in the fabric. Modifications to the let-off mechanism were also made in such a way that each yarn is controlled independently. Figure 10.9 shows a continuous 3-D shell fabric produced using this method. Chen and Tayyar (2003) employed an easy-to-use add-on device to the conventional loom for making 3-D shell fabrics. They used a profiled takeup roller to take up the fabric for the creation of the dome shape. Changes in the let-off were made to accommodate

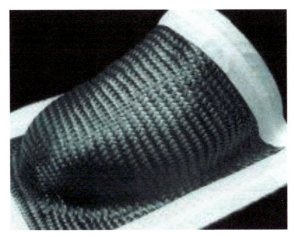

Figure 10.9 3-D shell fabric made by direct weaving.
Busgen (1999).

the takeup discrepancies as in the previous case. When the add-on device is removed, the loom resumes its original setup for making ordinary fabrics.

10.5.2   Use of combined weaves

In making everyday woven fabrics, the aim is to create flat fabrics with even material distribution. When weaves with different float lengths are used side-by-side in a fabric, care must be taken that the tension difference caused by different weaves are evened out in order not to affect the flatness of the fabric. However, in some cases, the effect of an uneven fabric is desired; one notable example is the honeycomb weave, which is a combination of short-float weaves (typically the plain weave) and long-float weaves. When short-float weaves are used, more intersections between the warp and weft yarns take place than in the fabrics with long-float weaves. This leads the short-float fabric section to take up more areal space than the long-float fabric section. In other words, the fabric section made from short-float weave tends to expand whereas the fabric section made from long-float weave tends to shrink. This is especially the case when the fabric is removed from the loom and is free of tension. This creates the bubbled effect of the fabric. This same principle was used to create 3-D shell fabrics (Chen and Tayyar, 2003). Concentric rings were planned, and the gradient change of weaves was employed in the areas between two adjacent rings. See Figure 10.10 for a domed fabric produced using different weaves.

10.5.3   Shell fabrics by moulding

Because of the extensibility of yarns and fabrics and the allowance of shear, most flat woven fabrics can to some extent be moulded into doubly curved surfaces. One example is fabric that is moulded into 3-D shapes as lining material for the inside of vehicle doors. However, the extent that a woven fabric can be moulded is quite

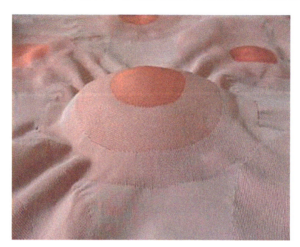

Figure 10.10 Domed fabric using weave combination.
Chen and Tayyar (2003).

limited, mainly because of the friction between the warp and weft yarns at the cross-over points and space limitation of the fabric's structure (Roedel and Chen, 2007). It is reported (Chen et al., 1992) that the 3-D angle interlock fabrics have notably lower resistance against shear than other 2-D and 3-D fabrics under the same circumstances. This finding has helped the development of helmet shells and female body armour. Refer to Figure 10.11 to see a helmet shell made from a single piece of angle-interlock fabrics with the optimal structural parameters (Roedel and Chen, 2007).

Figure 10.11 Helmet shell made from moulding a woven fabric.
Roedel and Chen (2007).

10.5.4 Use of origami principle for box shells

Shell shapes may also be in the form of an open box. For this type of 3-D shell structures, the fibres are continuous throughout the box shell. Chen and Tsai (2009) worked on such shell structures and proposed that the origami (paper-folding) principle can be used to fold the box shell as the first step. For a given box shape that is to be woven, there are usually different ways to get it folded in such a way that is suitable for weaving. The optimal solution for folding should be the one that gives the best fibre orientation and, hence, the most appropriate performance and the one that can be woven most conveniently. See Figure 10.12, which illustrates the use of an open box from paper, one of the ways that an open box can be folded down.

Once an open box has been folded flat, regions can be classified according to the structural features of the folded geometry. In the example of folded open box, the folded geometry is divided into regions A to E along the width, and I to IV in the length direction. This is shown in Figure 10.13a. When warp yarns go in the length direction, weaves can then be worked out for the individual regions I to IV. Figure 10.13b indicates the weaving of the fabric sections corresponding to region II across the width.

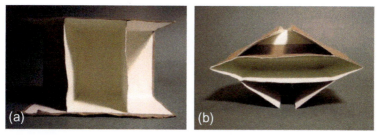

Figure 10.12 Folding of a box shape. (a) Before folding. (b) After folding.
Chen and Tsai (2009).

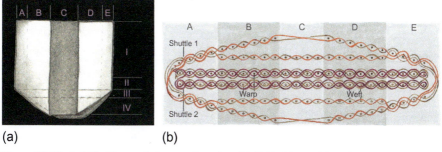

Figure 10.13 A folded box and the weaving plan. (a) Folded box with regions. (b) Weaving plan for region II (cross-section).
Chen and Tsai (2009).

10.6   3-D nodal woven fabrics

The term *3-D nodal fabric* refers to a fabric that facilitates a network formed by different tubular or solid members joining together. See Figure 10.14 for one type of the nodal fabrics (Smith, 2009). In such a case, each member is a tube whose wall may be of a single layer fabric or any form of 3-D solid fabric (Taylor, 2007). The fabric is woven flat and when taken off the loom, it is pulled into shape.

Many have worked on the design, manufacture, and application of the 3-D nodal fabrics for different end uses. Lowe (1987) developed woven nodal structures from the plain weave using thermal reactive fibres that enable the tubular members to have adaptable dimensions. The nodal structures described in this work are illustrated in Figure 10.15. Day et al. (1990) created a multitude of possible nodal structures based on the use of a jacquard loom, and sketches of some of the nodal structures are depicted in Figure 10.16. The woven flanges are required to have an area for defining and joining the two cloths together by stitching to obtain the tubular structures. Chen and Zhang (2006) reported that tubular members can be formed within a 3-D fabric

Figure 10.14 Illustration of a 3-D nodal fabric.
Smith (2009).

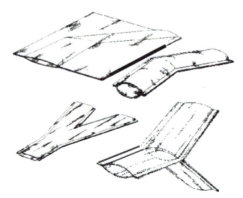

Figure 10.15 Integrated nodal structures.
Lowe (1987).

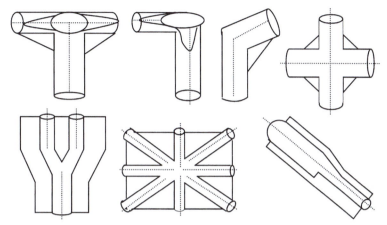

Figure 10.16 Nodal structures with integrally woven flanges.
Day et al. (1990).

Figure 10.17 Nodal structures made from various fibres. (a) Polyester. (b) Glass. (c) Carbon.
Taylor (2007).

and that the relationships of these tubular members can be one above the other or intersectional.

Taylor (2007) worked on the design and creation of nodal structures and developed a procedure for designing generic nodal fabrics. The design process involves the nodal structure specification, structure flattening, pattern segmentation, weave assignment, and weave combination. One important issue in manufacturing nodal fabrics is the smooth intersection of two tubes with or without the same diameter. The algorithm led to a solution that ensures the high quality of the intersection between tubes. Based on this achievement, Smith (2009) computerised the algorithms and produced software to assist the design and manufacture of 3-D nodal fabrics. Figure 10.17 shows nodal structures in various forms made from polyester, glass, and carbon fibres, respectively (Taylor, 2007).

10.7   3-D woven architecture from specially made devices

The conventional weaving machine is basically a device for making 2-D fabrics, primarily for domestic applications. As has been demonstrated, it also can be used for making various types of 3-D fabrics but would encounter problems when the thickness

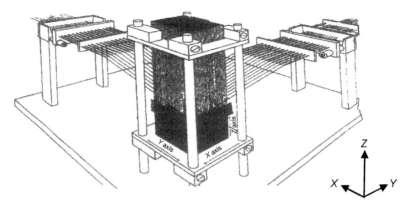

Figure 10.18 The non-interlacing weaving device.
King (1976).

of the 3-D fabric becomes substantial. In addition, the conventional weaving machine allows the weft yarns to be inserted only in one direction, which poses a limitation on the creation of 3-D fabrics. In parallel to the use of conventional weaving machines, different forms of weaving devices have been developed for making 3-D fabrics.

King (1976) developed a 3-D textile assembly using a purpose-made weaving device from static yarns oriented in the Z direction. The X yarns are inserted first and then beaten into place, and then the Y yarns are inserted and beaten. This is repeated to produce a compact structure until the desired height is achieved, producing a 3-D rectangular cross-sectional configuration. No shedding is involved in this technology, and the final product is an orthogonal structure that demands no interlacements. Figure 10.18 illustrates the principle of this non-interlacing weaving device.

Mohamed and Zhang (1992) patented a method for creating variable cross-sectional shaped 3-D fabrics based on the orthogonal structure. Except for the binding warp yarns, the straight warp yarns are made into several layers, and the weft yarns are inserted at different levels into the fabric simultaneously from one side or both sides of the fabric. If inserted from both sides, the weft yarns may be inserted simultaneously or alternately from each side of the warp. The binding warp yarns bind the structure together because of shedding. Depending on shedding, the 3-D fabric could be orthogonal, angle interlocked, warp interlocked, or a combination of all three in the same piece. Figure 10.19 shows the principle of the weaving machine and some of the products (Mohamed, 2008).

Based on the belief that weaving must involve yarns interlacing with one another, Khokar (2001) developed a weaving device that is capable of shedding in two directions. The linear–linear dual-directional shedding operation alternately displaces the gridlike arranged warp yarns Z to enable creation of sheds in the fabric thickness and fabric width directions. Consequently, two mutually perpendicular sets of corresponding weft yarns Y and X can be inserted into the created sheds. This enables the warp yarns to interlace with the vertical and horizontal weft yarns to create an interlaced woven 3-D fabric. The principle of the dual-directional shedding method is illustrated

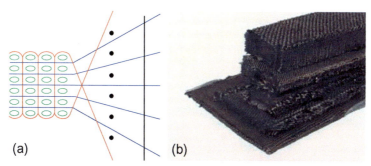

Figure 10.19 Weaving machine and products. (a) Sketch of 3Tex weaving principle.
(b) 3Tex fabrics.
Mohamed (2008).

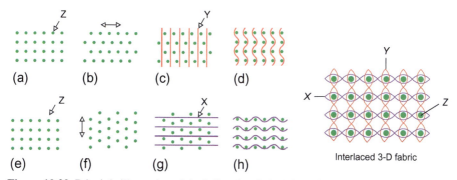

Figure 10.20 Principle illustration of dual-directional shedding with the dots being the
warp ends. (a) Warp level. (b) H. Shedding. (c) V. Picking. (d) Interlacing. (e) Warp level.
(f) V. Shedding. (g) H. Picking. (h) Interlacing.
Khokar (2001).

in Figure 10.20. Varieties of 3-D fabric structures can be made using this method.
Those 3-D fabrics made in this way offer ultimate structural integrity even when the
fabric is cut or damaged.

10.8   Conclusions

Textile technology is the only approach that can organise unidirectional materials into
2-D and 3-D assemblies with the desired shape, dimension, and, to some extent, material orientation. The use of 3-D fabrics has been widely discussed for preforms for
advanced composites because of their many advantages, including the structural integrity, light weight, and high performance. This chapter has reviewed different types of
3-D woven fabrics and routes for manufacturing them.

In the discussion of 3-D fabric structures, this chapter reviewed the 3-D woven fabrics in the categories of 3-D solid, 3-D hollow, 3-D shell, and 3-D nodal fabrics. The 3-D solid fabrics, which can be made using different weave geometries, can be used not only as reinforcements to board-shaped and tapered composites but also as constituent sections in other 3-D fabrics such as 3-D nodal and shell structures. The 3-D hollow fabrics lead to lightweight composites that are good for energy absorption and impact protection. The 3-D shell fabrics can be used to reinforce composites that have doubly curved shapes with structural integrity. The origami method is used to produce box-shaped fabrics that can be further refined for more complex components. 3-D nodal structures are particularly useful for making truss structures from textile composites.

Currently, 3-D woven fabrics are basically made through two routes, one by using the conventional weaving technology and the other by using specially developed weaving devices or heavily modified conventional looms. The chapter has highlighted two types of specially engineered weaving devices, one enabling simultaneous multi-level weft insertion and the other employing dual shedding. Although these are capable of making 3-D woven fabric in considerable sizes, the availability of such devices is limited in comparison to the prevalence of the conventional weaving machines.

References

Banister, M., Herszberg, I., 1995. Mechanical performance and modelling of 3D woven composites. In: Proceedings to the 4th International Conference on Automated Composites (ICAC95), Nottingham, UK.

Busgen, A., 1999. Woven fabric having a bulging zone and method and apparatus of forming same, Patent No. US6000442.

Chen, X., 2007. Technical aspect: 3D woven architectures. In: NWTexNet 2007 Conference, Blackburn, UK.

Chen, X., Potiyaraj, P., 1999. CAD/CAM of the orthogonal and angle-interlock woven structures for industrial applications. Text. Res. J. 69 (9), 648–655.

Chen, X., Tayyar, A.E., 2003. Engineering, manufacture and measurement of 3D domed woven fabrics. Text. Res. J. 73, 375–380.

Chen, X., Wang, H., 2006. Modelling and computer aided design of 3D hollow woven fabrics. J. Text. Inst. 97 (1), 79–87.

Chen, X., Zhang, H., 2006. Woven textile structures, Patent No. GB2404669.

Chen, X., Hearle, J.W.S., 2008. Developments in design, manufacture and use of 3D woven fabrics. In: TEXCOMP9 (International Conference on Textile Composites), University of Delaware, USA.

Chen, X., Sun, D., 2009. Textile materials in personal protection equipment for the police and military personnel. J. Xi'an Polytech. Univ. 23 (2), 67–74.

Chen, X., Tsai, L.-J., 2009. Weaving of complex composite preforms of 3D shapes based on the origami principles. In: Proceedings to the 3rd Annual Conference of Northwest Composites Centre, Manchester, UK.

Chen, X., Yang, D., 2010a. Use of 3D angle-interlock woven fabric for seamless female body armour, Part I: ballistic evaluation. Text. Res. J. 80 (15), 1581–1588. http://dx.doi.org/10.1177/0040517510363187.

Chen, X., Yang, D., 2010b. Use of 3D angle-interlock woven fabric for seamless female body armour, Part II: mathematical modelling. Text. Res. J. 80 (15), 1589–1601. http://dx.doi.org/10.1177/0040517510363188.

Chen, X., Knox, R.T., McKenna, D.F., Mather, R.R., 1992. Relationship between layer linkage and mechanical properties of 3D woven textile structures. In: Proceedings to the International Symposium on Textile and Composite Materials for High Functions, Tampere, Finland.

Chen, X., Ma, Y., Zhang, H., 2004. CAD/CAM for cellular woven structures. J. Text. Inst. 95 (1–6), 229–241.

Chen, X., Sun, Y., Gong, X., 2008. Design, manufacture, and experimental analysis of 3D honeycomb textile composites, Part I: design and manufacture. Text. Res. J. 78, 771–781.

Day, G.F., Robinson, F., Williams, D.J., 1990. Composite articles, Patent No. US4923724.

Diesselback, D., Stahl, D., 1983. Dimensionally stable composite material and process for manufacture thereof, Patent No. US4389447.

Fukuta, K., Aoki, E., 1986. 3D fabrics for structural composites. In: Proceedings to the 15th Textile Research Symposium, Philadelphia, PA, USA.

Goerner, D., 1989. Woven Structure and Design—Part: Compound Structures. British Textile Technology Group, Leeds, UK.

Grosicki, Z., 1977. Watson's Advanced Textile Design: Compound Woven Structures, fourth ed. Newnes-Butterworths, London, UK.

Han, Z., 1999. Production and tensile properties of 3D woven structures as composite reinforcements. MPhil Thesis, Department of Textiles, University of Manchester Institute of Science and Technology, UK.

Khokar, N., 2001. 3D-weaving: theory and practice. J. Text. Inst. 92 (Part 1 (2)), 193–207.

Khokar, N., 2008. Second-generation woven profiled 3D fabrics from 3D-weaving. In: Proceedings to the 1st World Conferences in 3D Fabrics and Their Applications, Manchester, UK.

King, R.W., 1976. Apparatus for fabricating three-dimensional fabric material, Patent No. US3955602.

Ko, F.K., 1999. 3D textile reinforcements in composite materials. In: Miravete, A. (Ed.), 3D Textile Reinforcements in Composite Materials. Woodhead Publishing Ltd, Cambridge, England.

Koppelman, E., Edward, A.R., 1963. Woven panel and method of making same, US Patent No. 3090406.

Long, A. (Ed.), 2005. Design and Manufacture of Textile Composites. Woodhead Publishing Ltd, Cambridge, England.

Lowe, F.J., 1987. Articles comprising shaped woven fabrics, Patent No. 4668545.

Miravete, A. (Ed.), 1999. 3D Textile Reinforcements in Composite Materials. Woodhead Publishing Ltd, Cambridge, England.

Mohamed, M.H., 2008. Recent advances in 3D weaving. In: Proceedings to the 1st World Conferences in 3D Fabrics and Their Applications, Manchester, UK.

Mohamed, M.H., Zhang, Z.-H., 1992. Method of forming variable cross-sectional shaped three-dimensional fabrics, Patent No. US5085252.

Newton, N., Georgallides, C., Ansell, P., 1996. A geometrical model for a two-layer woven composite reinforcement fabric. Compos. Sci. Technol. 56, 329–337.

Rheaume, J.A., 1970. Three-dimensional woven fabric, Patent No. US3538957.

Rheaume, J.A., 1976. Multi-ply woven article having double ribs, Patent No. US3943980.

Roedel, C., Chen, X., 2007. Innovation and analysis of police riot helmets with continuous textile reinforcement for improved protection. J. Inf. Comput. Sci. 2 (2), 127–136.

Scardino, F.L., 1989. Introduction to textile structures. In: Chou, T.W., Ko, F.K. (Eds.), Textile Structural Composites. Elsevier, Covina, CA, USA.

Smith, M.A., 2009. CAD/CAM and geometric modelling algorithms for 3D woven multi-layer nodal textile structures. PhD Thesis, School of Materials, The University of Manchester, UK.

Soden, J.A., Hill, B.J., 1998. Conventional weaving of shaped preforms for engineering composites. Compos. Part A Appl. Sci. Manuf. 29 (7), 757–762.

Takenaka, K., Eiji, S., 1988. Woven fabric having multilayer structure and composite material comprising the woven fabric, Patent No. US5021283.

Tan, X., Chen, X., 2005. Parameters affecting energy absorption and deformation in textile composite cellular structures. Mater. Des. 26, 424–438.

Taylor, L.W., 2007. Design and manufacture of 3D nodal structures for advanced textile composites. PhD Thesis, School of Materials, The University of Manchester, UK.

Van de Wiele website, 2010. http://www.vandewiele.com/src/.

Yip, J., Ng, S.-P., 2008. Study of three-dimensional spacer fabrics: physical and mechanical properties. J. Mater. Process. Technol. 206, 359–364.

Three-dimensional fabric structures. Part 2 – Three-dimensional knitted structures for technical textiles applications

10

S.C. Anand

Institute of Materials Research and Innovation, The University of Bolton, Bolton, UK

Chapter Outline

10.9 Introduction

The main object of this chapter is to discuss the structure, properties, and end-use applications of warp and weft knitted three-dimensional (3-D) fabrics, otherwise known as "spacer structures". During the last decade or so, significant research and development work has been carried out by machinery manufacturers as well as the manufacturers of technical textile products for a wide range of product applications.

Spacer structures offer many attributes and characteristics that have not been possible to achieve by using other technologies. They offer a wide spectrum of properties, such as low area densities, low bulk densities, wide range of thicknesses from 2 to 60 mm, high-comfort or thermophysiological properties and tailor-made tensile,

Handbook of Technical Textiles. http://dx.doi.org/10.1016/B978-1-78242-458-1.00014-5

elastic, compression, and permeability properties without using foam, neoprene, rubber, latex, and other coating or laminating techniques.

The advantages and limitations of warp and weft knitting spacer technologies, fabric structures and their properties are discussed in detail in relation to their current and potential product applications.

10.10   Knitted spacer fabrics

10.10.1   General

Warp and weft knitted spacer fabrics continue to find new and novel product applications and it is generally recognised that spacer fabrics will be extensively used in the future in a wide range of products, mainly because an extremely wide range of possibilities is available to tailor make their aesthetical, functional, and technical properties for niche applications.

10.11   Warp knitted spacer fabrics

Warp knitted spacer fabrics are structures that consist of two separately produced fabric layers joined back-to-back. The two layers can be produced from different materials and can have completely different structures. The yarns that join the two face fabrics can either fix the layers directly or space them apart. This 3-D space is the special feature of these structures. Typically, spacer fabrics can be from 1 to 15 mm thick with the two faces being from 0.4 to 1 mm thick. The major single feature of warp knitted spacer fabrics is that virtually any thickness can be obtained, depending on the type of machinery used and the type of yarns and structures used. The author has seen a warp knitted spacer fabric with a thickness of more than 100 mm (4 in.) for use as seating fabric for sports cars.

Figure 10.21 shows the basic set-up of a Karl Mayer spacer machine RD6N whose guide bars 1 and 2 knitted the front base fabric on the front needle bar only and guide bars 5 and 6 knitted the other separate base fabric on the back needle bar only. Guide bars 3 and 4 carry the spacer threads knitted on both needle bars in succession. The thickness of the spacer depends on the distance between the two needle bars and can be varied between 1 and 15 mm. In theory, the material used in guide bars 1 and 2, 3 and 4, and 5 and 6 can be different, and the structure of the two base fabrics can be completely different. It is possible to vary the structure from an inelastic, elastic, solid, net, or specific textured surface independently in each face fabric. Furthermore, the compression and resilience properties of the spacer can be altered at will, depending on the material and the pattern chains of the threads in guide bars 3 and 4. Figure 10.22 shows the structure of a spacer material of the end cross-sectional view of a spacer produced on RD6N; its compressional capability is illustrated in Figures 10.23 and 10.24. One of the major characteristics of these materials is that the spacer yarn is knitted-in the two base structures and cannot be pulled out, frayed, or laddered when cut during shaping or moulding the product (see Figures 10.25–10.27).

6 5 4 3 2 1

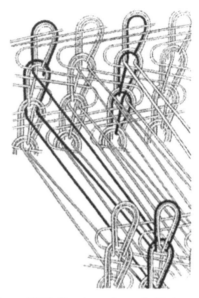

Figure 10.21 Knitting elements of Karl Mayer RD6N.

Figure 10.22 Structure of warp knitted spacer fabric.

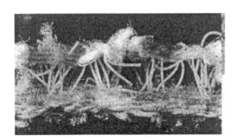

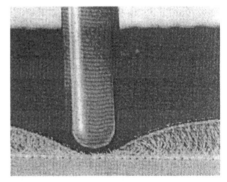

Figure 10.23 Cross-sectional view of spacer fabric.

Figure 10.24 Compression property of spacer fabric.

10.12 Knitting constructions

See Figure 10.25 for the lapping movements, threading order, and pattern chains of all six guide bars of a dimensionally stable and solid on both faces spacer structure knitted on a Karl Mayer RD6 DPLM machine. This particular structure either can be used as a spacer fabric or can be subsequently cut and separated by processing on a pile-shearing machine (PSM), which slits the fabric in the middle to produce two plush or velvet fabrics. The general rule is that the widthwise stability of individual face fabric is dictated by the threading and lapping movements of guide bars 1 and

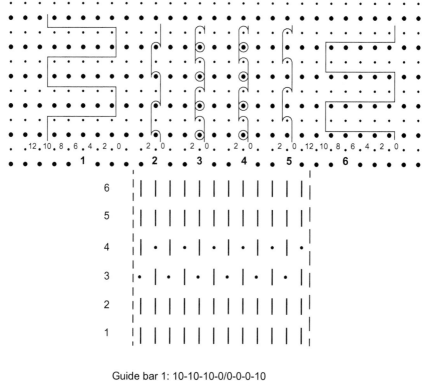

Guide bar 1: 10-10-10-0/0-0-0-10

Guide bar 2: 0-2-2-2/2-0-0-0

Guide bar 3: 0-2-0-2/2-0-2-0

Guide bar 4: 0-2-0-2/2-0-2-0

Guide bar 5: 0-0-0-2/2-2-2-0

Guide bar 6: 0-10-10-10/10-0-0-0

Figure 10.25 Lapping movements, threading order, and pattern chains of a stable and solid spacer structure knitted on an RD6 DPLM machine.

6, whereas the lengthwise stability is largely dictated by the threading and lapping movements of guide bars 2 and 5. Of course, the yarn types, their dtex, and so on also influence the overall appearance and properties of the spacer fabric.

The full technical specifications of two warp knitted spacer fabrics, fabric 1 and fabric 2, are given in Figures 10.26 and 10.27, respectively. Fabric 1 is a semistable spacer whereas fabric 2 is an elastic spacer fabric. The author leaves it to the individual reader to study the two structures and determine the main differences in the fabric appearance and properties of these two examples.

The area density of each individual guide bar fabric and, hence, percentage of yarn used in each guide bar and, of course, the area density of the whole fabric are calculated by using the following standard formula:

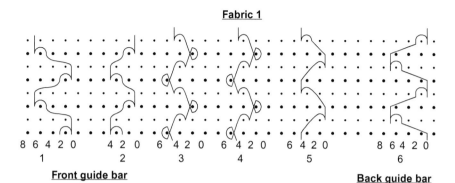

Fabric 1

8 6 4 2 0   4 2 0   6 4 2 0   6 4 2 0   6 4 2 0   8 6 4 2 0
   1   2   3   4   5   6

Front guide bar   **Back guide bar**

All guide bars: full set

Pattern chains:

Guide bar 1: 2-0-0-0/4-6-6-6
Guide bar 2: 2-4-4-4/2-0-0-0
Guide bar 3: 4-6-4-2/2-0-2-4
Guide bar 4: 4-6-4-2/2-0-2-4
Guide bar 5: 4-4-2-0/0-0-2-4
Guide bar 6: 0-0-4-6/6-6-2-0

Yarns		**%Age**	**Run-in rack**
GB1	44d tex/12 PA6 Bright	13.0	1700 mm
GB2	44d tex/12 PA6 Bright	9.2	1210 mm
GB3	22d tex/1 PA6 (monofil)	27.8	7300 mm
GB4	22d tex/1 PA6 (monofil)	27.8	7300 mm
GB5	44d tex/12 PA6 Bright	9.2	1210 mm
GB6	44d tex/12 PA6 Bright	13.0	1700 mm

Machine type	RD6N
Gauge	E32
Numbering of guide bars	6
Machine speed	850 rpm
Production	12.75 m h^{-1}
Finished courses per cm	21.4
Finished wales per cm	14.17
Finished area density	378.32 g m^{-2}

Figure 10.26 Full technical specifications of a semistable solid warp knitted spacer (fabric 1).

$$\text{Area density of each guide bar fabric:} \left(\text{c.p.c.} \times \text{w.p.c.} \times \ell(\text{mm}) \times \text{tex} \times 10^{-2}\right) \text{gm}^{-2}$$

where c.p.c. = courses per cm on one face; w.p.c. = wales per cm on one face; ℓ (mm) = stitch length of the guide bar yarn (run-in per rack [mm] ÷ 480), and tex = linear density of the yarn.

Warp knitting is the most versatile and flexible technology for producing spacer materials. The surface texture, opacity, thickness, dimensional stability, and properties of the fabric can be varied at will and the end-use applications of these materials are increasing all the time.

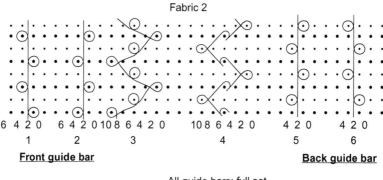

Fabric 2

6 4 2 0   6 4 2 0 10 8 6 4 2 0   10 8 6 4 2 0   4 2 0   4 2 0
　　1　　　　　2　　　　　3　　　　　　4　　　　　5　　　6

Front guide bar   **Back guide bar**

Pattern chains   All guide bars: full set

Guide bar 1   2-0-2-2/2-4-2-2
Guide bar 2   2-4-2-2/2-0-2-2
Guide bar 3   8-10-4-6/2-0-6-4
Guide bar 4   4-6-8-10/6-4-2-0
Guide bar 5   2-2-2-4/2-2-2-0
Guide bar 6   2-2-2-0/2-2-2-4

Yarns	**%Age**	**Run-in/rack**
GB1 167d tex/48 PES textured white	30.0	2300 mm
*GB2 .44d tex/1 Lycra (@ 40% extension)	3.0	880 mm
GB3 30d tex/1 Monofil PES white	17.0	7330 mm
GB4 30d tex/1 Monofil PES white	17.0	7330 mm
*GB5 44d tex/1 Lycra (@ 40% extension)	3.0	880 mm
GB6 167d tex/48 PES textured white	30.0	2300 mm

Machine type	RD6DPLM 12/3; E22; 6 guide bar
Machine speed	700 rpm.; production 14 m h^{-1}
Finished courses per cm	15.0 Finished wales per cm: 11.02
Finished area density	442.52 g m^{-2}

Figure 10.27 Full technical specifications of an elastic solid warp knitted spacer (fabric 2).

10.13 Fabric properties/product applications

The major benefit of using spacer material is to replace polyurethane (PU), neo-prene, and other types of foams that are laminated to textile fabrics for creating bulk, softness, flexibility, resilience, and so on. These foams, however, have some serious drawbacks. For instance, foams are generally flammable and are extremely uncom-fortable because of their extremely small cavities. Their thermophysiological prop-erties are poor, their compression and resilience properties deteriorate with time, and their mouldability, delamination, maintenance of original thickness when moulded into complex 3-D shapes, and washing and drying properties are often poor and do not meet the standard required. Relatively stiff monofilaments generally used as spacer material more or less overcome these drawbacks associated with laminated structures.

The major product applications for warp knitted spacer materials are vehicle seat covers (both solid or net structures in the face or back or both surfaces); automotive interiors (lining for doors, roofs, convertible hoods, and so on); seat heating systems for cars; mud flaps for trucks and buses; insoles and face fabric for sports and other shoes; lining for rubber and other boots; protective inner lining; mattress underlays and mattress covers for prevention and management of incontinence and pressure sores; for children's beds; diving and surfing suits; sports equipment; high-performance sportswear; reinforcement for composite structures; bras; underwear; swimwear; shoulder pads; fluid filters; geotextiles; bandages; plaster casts; braces; controlled release of drugs, antimicrobials, cosmetics, and so on; and finally heat and moisture regulation fabrics.

10.14   Knitting equipment

The spacer material is normally produced on Karl Mayer RD4N and RD6N machines. These 4- and 6-guide bar machines are normally used in gauges from E18 to E32 and with gaps up to 9 mm. There has been a big demand for coarser gauges and wider gaps and to satisfy the consumer demands, Karl Mayer recently introduced its latest RD7 machine.

The machine is currently available in widths of 1960 mm (77 in.) and 3500 mm (138 in.), gauges of E12 and E16 although other gauges are available on request, and fabric thickness of up to 15 mm. Special lappings inside the gap also can offer thicknesses higher than the distance set between the two trick plates. The machine is extremely versatile with two guide bars available for each base fabric, and three middle guide bars can be used to create the 3-D spacer design and required pattern, depending on the application.

The distance between the two trick plates can be altered centrally and synchronously within a few minutes with the two web holder bars and the position of the latch wire guards being adjusted at the same time.

Karl Mayer also offers their standard and up-to-date equipment as optional if requested, such as their electronic warp-let-off motions EBC; pattern drives either mechanical N or PN or electronic pattern drive EL. This is a truly versatile machine that can be used to produce an extremely wide spectrum of 3-D technical textiles. Spacer materials can also be produced on Karl Mayer type RD DPLM equipment in a wide range of thicknesses and gauges.

To create complex patterns and shapes for both fashion and performance products, Karl Mayer has also developed and offers a double-bar Raschel machine RDPJ4/1 with a piezo jacquard mechanism for producing jacquard-patterned spacer fabrics.

Karl Mayer presented for the first time its latest high-distance double-needle bar machine at ITMA in Birmingham, United Kingdom, in October 2003 for the production of spacer fabrics in thicknesses of between 25 and 60 mm, which have not been possible until then. This new machine is equipped with electronic guide bar control (EL), electronic sequential let-off motion (EBC), electronic fabric take up motion (EAC), and electronic fabric batching motion. The completely new feature is the quick and accurate electronic adjustment of the knitting elements across the whole width of the machine (1).

10.15   Weft knitted spacer fabrics

Weft knitted spacer fabrics can be produced on circular double-jersey machines as well as electronically controlled flat machines. The major advantages of these structures are:

(a) Plain as well as colour and design, and surface texture effects can be produced on the face of the fabric knitted by the cylinder needles.
(b) Shaped and true 3-D structures can be produced on electronically controlled flat machines.

The major limitations of weft knitted spacer fabrics are:

(a) The thickness of the spacer is normally limited to between 2 and 10 mm.
(b) The basic structure of the spacer fabric is limited to either knitting the spacer threads on the dial and tucking on the cylinder or tucking the spacer threads on the dial and cylinder needles.

It is obviously more practical to use tuck stitches with spacer monofilament yarns to ensure that the spacer yarns lie correctly inside the knitted fabric and prevent the face and back of the fabric from having a rough or harsh feel.

The areas of application of these structures are more or less the same as those of warp knitted spacer material and, hence, will not be repeated here. A number of products currently made from warp and weft knitted spacer fabrics are shown at the end of this chapter (see Appendix).

10.16   Knitting constructions

The structure of a circular knitted monofilament spacer fabric is illustrated in Figure 10.28, and two examples of spacer fabrics produced on circular interlock-gaited machines are shown in Figures 10.29 and 10.30. Three different yarns are required for each course: (a) yarn for the dial needles, (b) yarn for the cylinder needles, and (c) spacer yarn, normally monofilament yarn.

Two different techniques for knitting these three yarns are described below (2):

(1) Tucking on dial and cylinder needles at the same feeder (Figure 10.29):
  (a) Tucking on the dial and cylinder needles on feeders 1 and 4 (monofilament spacer yarn) on low and high butt needles, respectively.
  (b) Knitting dial needles with dial yarn at feeders 2 and 5 on low and high butt needles, respectively.
  (c) Knitting cylinder needles with cylinder yarn at feeders 3 and 6 on low and high butt needles, respectively.

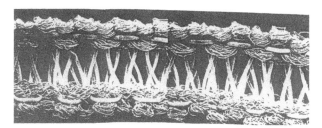

Figure 10.28 Circular knitted monofilament spacer fabric.

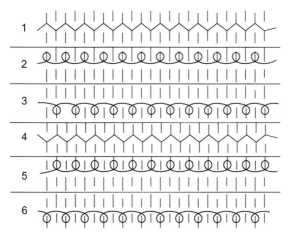

Figure 10.29 Tucking on dial and cylinder needles.

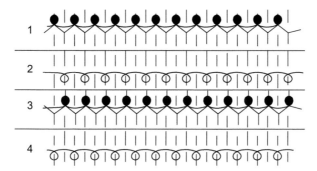

Figure 10.30 Knitting/plating on dial and tuck on cylinder.

(2) Knitting/plating on the dial needles and knitting on cylinder needles (Figure 10.30): a special yarn feeder is required with two holes to enable two yarns to be knitted at the same feeder (feeders 1 and 3):

  (a) Dial yarn knitted on the dial needles on low and high butt needles on feeders 1 and 3, respectively; and spacer yarn knitted on dial needles and tucked on cylinder needles, as shown in feeders 1 and 3.

  (b) Cylinder yarn knitted on cylinder needles at feeders 2 and 4 on low and high butt needles, respectively.

It was mentioned earlier that weft knitting equipment for producing some spacer materials is more versatile and cheaper. For instance, it is more convenient to produce textured, openwork, and jacquard effects on electronically controlled circular or flat machines. By altering the stitch structure (i.e. knitted, tuck, and miss stitches of the face, back, and spacer threads) on an individual needle selection basis, it is feasible to produce a very wide range of designs and effects in the spacer fabrics. It is also more convenient to process almost any yarn type on circular and flat machines. Figure 10.31 illustrates the structure of a circular weft knitted spacer fabric with a plain back and a micromesh structure on the face of the fabric (3). It must be stressed, however, that for the majority of high performance, high-tech applications, warp knitted spacers are often more suitable and

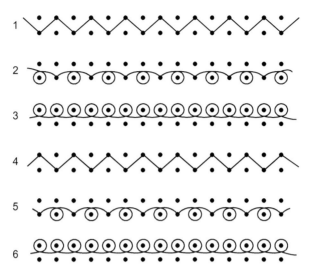

Figure 10.31 Weft knitted spacer with micromesh structure on face and plain back.

are generally preferred as opposed to weft knitted spacer fabrics. Both technologies have advantages and limitations and the ultimate choice depends on a number of factors. Both interlock and rib-gaited double-jersey machines are used to produce spacer structures.

10.17   Knitting equipment

It is generally agreed that weft knitted spacer fabrics, both circular and flat knitted, could be successfully used in a large number of technical textiles and products and as stated earlier, they are likely to compete with Raschel knitted structures for a number of applications. A number of knitting machine builders, such as Monarch, Terrot, Vignoni, Orizio, Mayer and Cie, Pai Lung, and so on, are already concentrating on developing specific models for producing spacer fabrics for a wide range of products. This area of development is considered to be the next big double-jersey boom (3,4). It was noted earlier that the major limitation of weft knitted spacer materials is that the maximum fabric thickness possible on standard equipment is 10 mm. Some machine builders are studying the possibility of increasing this thickness by reducing the diameter of the dial.

10.18   Comparison of properties of warp and
weft knitted spacer fabrics

The test results of the dimensional, tensile, and comfort properties of a warp knitted spacer fabric (Quality 5870) and a weft knitted spacer (Quality 95089) fabric are given in Table 10.5. Both fabrics were produced from polyester and Lycra yarns for face and back fabrics, and monofilament polyester yarn was used for spacer yarn in both cases.

Table 10.5 **Comparison of properties of warp and weft knitted spacer fabrics**

Dimensional properties	5870 (warp knitted spacer)	95089 (weft knitted spacer)
Area density (g m⁻²)	461.3	427.23
Thickness (mm)	2.82	2.58
Bulk density (g cm⁻³)	0.163	0.166
Tensile properties		
Breaking load (N)		
Machine direction	647.0	702.7
Cross-direction	653.7	455.7
45° direction	594.6	458.7
Tenacity (N tex⁻¹)		
Machine direction	0.029	0.033
Cross-direction	0.029	0.021
45° direction	0.027	0.021
Breaking extension (%)		
Machine direction	221.1	291.1
Cross-direction	201.3	338.2
45° direction	173.3	236.0
Modulus (N)		
Machine direction	60.0	–
Cross-direction	40.0	–
45° direction	57.1	–
Specific modulus (N tex⁻¹)		
Machine direction	0.003	0.00053
Cross-direction	0.002	0.00033
45° direction	0.003	0.00046
Comfort properties		
Alambeta		
Thermal resistance (W⁻¹ K m² × 10⁻³)		
Dry Back down	61.10	52.30
Face down	60.80	52.80
Wet (1 min) Back down	22.15	27.40
Face down	28.03	26.20
Wet (4 min) Back down	26.43	28.60
Face down	27.93	29.00
Recovery after 4 min of wetting (%)	45	45
Thermal absorptivity (W m⁻² s¹ᐟ² K⁻¹)		
Dry Back down	101.00	97.20
Face down	90.40	98.70
Wet (1 min) Back down	435.50	185.75
Face down	264.75	186.00

Table 10.5   Continued

Wet (4 min)	Back down	263.50	167.20
	Face down	219.00	171.25
Loss in warmth to touch from dry to 4 min wetting (%)		152.00	72.74
Permetest			
Water vapour permeability (%)			
	Back down	12.13	18.8
	Face down	12.24	17.6
Resistance to evaporative heat loss (m² Pa W⁻¹)			
	Back down	13.30	5.20
	Face down	16.00	5.20
Absorption (g g⁻¹)		1.27	3.05
Wicking (g cm)			
Machine direction		17.30	7.66
Cross-direction		14.10	8.04
Wicking height (mm)			
Machine direction		113.8	72.0
Cross-direction		103.2	71.0

5870 Warp knitted polyester/Lycra – solid spacer.
95089 Weft knitted polyester/Lycra – solid spacer.

It should be noticed that the weft knitted spacer fabric is somewhat lighter and thinner than the warp knitted fabric. Their bulk densities are more or less the same.

The most interesting results from the tensile testing are that both spacer fabrics are more or less isotropic in their breaking load, tenacity, breaking extension, and initial modulus properties. The majority of warp and weft knitted as well as woven fabrics used for technical textile applications have a much lower tenacity, much higher breaking extension and much lower initial modulus in the bias or 45° direction because of the scissoring effect of the structure. The near isotropic characteristics of spacer fabrics could be extremely beneficial in many technical textile products, such as composites, shoe fabrics, automotive seating fabrics, sportswear, and so on. It is obvious that this particular characteristic is a result of the spacer structure, which minimises the scissoring tendency of the structure when extended in the bias direction (45° to the course or wale direction). Note that most spacer structures also exhibit high shear modulus, which also is a significant asset in many medical devices, such as orthopaedic supports and mattresses for pressure sores. It is also interesting to observe from the tensile test results that weft knitted spacer fabrics can be engineered with similar tenacity and breaking extension values when compared with those of warp knitted spacer structures. In this particular case, the specific modulus of warp knitted spacer structure is of the order of 10 times higher than the weft knitted fabric.

The thermal insulation properties of both spacer fabrics are quite good with warp knitted spacer fabric showing a higher thermal insulation in the dry state. It is interesting to observe from thermal resistance values that both spacer structures recover 45% of their dry values after wetting and retesting after 4 min. The thermal absorptivity values of weft knitted fabric are lower than those of warp knitted fabric, which signifies that the former is relatively warmer to touch as compared with the warp knitted fabric, particularly when the fabrics have been wet with water. Even after 4 min from wetting time, weft knitted spacer fabric feels warmer to touch as compared with the warp knitted spacer fabric.

The water vapour permeability and resistance to evaporative heat loss properties of weft knitted spacer fabric are superior to those of warp knitted spacer material; therefore, the former is relatively more comfortable to wear next to the skin, during a strenuous activity. What is even more significant is that the absorption capacity of weft knitted spacer fabric is substantially superior to that of the warp knitted fabric.

The wicking characteristics of the warp knitted spacer fabrics are much superior to the weft knitted spacer fabric, in spite of the fact that both fabrics were treated with an appropriate hydrophilic softener during the finishing process. On the whole, the thermophysiological properties of both spacer fabrics were found to be satisfactory, and the wearer would feel relatively comfortable when wearing both fabrics.

The compression and recovery properties of the two spacer materials were also investigated over 10 cycles of compression and recovery. Both spacer fabrics showed a maximum compression of 20% at each cycle, and both fully recovered their thickness after relaxation.

This author has designed, developed, and characterised a wide range of both warp and weft knitted spacer structures, and it has been established that the properties of spacer materials can be engineered and fine-tuned to suit the specific end-use requirements.

For instance, a very wide range of comfort properties can be achieved by altering the fibre type and structure of the face, back, and spacer individually and independently.

10.19   Research in spacer structures

Ye et al. studied a number of relevant properties of warp knitted spacer fabrics for their suitability for use as cushion or padding in car seats. They investigated the compression behaviour, thickness variation, pressure relief, thermal properties, and air permeability of warp knitted spacer fabrics and compared them with alternative solutions for car seats, especially PU foams, which are currently used as a three-layer laminate with seating fabric as the face and a knitted fabric as the back, both fabrics being flame bonded to the PU foam as the thick middle layer.

The main conclusions drawn from this research are summarised below (5):

(a) Warp knitted spacer fabrics show very good linear elastic compressibility in the first compression phase, which is of great interest for seating fabrics.
(b) Warp knitted spacer fabrics are better at reducing peak pressure as compared to PU foam. The use of several layers of spacer material is suggested for achieving increased pressure relief, which makes drivers feel more comfortable during driving.
(c) Warp knitted spacers are more breathable substrates for car seats than foam or other nonwoven alternatives.
(d) Warp knitted spacer fabrics have higher thermal conductivity or lower thermal resistance and are capable of transferring heat more effectively away from the driver's body and thus possess better thermoregulation properties than PU foam for warm climatic conditions.
(e) Warp knitted spacer fabrics are easier to recycle than PU foam or other nonwoven alternatives used for car seating.

The compression and recovery curves for different fabrics investigated by Ye et al. are illustrated in Figure 10.32 (5). It can be observed in Figure 10.32 that the hysteresis loop of the PU foam is much larger than the ones shown for other materials, which indicates a greater dissipation of energy during the compression/recovery cycle. The compression modulus of warp knitted spacer 1 is much higher than that of the PU foam; even at 50 kPa pressure, it is still in compression phase 1. Nonwoven wadding is crushed by a pressure of less than 10 kPa.

One of the interesting patents granted during the last few years concerns the production of a wide range of both aesthetical and functional effects and properties by modifying and/or enhancing the fabric structure upon dyeing and finishing after knitting (6,7).

A spacer or 3-D structure is produced on a double-needle bar Raschel machine, which is slit in the middle to produce two separate fabrics. One of the major features of this particular technique of knitting is that the pile threads (or loops) from the middle two or three guide bars always plate on the technical face of the two separated individual fabrics. This specific feature of these structures is used to produce fleece on the technical face of the fabric, thus producing a double-sided velour fabric in which the technical back produces a velour or velvet effect from the threads that knitted on both needle bars during knitting; a fleece or nap is produced on the technical face of the fabric after raising the technical face of the individual fabric during finishing (Figure 10.33). The pile yarns, which plate on the technical face of each fabric, are raised into a fleece using wire cloth rollers that rotate in directions opposite to the ones situated on either side.

It is also feasible to apply a coating or a binder to the technical face in a certain pattern or design that will resist raising during the napping process to provide the finished technical face with a pattern of raised and unraised areas. The technical face can also be printed by using any of the standard techniques to enhance the colour and design effect in the finished fabric (7).

Tytex Group, a multinational healthcare company with its headquarters in Ikast, Denmark, has been actively engaged in research and development programmes in 3-D spacer structures for a number of years. The company has commercially designed and innovated a wide range of 3-D products using warp and weft knitted spacer fabrics. One of

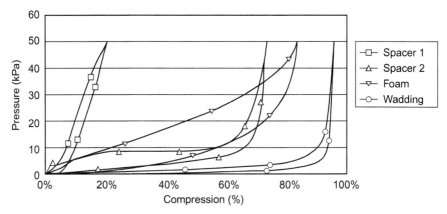

Figure 10.32 Compression/recovery curves of different materials.

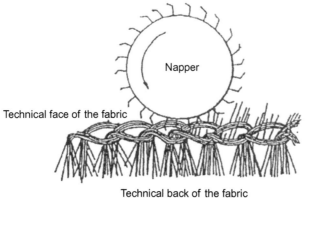

Raising machine

Napper

Technical face of the fabric

Technical back of the fabric

Before napping

After napping

Technical face

Technical back

Figure 10.33 Production of fleece on technical face of fabric.

its range of products, called AirX™ Orthocare, is designed for soft, semirigid, and rigid orthopaedic supports and braces for different parts or joints in the body. These supports for shoulder, elbow, back, hand, knee, and ankle are claimed to possess a wide range of attributes, which includes high breathability and good heat and moisture management, low shear and friction, and provide cushioning. They also are latex free, heat mouldable in complex shapes, and isotropic in mechanical properties, and reduce pressure buildup by uniform pressure distribution in all directions. For further information on AirX™ products and other healthcare products marketed by Tytex Group, visit its Web sites (8,9).

A number of these AirX™ Orthocare products are illustrated in Figure 10.34. All these products use either warp or weft knitted spacer materials, and both are sometimes combined into one product.

In a major research programme that began in 2004 at the University of Bolton, United Kingdom, novel warp and weft knitted spacer fabrics have been designed, developed, fully characterised, and compared with a number of commercially available knee braces made from different materials and by using different manufacturing techniques. The dimensional, mechanical, thermophysiological, compressional, and elastic properties of spacer materials and a number of commercial structures were tested and analysed in terms of their raw material, structure, and finishing treatments applied to them (10–12). The programme has established the fact that the knitted spacer fabrics are generally lighter, thinner, and more voluminous than commercially available knee brace materials (11).

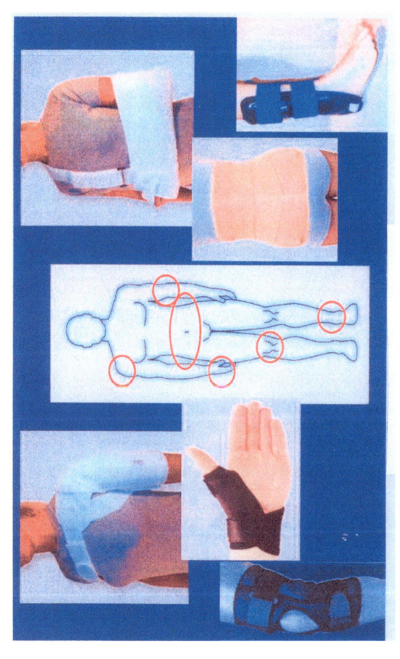

Figure 10.34 Orthopaedic supports and braces for different body parts.

One of the major attributes of spacer materials is that their tensile and mechanical properties are more or less isotropic (i.e. similar in all directions). This property is crucially important to an even distribution of pressure in many medical devices, such as mattresses for the management of pressure sores (12).

The comfort properties of the novel spacer structures developed during this research were substantially superior to neoprene- and foam-containing commercial products (11). For example, the water vapour permeability, resistance to evaporative heat loss, absorption of water, and wicking characteristics of spacer structures developed during this research programme were much superior than the commercially available products (11).

The empirical modelling of elastic and compressional properties of warp and weft knitted spacer fabrics was presented recently at an international conference and exhibition (12).

It has been established that knee braces are required to provide a support of around 15.5 mm Hg to the patient's limb. For testing purposes, these braces have been compared with a type 2 bandage conforming to the BS 7505. A knee brace in its basic form has a tubular shape; therefore, the cross-direction of the material has been analysed. It was found that both warp and weft knitted spacer fabrics developed in this work offer a quasilinear relationship between load and displacement when tested for their tensile properties. The two novel spacers (one warp knitted and one weft knitted) also offer very similar curves for the relationship between limb circumference and percentage reduction in material's width during manufacture in order to obtain optimum compression of 15.5 mm Hg around the knee (12). Figure 10.35 illustrates the individual as well as combined curves obtained for warp knitted spacer (quality 5870) and weft knitted spacer (quality 3500C). The best-fit second-order polynomial regression curve for both combined structures is shown in solid bold line. The equation and correlation coefficient for this combined curve is $y = -0.0175x^2 + 1.5006x - 10.343$ and $R^2 = 0.9599$, respectively (12).

Pressure ulcers, also known as *decubitus ulcers*, *pressure sores*, or *bedsores*, are injuries that break down the skin and underlying tissue. They are caused when an area of skin is placed under pressure. The severity of pressure ulcers can vary from grade 1, which is the most superficial type with the skin remaining intact but may hurt or itch. The skin may also feel either warm and spongy, or hard, to grade 4 pressure ulcer. A grade 4 pressure ulcer is the most severe type, in which the skin is severely damaged and the surrounding tissue begins to die (tissue necrosis). The underlying muscles or bone may also be damaged. Pressure ulcers can develop when a large amount of pressure is applied to an area of skin over a short period of time or when less pressure is applied over a long period of time. This extra pressure disrupts the flow of blood through the skin. Without a blood supply, the affected skin becomes starved of oxygen and nutrients and begins to break down, leading to an ulcer formation. Pressure ulcers tend to affect people with health conditions that make it difficult to move, especially those confined to lying in bed or sitting for prolonged periods of time. Pressure or decubitus ulcers are considered to be a major disease that costs the NHS in the United Kingdom between £1.8 and 2.6 bn per year (19).

PU foam is commonly used for seats and cushions in many applications in vehicles, chairs, wheelchairs, sofas, mattresses, and furniture. It is well known the PU foam cushions and seats are not breathable and lack thermophysiological comfort, are not washable at high temperatures, and are environmentally hazardous both in terms of flammability and recycling.

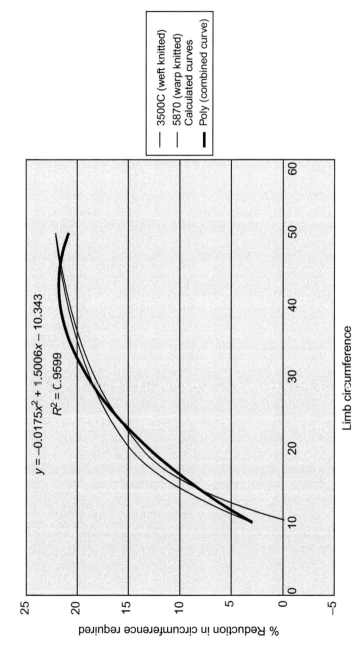

Figure 10.35 Combined plot of different curves.

Warp knitted spacer fabrics are extremely versatile in terms of design, thickness, comfort, washability, compression, and resilience; above all, they are extremely efficient in pressure relief. These unique 3-D structures can have between 2 and 60 mm thickness with good compression, resilience, and breathability. The two independent fabric faces can be knitted into any kind of mesh or solid structures, and to sustain the space between the two faces and to obtain the required compression and resilience properties in the thickness direction, monofilament yarns are normally used as the space yarn.

A systematic research and development project has been performed on the design, development, and characterisation of pressure-relieving cushions by combining a series of 3-D warp knitted spacer structures to achieve the desirable properties, such as good compression resistance and resilience, reduction of peak pressure, and pressure distribution over a much larger area of the cushion (20). The fundamental research work was carried out at the University of Bolton, United Kingdom. The Airospring® cushions are covered by a full granted patent and are being commercially developed and marketed by Baltex Ltd, Ilkeston, Derbyshire, United Kingdom.

The various innovative and unique features of the Airospring® cushions are as follows:

(1) They are much better at reducing peak pressures under the bottom than PU foam cushions.
(2) They distribute the pressure evenly over much larger areas of the cushions or seats than PU foam cushions.
(3) They conform to the shape of the body and have sufficient compression resistance to support the person without "bottoming out".
(4) They provide a well-ventilated, comfortable surface that does not unduly restrict movement.
(5) They can be laundered in the washing machine, are nonflammable, and can be easily recycled.
(6) They are particularly beneficial to individuals who are prone to developing pressure sores because they spend prolonged periods sitting in a wheelchair and other support systems.

Improving the quality of life for an ageing population is an important issue for healthcare and medical research. It has been recognised that venous leg ulcers are an increasing problem for the elderly, and about 1–2% of the adult population in the United Kingdom suffers from active ulceration during their lifetime with evidence that similar percentages are reported internationally (13,14). The total cost to the NHS in the United Kingdom for venous leg ulcer treatment is about £650 million per annum, which is 1–2% of the total annual healthcare expenditure (15). These figures could be underestimated, however, because of underreported, inadequate assessment, and restricted access to the latest statistics (14,16).

It is widely accepted that the only efficient treatment for venous leg ulcers is compression therapy, which treats the underlying venous insufficiency. The principle of compression therapy relies on external pressure applied to the lower limb, which assists the interstitial pressure and improves venous return and the reduction of superficial venous hypertension. Facilitating effective venous return of blood and the vital blood circulation is crucial for the healing of ulceration (14). In the United Kingdom, a four-layer compression system is widely used; in other European Union countries, a two-layer bandaging system is often preferred. However, these techniques have raised many issues with regard to the quality of life and comfort of patients.

A new single-layer 3-D bandage system for the treatment of venous leg ulcers has been designed and developed at the University of Bolton. This 3-D knitted spacer fabric structure has been designed by using mathematical modelling and Laplace's law. The sustained graduated compression of the developed 3-D knitted spacer bandages were tested, characterised, and compared with that of commercially available compression bandages. The developed 3-D single-layer bandage meets the ideal criteria stipulated for compression therapy. The laboratory results were verified by a pilot user study incorporating volunteers from different age groups (17,18).

The results from the pilot user study (Figure 10.36) show the comparison of the novel single-layer 3-D weft knitted spacer bandage and a two-layer compression bandage system (Surepress® ConvaTec). The desired trend of sustained graduated pressure from the ankle to the knee is evident in the pilot user study. The sub-bandage pressure readings at the ankle are 40–50 mm Hg, decreasing to 11–18 just below the knee. Therefore, pressure readings for both the novel single-layer 3-D knitted spacer and the commercial two-layer compression bandage system tested are slightly higher than that recommended at the ankle, but both have very similar if not the same results (18).

The harvesting of waste energy from the ambient environment and human movement has long been considered an attractive alternative to traditional rechargeable batteries for providing electric power to low-energy consumption devices such as wireless body-worn sensors and wearable consumer electronics.

In an ongoing research project, the possibility of having highly flexible, efficient, and durable piezoelectric generators based on the simple processing and low-cost strategies has been successfully achieved and fully characterised. In this novel and futuristic research project, all polymeric fibre-based weft knitted 3-D structures piezoelectric generators comprising of PVDF monofilament spacer yarn and a silver-coated polyamide 66 (PA66) yarn as the top and bottom conducting fabric faces have been designed, developed, and fully characterised. The generator is shown to be capable of efficiently converting mechanical energy into electrical energy; under a peak compressive pressure

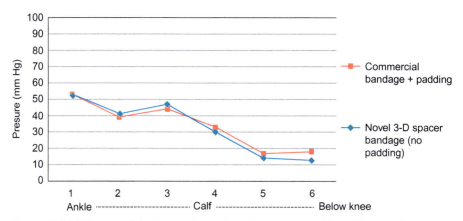

Figure 10.36 Pilot user trial of comparison of two-layer system with padding and single-layer novel 3-D spacer bandage without padding.

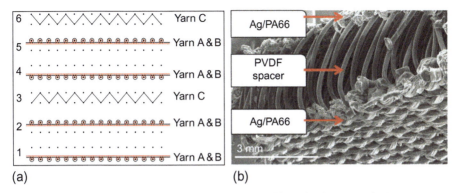

Figure 10.37 (a) Schematic of fabric structure with the position of various yarns in the structure. (b) Cross-sectional SEM image showing the position of piezoelectric and conductive yarns.

of 0.10 MPa, the fabric consistently produced a voltage of ~14 V and a peak current of ~30 μA, corresponding to a peak power density of 5.10 μW cm^{-2}, thus demonstrating a significantly higher power output under similar experimental conditions over existing 2-D woven and nonwoven piezoelectric structures (21).

10.20 Knitting of 3-D spacer piezoelectric fabric

The fabrics were knitted on an E20 (20 needles per inch) circular weft knitting double-jersey machine with a 30-in. diameter at a machine speed of 30 rpm. See Figure 10.37 for the specific knitted structure produced in this project by using three different yarns: conductive yarn A, insulating yarn B, and piezoelectric yarn C (22). The conductive yarn A was Ag-coated PA66, a 143/34 dtex yarn, and was plated on the outside of each fabric face; the insulating yarn B was a 84 dtex false-twist texturised polyester yarn and was plated inside the structure in such a manner that it shows on the inside of the two fabric faces; finally, the piezoelectric monofilament spacer yarn C, 300 dtex, was tucked inside the two fabric faces. It should be noted that yarn C is tucked on both cylinder and dial needles in such a manner that is does not protrude through either of the fabric faces and always remains inside the two faces keeping them apart. The total thickness of the fabric structures developed in this work was approximately 3.5 mm; however, it should be noted that this thickness can be varied between 2 and 60 mm, depending on the type of knitting machine (warp or weft knitting) used and the end-use application.

10.21 Conclusions

The spacer technology in relation to both warp and weft knitting has been explained in depth, and its salient features explored for a number of existing and future areas of application.

There is little doubt that warp and weft knitted spacer fabrics will become even more important and popular as fabric manufacturers and consumers become aware of their attributes and appreciate and discover their unique features and properties for specific niche applications today and in the future.

Acknowledgements

The figures and tables used in this chapter are taken from my publication, in the book: *Progress in Textile: Science and Technology,* vol. 3, *Technical Textiles: Technology, Developments and Applications,* published in 2008. I have been granted an official permission for using these tables and figures in this chapter by IAFL Publications, HS27 (first floor), Kailash Colony Market, New Delhi 110048, India. I am grateful for their cooperation and support in this matter.

Appendix

Some examples of products using spacer material (see Figures 10.A1–10.A10).

Figure 10.A1 End cross-sectional view of spacer material.

Figure 10.A2 Upper and lining spacer material for trainers.

Figure 10.A3 Footwear using spacer net structure.

Figure 10.A4 Sportswear using spacer material.

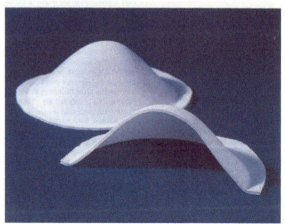

Figure 10.A5 Bra and brief spacer material.

Figure 10.A6 Moisture and heat regulation in spacer material.

Figure 10.A7 Lining fabric for body armour and fire fighter's suits.

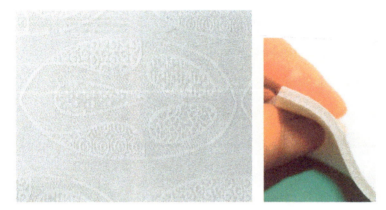

Figure 10.A8 Jacquard spacer material for fashionwear.

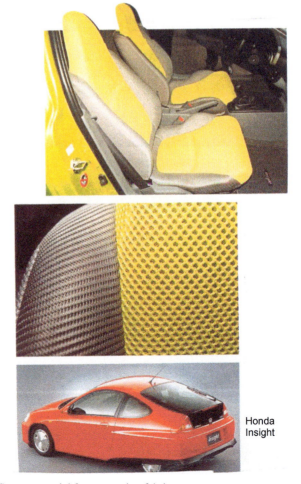

Figure 10.A9 Spacer material for car seating fabric.

Figure 10.A10 3-D pressure-relieving cushion for prevention of pressure sores for patients in wheelchairs.

References

[1] Anand SC. Customer oriented production. Knitting International, April 2004:45.
[2] Double-layer, circular weft knitted fabrics with monofilaments (spacer fabrics), Knitting Technique, 1994;16(5):306.
[3] Bremner N. Spacers about to take-off? Knitting International, June 2004:40.
[4] Millington J. Do we have lift off? Knitting International, October 2002:3.
[5] Ye X, Fangueiro R, Hu H, de Araujo M. Application of warp-knitted spacer fabrics in car seats. JTI 2007;98(4):337–43.
[6] Rock M., et al. Knit fabric with face-to-back differentiation, U.S. Patent No. 6837, 078, B1, Published 4 January 2005.
[7] Anand SC. Double-sided pile fabrics for apparel. Knitting International, July 2005:28.
[8] www.air-x.net
[9] www.tytex.com
[10] Pereira S, Anand SC, Rajendran S, Wood C. Novel 3D knits for knee braces. Knitting International, November 2006:32.
[11] Pereira S, Anand SC, Rajendran S, Wood C. A study of the structure and properties of novel fabrics for knee braces. J Ind Text April 2007;36(4):279.
[12] Pereira S, Anand SC, Rajendran S, Wood C. Empirical modelling of elastic properties of pressure garments for healthcare. In: Proceedings of MEDTEX07 Conference, Bolton, UK, 17 and 18 July 2007.
[13] Valencia IC, Falabella A, Kirsner RS. Chronic venous insufficiency and venous leg ulceration. J Am Acad Dermatol 2001;44(3):401–24.
[14] Palfreyman S, King B, Walsh B. A review of the treatment for venous leg ulcers. Br J Nurs 2007;16(15):S6–14.
[15] Smith PD. Venous ulcers. Br J Surg 1994;81(10):1404–5.
[16] Hildegard C, Ellie L. Principles of leg ulcer management and prevention. In: White R, editor. Wound Care Society Supplement. Huntingdon: Wound Care Society; 2003.
[17] Lee G, Rajendran S, Anand SC. New single-layer compression bandage system for chronic venous leg ulcers. Br J Nurs 2009;18(15):S4–18.
[18] Anand SC, Rajendran S. Development of 3D structures for venous leg ulcer management, Fiber Med 11, Tampere, Finland, 28–30 June 2011.
[19] http://www.nursingtimes.net/journals/2012/01/19/i/q/p/210124-Disc-guy.pdf

[20] Anand SC, Wood C, McArdle B. Development of three dimensional pressure relieving cushions for prevention of pressure sores. In: 5th International Conference: 3-D Fabrics and Their Applications, I.I.T. Delhi, India, 16 and 17 December 2013.

[21] Soin N, et al. Novel "3-D spacer" all fibre piezoelectric textiles for energy harvesting applications. Energy Environ Sci 2014;7:1670–9.

[22] Soin N, Shah TH, Anand SC, Siores E. U.K. Patent Applied for, Application No. 1313911.8, The University of Bolton, UK.

Three-dimensional fabric structures. Part 3 – Three-dimensional nonwoven fabrics and their applications

10

S.C. Anand
Institute of Materials Research and Innovation, The University of Bolton, Bolton, UK

Chapter Outline

10.22   Introduction

Nonwovens are unique engineered fabrics offering cost-effective and often superior alternative solutions for an increasingly wide variety of applications. Individuals and businesses use nonwoven fabrics in every day without knowing it or seeing them because they are often hidden from view. Nonwovens created by a modern and innovative industry are a product for our time. They are often combined with other materials and are often called *composite materials*.

Their end uses range from apparel, household, and soft furnishing to an extremely wide spectrum of technical textiles. This chapter presents evidence that nonwovens are the fastest growing area of technical textiles in Europe and many other countries. Think of tea and coffee bags; personal hygiene products for babies, feminine products, and adult incontinence; infection control garments; geotextiles; fire and heat protection; environmental protection; shroud and coffin linings; and a wide range of textiles for automobiles, aerospace uses, and satellites, just to name a few nonwoven textile uses we encounter everyday.

The world's nonwovens market value has more than doubled in the last 10 years from 3.19 million tonnes in 1999 to an estimated 6.32 million tonnes in 2009. This market was estimated to be US$22.4 billion in value in 2009 and 2.67 kg per capita annual consumption

Handbook of Technical Textiles. http://dx.doi.org/10.1016/B978-1-78242-458-1.00015-7

in Europe [1]. The value of the global nonwoven market was predicted to be \$33.1 billion in 2013 and is projected to grow 7.5% per year to reach \$47 billion by 2018 [2].

This chapter discusses the recent advances in three-dimensional (3-D) nonwoven technologies and products and demonstrates that nonwovens are the most exciting and expanding market worldwide and make a crucial contribution to the enhancement of quality of life in the modern world and in the foreseeable future.

10.23   Definition

A number of definitions have been proposed to describe modern nonwoven fabrics, including (ISO) 9092:1988 and (ASTM D) 1117-95. In its *Textile Terms and Definition* (11th edition), the Textile Institute defined nonwoven fabric as "Fabrics normally made from continuous filaments or from staple fibre webs or batts strengthened by bonding using various techniques; these include adhesive bonding, mechanical interlocking by needling or fluid jet entanglement, thermal bonding, and stitch bonding". Controversial areas are (a) wetlaid fabrics containing wood pulp for which the boundary with paper is not clear, (b) stitch-bonded fabrics that contain yarn for bonding purposes, and (c) needled fabrics containing reinforcing fabric [3].

10.24   Classification of nonwovens

The production of nonwovens can be described as taking place in the three stages of web formation, web bonding, and finishing treatments although modern technology allows an overlapping of these three stages, and in some cases, all three can take place at the same time.

The versatility of each process enables the nonwoven manufacturer to produce different types of fabrics with a variety of properties suitable for the final end use. The various developments in nonwoven technology also enable specialist properties and attributes such as absorbency, capillary wicking, filtration, breathability, and protection (barrier) to be achieved. Figure 10.38 highlights the nonwoven fabric formation techniques available [4].

10.25   Fibre trends

Virtually all types of fibrous material can be used to make nonwoven fabric with the choice being depending on:

a) The required profile of the fabric
b) The cost/use ratio (cost effectiveness)
c) The demands of further processing

The range of staple fibres that can be used in nonwovens can be between 5 and 150 mm at the two extremes of between 0.6 and 300 dtex in thickness. The bulk of the

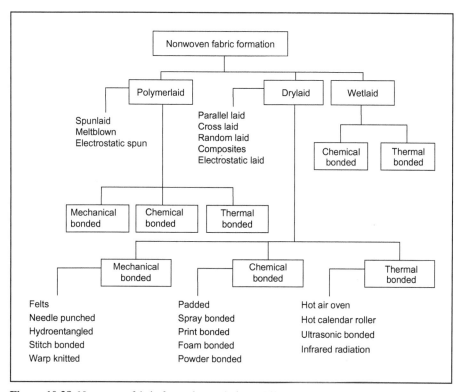

Figure 10.38 Nonwoven fabric formation techniques [4].

fibres used is from 1.5 to 20 dtex and 20 to 100 mm. All spunlaid fabrics are made from continuous filament yarns, which are also incorporated in many types of nonwovens during processing for strengthening the final structure.

The most common materials with the highest growth rate since 1996 have been polypropylene polymer, polypropylene fibre, polyester fibre, wood pulp, viscose fibre, and polyester polymer.

10.26   Three-dimensional nonwovens

In the majority of nonwoven fabrics produced currently, the orientation of fibres in the web just before consolidation or bonding are in the main in two dimensions as shown in Figure 10.39. Drylaid webs can be parallel laid (longitudinal orientation), parallel laid (transverse orientation), cross laid (longitudinal and transverse orientation), and random laid (not oriented or isotropic). Although a small portion of fibres is bound to be oriented in the third dimension (through thickness), the mass of the web is generally built up by depositing many single layers of the web on top of one another [5].

During the last few years, a number of different technologies have been developed and commercialised for producing genuine 3-D nonwoven fabrics. The structure,

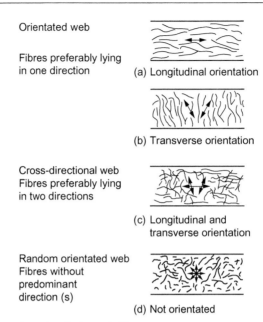

Figure 10.39 Fibre orientation in the web (5).

production, properties, and end-use applications of different 3-D nonwoven technologies are discussed in detail in this chapter.

Struto technology was developed at the Technical University of Liberec, Czech Republic, during 1988–1992. Because of the position of fibres that are predominantly oriented perpendicularly to the area of the fabric, the fabrics show high resistance to compression and excellent elastic recovery after repeated loading. One device for perpendicular layering of fibres in the web before bonding is shown in Figure 10.40. A reciprocating comb (2) pulls the carded web (1) towards conveyor belt (3). A fold is pushed to the fibre layer that is created and moved between the conveyor belt (3) and a wire grid (4). The 3-D web is then either thermally bonded by melting low-melt fibres present in the fibre blend or bonded to another fabric or sprayed with adhesive. These fabrics have found application in thermal insulation, sound insulation, vibration insulation, furniture manufacturing, automotive industry, sleeping bags, acoustic blankets, mattresses, and so on. In another version of this technology, the reciprocating comb is replaced by a worker roller similar to the one used in a carding machine [6]. Further information on Struto technology can be obtained in *Handbook of nonwovens* [10].

V-Lap Pty Ltd was established in 2004 by the CTS Group to produce a very wide range of vertically lapped products by developing specifically designed and engineered nonwoven equipment. It was mentioned earlier that in conventional nonwoven production, fibres are processed and built up into a web structure mainly in two dimensions (i.e. parallel-laid, cross-laid, random-laid, or composite webs). The V-Lap vertical lapping system produces a unique, truly vertical structure that gives the products their special characteristics. Fibres are normally opened and blended using a combination of hoppers, openers, and a carding machine to produce the web ready for vertical lapping [7].

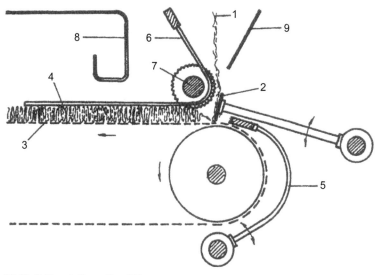

Figure 10.40 3-D web formation [6].

The V-Lap vertical lapping system can process a very wide range of fibres, both natural and manufactured, that can be opened into a web and thermally bonded. Webs and products can be much lighter in weight than needlepunched fabrics. Webs normally contain between 5% and 30% of low-melt fibres and can be made from 100% bicomponent fibre if required [7].

The finished fabric area density can range from 120 to $3000\,\text{g}\,\text{m}^{-2}$ and thickness from 12 to 55 mm. Materials can be split to produce 5-mm-thick products or combined to produce up to 200-mm-thick products [7]. A typical product is illustrated in Figure 10.41. The V-Lap technology was exhibited at ITMA 2011 at Barcelona, Spain.

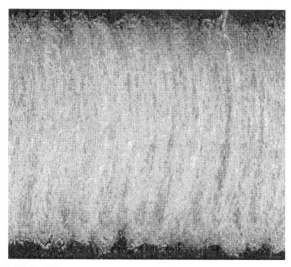

Figure 10.41 3-D fabric [7].

Unique V-Lap material has been successfully tested and approved by several of the largest car manufacturers in the world and is being tested for specific applications, such as door insulators, headliner pads and base materials, under-carpet insulators, and hood and truck liners. Other major industrial areas are filtration, building and construction, and furniture markets [8].

The compression and elastic recovery of these structures are largely influenced by the proportion of thermoplastic bicomponent fibres present and the fibre diameter, which governs fibre rigidity. Stiff, boardlike products are produced by using a high proportion of coarse bicomponent fibre (>5 dtex). Depending on composition and fabric structure, the fabrics have higher resistance to compression and elastic recovery than comparable cross-lapped and high-loft airlaid fabrics as shown in Figure 10.42 [10].

Kunit and Multiknit stitchbonding technologies are fairly recent and were developed to produce 3-D voluminous nonwoven materials to replace foam, neoprene, and other such materials that have certain inherent limitations for use in many technical textiles. In the Kunit process, the fibres are fed to the stitchbonding head either as webs or fleeces. Voluminous textiles with a variable thickness and density that either can have a dense appearance or be in the form of a structure having pile folds in the third dimension can be produced. Such materials have been used as lining, soft-toy fabrics, filters, covering materials for polishing discs, coating substrates, and upholstery materials for car interiors.

The Kunit machine uses a compound needle having a round head. It also uses a brushing bar in conjunction with the stitch-forming elements whose oscillating path can be varied between 6 and 51 mm, depending on the required pile fold height. The

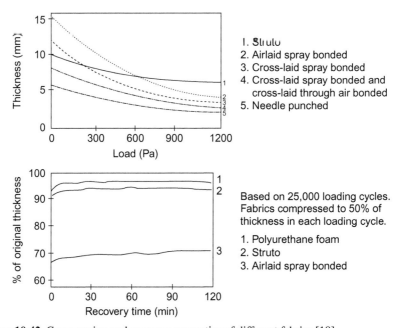

Figure 10.42 Compression and recovery properties of different fabrics [10].

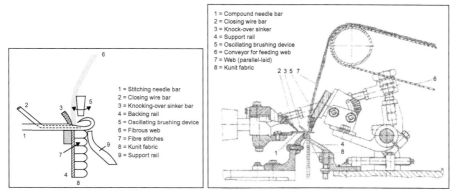

Figure 10.43 (a) Close-up of main element. (b) Main elements of Kunit stitch bonding machine.

elements of a Kunit machine and their relative positions are shown in Figure 10.43. The flat, oscillating brush compacts the lightweight web whose fibres are mainly oriented in the lengthwise direction so that the fibres are pressed into the needle hooks and formed into stitches on one face of the fabric whereas the proportion of the fibres that are not knitted are arranged as crosswise pile folds on the other face. The structure of a Kunit fabric is shown in Figure 10.44.

The Multiknit technique forms both sides of the 3-D nonwoven into a knitted structure, and the two faces are joined by vertical or perpendicular fibres. The starting material is usually a nonwoven that has been stitch bonded on one side by the Kunit technique to form the pile folds. The fibres in the pile folds are then stitch bonded on the Multiknit machine to produce a double-sided, 3-D, stitchbonded nonwoven. Other substrates, such as fabrics, webs, films, powders, and so forth, can be incorporated within the base web and covered by stitch bonding the multilayer material to produce a composite material. The machine uses a pointed-head compound needle.

A line for producing a Multiknit nonwoven would include, for example, a web/fleece-forming unit (card). In the first stage, the web/fleece, which is supplied continuously, is stitch bonded on one side on a Kunit machine and, in the second stage, it is fed continuously to the Multiknit machine where it is stitch bonded on the other side. The main elements of Multiknit stitchbonding machine are illustrated in Figure 10.45. Observe there that two Kunit fabrics are combined face-to-face to produce a double-thick

Figure 10.44 (a) Close-up of fabric structure. (b) Structure on Kunit fabric.

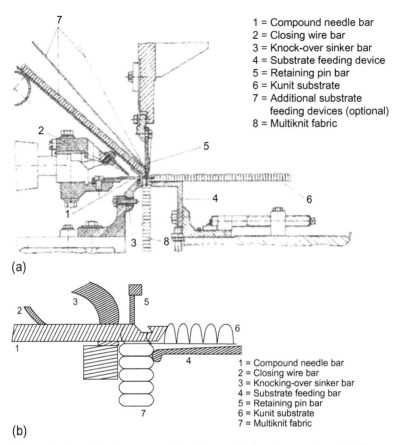

1 = Compound needle bar
2 = Closing wire bar
3 = Knock-over sinker bar
4 = Substrate feeding device
5 = Retaining pin bar
6 = Kunit substrate
7 = Additional substrate
     feeding devices (optional)
8 = Multiknit fabric

(a)

1 = Compound needle bar
2 = Closing wire bar
3 = Knocking-over sinker bar
4 = Substrate feeding bar
5 = Retaining pin bar
6 = Kunit substrate
7 = Multiknit fabric

(b)

Figure 10.45 (a) Main elements of multiknit stitch bonding machine. (b) Close-up of main elements.

Multiknit structure with a knitted structure on both faces. Other substrates 7, shown in Figure 10.45(a), which could be woven, knitted, or other nonwoven or other types of substrates can be incorporated during processing. Figure 10.46 shows the structure of a Multiknit fabric and a complete line from fibre opening, carding, Kunit machine, Multiknit machine, and the batching device is shown in Figure 10.47. Possible fabric

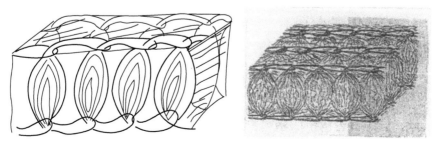

Figure 10.46 (a) Close-up of fabric structure. (b) Structure of multiknit fabrics.

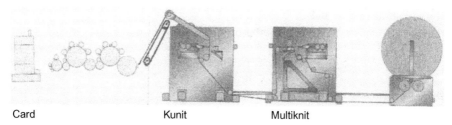

Card   Kunit   Multiknit

Figure 10.47 A continuous line to produce Kunit or multiknit fabrics.

area densities currently range from around 120 to 800 g m^{-2} for single-layer Multiknit nonwovens and from about 150 to 1500 g m^{-2} for multi-layer structures.

Both Kunit and Multiknit machines are currently offered in widths of either 114 in. (2900 mm) or 149 in. (3800 mm) and in gauges in widths ranging from 10 to 22 needles per 25 mm. The working width can be reduced at will, and other machine gauges are supplied upon request.

Both systems are also supplied with Karl Mayer Command System (KAMCOS) with a 12.1-in. touch screen, operator interface – IPC motion control, single speed for the control of the basic functions, and Ethernet interface for interconnection of machines via network and connection to relevant MDE systems, Teleservice [9].

The full production of Malimo stitchbonding machines was transferred to Pinkert Machines-Malimo in Germany in September 2012. The company offers the complete range of machines, including Maliwatt, Malivlies, Kunit, and Multiknit machines [11].

More than 65 million cars are produced worldwide each year; each one uses between 10 and 12 m^2 seating fabric. Seating fabric is a laminate consisting of a face fabric, which is normally produced from either air-textured or false-twist polyester yarn, which is flame bonded to PU foam and a warp knitted scrim, in such a manner that the foam is sandwiched between the polyester face fabric and polyester warp knitted scrim. The PU foam has excellent elastic compression recovery and imparts softness and bulk to the seating fabric, which in turn makes the seat comfortable. The foam, however, is negatively associated with emissions from additives, including flame-retardant chemicals and full life cycle problems relating to disposal and recycling after use. Other problems are that during the flame-bonding process, singed substances cause strong fogging. Foam seating materials also exhibit poor air permeability and moisture and water vapour permeability and facilitate poor moisture and temperature regulation. Caliweb, the registered trademark of products manufactured using the Kalitherm technique, is used for mechanically and thermally bonded nonwovens and for lamination of nonwoven composites with fabrics such as those used for car seating. This system uses Kunit or Multiknit fabrics that have been thermally bonded to produce a foam substitute.

Lamination is preferably done by using flat-bed laminating systems. The nonwoven foam substitute requires a proportion of low-melt or biocomponent fibres (sheath-core type) to effect stabilisation in thermal bonding. If produced from 100% polyester, these materials can be completely recyclable and have high air, moisture, and water vapour permeability and good ageing behaviour. They have low emissions and do not cause fogging. This is only one example of the successful use of 3-D nonwoven structures made by the Kunit and Multiknit techniques. These lightweight, low-density, high-volume, 100% nonwoven structures with structured and smooth surfaces either on one face (Kunit) or on both faces (Multiknit) are successfully used as heat, noise, vibration, and sound insulation materials and as dust filters, adhesive tapes, shoe linings, antidecubitus mattings, and medical, hygiene, and sanitary materials [12]. These fabrics have been used as part of upholstered furniture, car seating, and other furniture as a replacement for PU foam. Other applications include filter fabrics; heat, noise, and vibration insulation; and excellent mouldability with smooth and uniform surface on both faces; these fabrics are weldable as long as the correct materials are used.

The Struto, V-Lap vertical lapping system, and the Kunit and Multiknit technologies can be used for the following applications:

- Mattress and upholstery fillings
- Thermal insulations
- Acoustic insulations
- Clothing fillings
- Fatigue pads
- Shoe liners
- Underfelts
- Medical/hygiene products
- Sports underlays
- Horse blankets
- Automotive products
- PU foam replacement
- Filtration
- Grass/plant root reinforcement and stabilisation
- Vibration control

Other examples of recently developed 3-D nonwoven structures include Hydrospace fabrics developed and characterised by the Nonwoven Research Group at the University of Leeds, United Kingdom. These fabrics are manufactured using a hydroentanglement process; one objective is to incorporate active ingredients, such as sustained drug delivery systems when the fabric is made rather than applying them after the fabric has been produced. One of the major attributes of these novel structures is that 3-D cavities are incorporated in the fabric structure during manufacture [13].

10.27 Summary

This chapter has provided evidence that the development in all aspects of nonwoven fabric manufacture is taking place at a rapid rate. We live in the age of relatively short-life textile products that fit current purposes and can be replaced or improved in a short timescale.

Recent advances in 3-D nonwoven structures and technologies have enabled the technical textiles manufacturers to produce technically advanced products that offer superior properties and can be easily recycled because they are made from 100% textile fibres that normally consist of only one fibre type.

Acknowledgements

I have been granted official permission to use the following tables and figures in this chapter:

Figure 10.40: Carolina Academic Press, 700 Keutst., Durham, NC 27701, USA.
Figure 10.41: CTS Group PTY LTD/VLAP PTY LTD., Australia.
Figures 10.43–10.47: Karl Mayer Textilmaschinenfabrik GmbH, Germany, Brühistrasse 25, 63179 Obertshausen, Germany.

I am truly grateful to the above companies for their kind cooperation and support.

References

[1] Wiertz, P. In: International conference on technical textiles and nonwovens. 11–13 November 2008, New Delhi, India; 2008.
[2] The future of global nonwovens markets to 2018, Smithers Apex Market Reports, publications@smithersapex-info.com.
[3] Textile terms and definitions, Textile Institute, UK, 11th ed.; 2002, p. 234.
[4] Rigby AJ, Anand SC. Technical Textiles International, September 1996, p. 22.
[5] Lunenschloss, J., Albrecht, W., 1985. Non-woven bonded fabrics. Chichester, UK: Ellis Harwood Ltd.; 1985, p. 120.
[6] Jirsak, O., Wadsworth, L.C., 1999. Nonwoven textiles. North Carolina, USA: Carolina Academic Press, 1999, pp. 58–9.
[7] www.v-lap.com
[8] The V-Lap advantage, Nonwovens Report International, Issue 4, 2011, p. 16.
[9] Karl Mayer Technical Bulletin on Kunit and Multiknit Technology.
[10] Russell, S.J. (Ed.), 2007. Handbook of nonwovens. Cambridge, UK: Woodhead Publishing Ltd, pp. 72–3.
[11] www.pinkert-machines.com
[12] Russell, S.J. (Ed.), 2007. Handbook of nonwovens. Cambridge, UK: Woodhead Publishing Ltd, p. 219.
[13] Wilson A. 2009. Textiles, Number 1, p. 22.

One-dimensional textiles: rope, cord, twine, webbing, and nets

J.W.S. Hearle
University of Manchester, Manchester, United Kingdom

Chapter Outline

11.1   Introduction

11.1.1   Tension textiles

The great majority of textile materials come in thin, planar two-dimensional (2D) sheets. In recent years, three-dimensional (3D) fabrics, either in overall form or in thick multiple layers, have grown in importance. This chapter is concerned with one-dimensional (1D) textiles, commonly referred to as *cordage*. Their use is almost always to apply tension, whether in ship moorings, ropes for climbing, tent guylines, parcel ties, a cut sewn by a surgeon or a host of other uses on land and sea. They range in size from large ropes to fine twines. The majority are made by twisting yarns together. For example, the most common form of rope was and is a three-strand rope with multiple twist levels, although there are now other forms that are described in Section 11.3.3.

Two 2D forms also are considered in this chapter. Narrow rectangular fabrics made by weaving or braiding have a similar function as tension textiles and are called *webbing*, *strapping*, or *tape*. Nets are essentially a 2D collection of holes usually formed by cordage.

11.1.2   Definitions of cordage: Rope, cord, twine, string, and thread

Textile Terms and Definitions (TT&D) (Denton and Daniels, 2002), which is produced by expert panels of the Textile Institute, contains many relevant definitions. *Cordage* is defined as "any product, regardless of size made by twisting or braiding together yarns

Handbook of Technical Textiles. http://dx.doi.org/10.1016/B978-1-78242-458-1.00011-X

which is generally round in cross-section and capable of sustaining loads". *Ropes* are defined as being "cordage more than approximately 4 [millimetre] mm in diameter". *Cable* and *hawser* are terms used in industry for the larger sizes. The definition of *cord* is "a variety of textile strands including: (i) **cabled yarns**; (ii) **plied yarns** and structures made by **braiding, knitting,** or **weaving**. See also **bullion cord, cable cord, case cord, crepe cord, upholstery cord,** and **welting cord**".[1] It is most commonly used with a specific descriptor such as tyre cord and sash cord. *Twine* has this generic definition: "Twisted cordage less than 4 mm in diameter". *String*, typically 1 or 2 mm in diameter, is a commonly used term that is not technically defined. *Thread* is defined as "(1) A textile yarn in general. (2) The result of twisting together in one or more operations two or more single or folded yarns. (3) A term related to silk yarn processing". It is commonly used in reference to a particular twine such as sewing thread. *Yarn* is defined as "a product of substantial length and relatively small cross-section consisting of fibres and/or filament(s) with or without twist"; it is not a term used in cordage, except as the starting material in manufacturing or more specifically as the synonymous terms *folded, doubled,* or *pled yarns*. Various professional and standards organisations have other definitions for cordage.

11.1.3   Outline of chapter

This chapter starts with a short account of the materials used to make cordage. It is followed by a section on ropes, which covers their history, the various types and structures, and the means of production. Smaller cordage, webbing and tapes, and nets are dealt with in successive sections. The conclusion introduces future trends and lists sources of information.

11.2   Materials

11.2.1   Traditional fibres

Most of the conventional fibres referred to in Chapter 2 have been or are widely used in cordage with the exception of wool and other hairs, acetate, glass, ceramic, and carbon fibres.

Until the mid-twentieth century, natural plant fibres were the dominant material, particularly the coarse fibres extracted from the leaves or stems of plants. Locally, a great many different types have been used. The most important for mainstream production are listed in Table 11.1. The amounts used in 1951 still reflect the relative importance of the various types, although they have lost market share to manufactured fibres.

Leaf and stem fibres are characterised by relatively high strength (~0.5 N/tex) and low break extension (~3%). Although largely replaced by manufactured fibres, there is still substantial usage of these natural fibres. Cotton is weaker and more extensible (~0.3 N/tex, 7%) and makes softer ropes. It was the first fibre to be used in tyre cords.

---

[1] Items in **bold font** refer to other definitions in *TT&D*.

Table 11.1 Principal vegetable fibres for cordage, with consumption in the United States in 1951, as listed by Himmelfarb (1957)

Fibre	Botanical name	Location in plant	Million pounds (tonnes)
Hard fibres			
Abaca (Manila hemp)	*Musa textilis*	Leaf	94 (43)
Sisal	*Agave sisalana*	Leaf	121 (66)
Henequen	*Agave fourcroydes*	Leaf	57 (26)
Soft fibres			
Jute	*Corchorus*	Stem	45 (20)
Hemp	*Cannabis sativa*	Stem	13 (6)
Flax	*Linum usitatissimum*	Stem	
Cotton	*Gossypium*	Attached to seed	13 (6)

Silk ropes were used in decorative situations but also in the earliest parachute cords because of its combination of strength and extensibility (tenacity ~0.4 N/tex, breaking extension ~20%), which gives a high work of rupture.

11.2.2 Cellulose fibres

Regenerated rayon fibres have been little used in cordage with one exception. For a period from the 1930s to the 1950s, successively improved forms of high-tenacity viscose rayon were the most important tyre cord fibre.

11.2.3 First wave of synthetic polymer fibres

Nylon 66, launched by DuPont in the late 1930s, by the 1940s had replaced silk in parachute cords and went on to become a major rope fibre. Polyethylene and, more importantly, polypropylene and polyester followed later. This group of manufactured fibres now dominates the mass market for cordage. They are characterised by relatively high strength and break extension (~0.6 N/tex and 15% in high-tenacity forms, respectively). The factors influencing choice in this group include:

- Nylon (6 or 66): good elastic recovery; high energy absorption; poor wet abrasion resistance.
- Polyester: higher modulus (stiffness) than nylon; good abrasion resistance.
- Polypropylene: poorer elastic recovery.
- Polyethylene; cheaper; weaker; poorer elastic recovery.

It is not only the material that is important. The finish has a major effect on abrasion resistance. In particular, yarns with special marine finishes should be used in water.

11.2.4 High-performance fibres

In the 1970s, a new group of high-modulus, high-tenacity fibres became available, characterised by high strength and low break extension (~3 N/tex; 3%). One group is made of stiff, interactive polymer molecules, which associate in highly oriented, chain-extended forms. They include:

* Aramid: *Kevlar* (Dupont), *Twaron* (AKZO); a variant form, *Technora* (Teijin).
* Aromatic copolyester: *Vectran* (Celanese).
* Polybenzoxazole (PBO): *Zylon* (Toyobo).

A similar structure can be made from flexible, inert polyethylene molecules, which can be highly drawn after extrusion as a gel. *Dyneema* and *Spectra* are high-modulus polyethylene (HMPE) fibres. These fibres are prone to creep, so they are less suitable for use in long periods under load, particularly in hot conditions.

11.2.5 Wire ropes

The development of iron technology in the nineteenth century led to the production of steel wire. Multiple wires, analogous to textile monofilaments, could be twisted together to make ropes. This led to a major change from natural fibre ropes to wire ropes in the most demanding applications such as bridge cables and hoists. However, these ropes have only an indirect relevance to technical textiles. A significant difference is that steel has a density of ~8 g/cm^3 whereas the polymer fibres are in the range of 0.9–1.5 g/cm^3. This means that a shorter length of steel breaks under its own weight, which has led, for example, to polyester ropes replacing wire ropes in deepwater moorings.

11.3 Ropes

11.3.1 History of ropes

The 1D structure, cordage, is probably the oldest form of textile material, predating the change from skins to 2D woven fabrics for clothing. Early mankind once used natural creepers as "tension textiles" but then found that they could twist natural fibres together into yarns and then assemble them into larger and stronger forms. The earliest record of the use of a rope (or a creeper) may be in a cave painting in Spain from 10,000 or more years ago, which shows a honey gatherer climbing down a cliff. An Egyptian tomb from circa 2400 BC has a drawing of ropes on a sailing boat. A clear view of three-strand ropes being used to move a huge statue of a bull is shown in a bas-relief from Nineveh, circa 700 BC (see Figure 11.1). By Greek and Roman times, making cordage was a well-established trade.

Prehistoric manufacture of rope is shown in Figure 11.2, which is a drawing from a tomb in Thebes, circa 1450 BC. It shows a rope being formed by leather strips being twisted together. A more tangible find was a three-strand papyrus rope made in 500 BC and found in a cave in Egypt in the 1940s (Figure 11.3). This

Figure 11.1 Detail of a bull colossus being pulled along by ropes, circa 700 BC; bas-relief from Neineveh.
From a drawing made on the spot by A.H. Layard during his second expedition to Assyria (Layard, 1853). Reproduced by courtesy of the Director and Librarian, the John Rylands Library at the University of Manchester.

Figure 11.2 Leather ropemakers from a tomb in Thebes, circa 1450 BC.
From Gilbert (1954).

remained the basic technology until the mid-twentieth century. The production was mechanised in rope walks (Figure 11.4) but remained manually operated. Twisting is by a rope jack (Figure 11.5), which runs on a track from the yarn supply end to the full length of rope being made. In stage 1, three sets of yarns are pulled off bobbins and pulled along the length of the ropewalk. In stage 2, an assistant turns the crank handle of the jack so that the yarns are twisted into strands by the rotation of the three hooks on the jack. Twist causes the lengths to contract so that the carriage has to move back along the ropewalk under the control of the ropemaker. In stage 3, the hook on the carriage rotates in order to twist the strands into the rope. Figure 11.6 shows the ropewalk at Chatham Dockyard, United Kingdom, which remained operational until the mid-1900s. Today ropewalks remain only in heritage

Figure 11.3 A papyrus rope made in 500 BC.
From Tyson (1966).

Figure 11.4 A ropewalk from Denis Diderot's Eighteenth Century *L'Encyclopedia, ou Dictionnaire Raisonne des Sciences, des Arts et des Metiers* [Trades].
Reprint edited by Gillispie (1950).

attractions, such as Hawes Ropemakers in Yorkshire; its workshop also contains more modern machinery and offers a variety of ropes for sale.

11.3.2   Three- and four-strand ropes

Three-strand ropes dominate the market for commodity ropes. Four-strand ropes are a less important form made in a similar way. For both types, the ropemaker starts with "textile yarns", either twisted, natural staple-fibre yarns, or continuous filament yarns

Figure 11.5 A ropemaker's jack.

Figure 11.6 The 344-m long ropewalk at Chatham dockyard.

with low-twist or interlacing from the yarn producer. These are then twisted in the following sequence: textile yarn → rope yarn → strand → rope. The components making up the rope are illustrated in Figure 11.7. The twist alternates between S and Z in order to minimise torque, which promotes overtwisting or undertwisting when the rope is under tension.

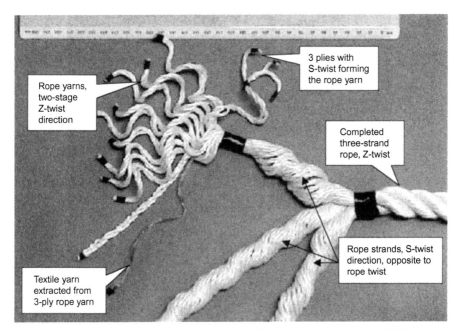

Figure 11.7 The components of a three-strand rope. Picture by McKenna et al. (2004).

11.3.3 Newer rope types

Traditional natural fibre ropes require high twist levels to prevent the slippage of short fibres past one another. The obliquity reduces strength. With continuous filament yarns, the high twist is no longer essential. New rope structures shown in Figure 11.8 were developed in the second half of the twentieth century. The advances were also driven by two other factors: Some uses demanded much larger ropes; balanced, torque-free ropes were needed to avoid failure mechanisms resulting from rope twisting.

Braiding had long been used for small cords, but it was not until machinery manufacturers produced large braiding machines that it became possible to make large ropes up to around 30 cm in diameter. A circular braid has two sets of identical yarns that interlace with one another as they follow helical paths in opposite senses. Both 8-strand (Figure 11.8a) and 12-strand (Figure 11.8b) braided ropes are made. Another form is braid-on-braid (Figure 11.8c), which has a core and a cover.

Wire ropes are made with multiple layers of stands, such as the 6-round-1 and the 36-strand forms (Figure 11.8e and f). These constructions are easier to make in large sizes than three-strand ropes and have been adopted for fibre ropes. They are usually encased in a braided jacket.

Another approach was to make smaller low-twist, three-strand ropes on existing machinery. They are rather loose structures, but a large rope with good coherence can be made by assembling a number of the three-strand ropes in a braided jacket (Figure 11.8g). A variant has long-lay braided ropes as the subropes. Somewhat confusingly, both these types are known as *parallel strand ropes*.

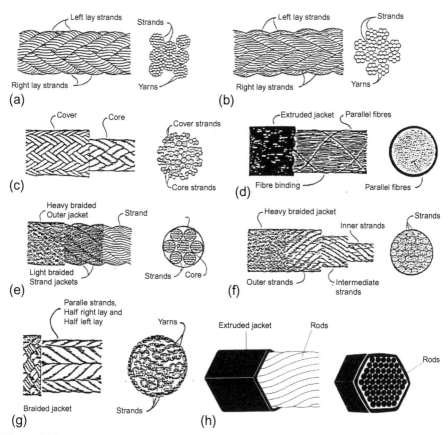

Figure 11.8 Modern rope types: (a) 8-strand braided, (b) 12-strand braided, (c) double braid or braid-on-braid, (d) parallel yarn, (e) 6-round-1 wire-rope construction, (f) 36-strand wire-rope construction $(18+12+6+1)$, (g) parallel strand, and (h) pultruded rod.

Ropes with a twisted or braided structure fit naturally into the ropemaking industry and are made by a number of suppliers. If required, the supply yarns can be given a special finish to reduce internal abrasion, and a polymer coating can be applied to give external protection.

In principle, a continuous filament tow or a zero-twist assembly of continuous filament yarns could act as a rope, much like Rapunzel's hair. This would eliminate the loss of strength when fibres lie at an angle in a structured rope. In practice, a parallel assembly lacks the coherence needed in a rope. The invention of parallel yarn ropes (Figure 11.8d), such as Parafil developed by ICI and now produced by Linear Composites, achieved coherence by encasing multifilament yarns in a plastic coating. These ropes are stiffer than structured ropes but are suited for applications where appreciable bending is not required.

Pultrusion is a way of making rigid composite rods that can be adapted to make pultruded ropes (Figure 11.8h) by pulling a filament tow through a die and coating with resin. These ropes do not have the flexibility of more conventional ropes and are not common. The technique could offer a way of making ropes of glass, carbon, or ceramic fibres.

11.3.4   Rope production

The preliminary stages of rope production follow normal textile operations. The "textile yarns" are either spun from staple fibres or purchased as continuous filament yarns from fibre producers. They are then converted into multiply "rope yarns" on uptwisters or downtwisters.

As described earlier, traditional production of three-strand ropes was a manual batch process. In the twentieth century, machines were developed for continuous production. The schematic diagram, Figure 11.9, shows how three strands can be continuously hauled off large bobbins, twisted, and wound up. The next development was to combine this with strand production in a two-stage machine such as the one shown in Figure 11.10. In the first half of the machine, rope yarns on a creel are fed to strand formation; the strands then pass to the rope formation stage.

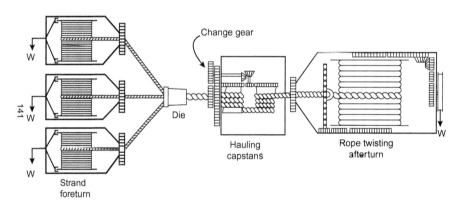

Figure 11.9 Ropelaying machine from Himmelfarb (1957).

Figure 11.10 A ropemaking machine form Roblon.

Large braiding machines for ropemaking operate in the same maypole fashion as the long-established small braiders. The bobbins containing the strands are mounted above horn gears that cause the strands to interlace as the bobbins rotate in opposite directions. Braid-on-braid is produced in two stages with the braid from the first stage being fed through the middle of the second-stage braider. Parallel strand ropes (subropes in a braided cover) and the application of braided jackets to other rope types are applied in a similar way.

For wire ropes, twisting machinery is available from the wire-rope industry.

For parallel yarn and pultruded ropes, pultrusion machinery, which is used in the plastics industry to make cylindrical, fibre-reinforced composites with uniform cross-section, can be modified.

11.3.5   After treatment

Ropes are produced under relatively low tension, and the structure is not fully compacted and will contain irregularities in local packing of fibres, yarns, and strands. On loading, the rope tightens and increases by a length that is not recovered on removing the load. For many purposes, it is therefore desirable to apply a number of moderate cyclic loads before use. In some cases, ropemakers apply stretching treatments to improve performance.

Plasma-braided HMPE ropes made by Cortland's patented recrystallisation process (Cortland, 2014), are a specific example. The process draws twisted strands through a hot fluid vessel. This enhances the properties of the HMPE fibre and improves strength efficiency.

11.4   Cordage

11.4.1   Production of cords

Singles, twisted, staple-fibre yarn untwists when hung freely under tension; the filaments in a multifilament yarn tends to spread out when free of tension. These effects make such yarns generally unsuitable as 1D tension textiles. Greater cohesion is needed.

Plied yarns in which two or more yarns are twisted together are stable structures and one of the common forms of cords, such as tyre cords, twine, string, and thread. The two ways of making twisted cords are (1) down-twisting on ring or flyer twisting machines in which the takeup package is rotated and (2) uptwistng in which the supply package is rotated.

Another way of making cords is braiding, also called *plaiting*, on small braiding machines.

11.5   Webbings, strappings, belts, and tapes

11.5.1   Narrow-fabric forms

There are occasions when 1D tension textiles need to be flat strips, which can also be regarded as 2D fabrics rather than the circular cordage described so far. These narrow fabrics range from lightweight *tapes* to heavy *webbing* used for hoisting loads and may also be referred to as *strappings* or *belts*. The rectangular cross-section provides a structure with a lower thickness and greater width than the equal thickness in all directions of a circular cross-section. This geometry is more suitable for some applications; for example, they are more easily fitted with buckles. When used as a belt or strap, the contact loading is more widely distributed. For conveyor belts, a circular cross-section would be useless.

The two structures that are used are weaves, which are described in Chapter 4, and braids, which are similar in their interlacing to the circular braids described in Section 11.3.2 except that the two sets of yarn move at opposite angles to and fro across the flat braid.

11.5.2 Production

Woven tapes and webbings are made on narrow-fabric looms. The warp is fed from a creel, and the weft is translated by a needle that grips the yarn.

Flat braiding machines operate with horn gears that guide the interlacing yarn paths.

11.6 Nets

11.6.1 2D cordage assemblies

Nets are a form of 2D fabric in which holes occupy more space than yarns. The definition of *net* in *TT&D* (Denton and Daniels, 2002) is "an open mesh fabric in which a firm structure is ensured by some form of twist, interlocking, or knitting of the yarn. It may be produced by gauze weaving, knitting, or knotting or on a lace machine". The essential feature is that the crossovers must be firmly locked. The "yarns" are commonly some form of cordage, but some horticultural nets are made of thermoplastic monofilaments bonded at crossovers.

Nets range from very light structures, such as a mosquito net, through coarser forms, such as fishing nets, to massive structures, such as riser protection nets for an oil rig where the "yarns" are heavy-duty ropes.

11.6.2 Production

Nets are another ancient technology. Archaeologists have found remains and pictures of fishing nets from thousands of years ago. Until recently, most technical nets were made by hand knotting. This process continues in the craft environment, as described by Brandon and Bunting (2004), who provide equipment and instructions for netmaking. Figure 11.11 shows a typical netmaking knot (a), a net being constructed on a pegboard (b), and an example of what can be made (c).

Industrial machinery is now available for making nets. An example, based on a sewing machine, is shown in Figure 11.12 (a) a full-size fishing net machine and (b) other machines are based on Raschel knitting.

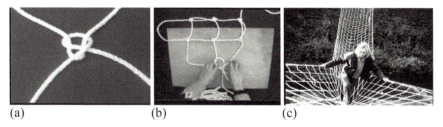

Figure 11.11 (a) Netting knot, (b) making a net on a pegboard, and (c) a net in use. From Brandon and Bunting (2004).

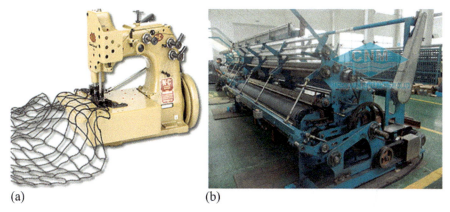

Figure 11.12 (a) Model 81200TF heavy-duty netting machine from Nettingmachinery. com, Netting Machinery, Newark, NJ, USA. (b) Double knot nylon multifilament fishing net making machine from Yangzhou Ocean Friend Import and Export Co., Ltd, Yangzhou, Jiangsu, China.

11.7 Cordage properties

11.7.1 Mechanical and other properties

The obviously important properties of ropes and cords are easily listed and follow those of other technical textiles with well-known test methods.

- Load-elongation, normalised as stress–strain, performance.
- Break load (tenacity), break extension, energy absorption, tensile stiffness (modulus).
- Bending and twisting performance.
- Internal and external friction.
- Melting point.
- Effects of moisture.
- Environmental stability (e.g. effect of UV exposure and chemical and biological attack).

One particular aspect requires more attention. Ropes are often used for long periods subject to changing loads. What determines their effective life?

11.7.2   Rope failure modes

Fibres can fail in a variety of ways, depending on their loading history. An *Atlas of Fibre Fracture and Damage to Textiles* (Hearle et al., 1998) shows the different forms of fibre failure. For example, breaks from tensile overload of fibres are shown in Figure 11.13a–d. The maximum force that a rope can stand without breaking is a basic

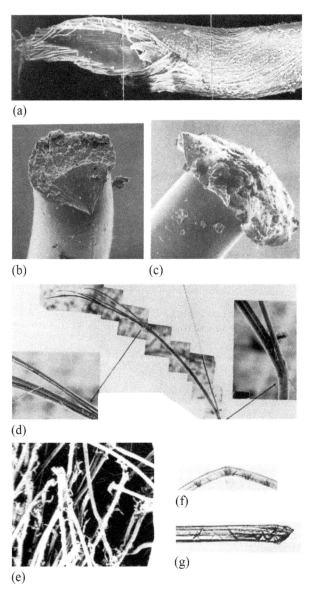

(a)

(b)   (c)

(d)

(e)   (f)

(g)

Figure 11.13 (a) Tensile break of cotton, (b) tensile break of nylon, similar for polyester, (c) break of nylon at high speed often observed at the final stage of a rope break as stored energy is released, (d) tensile break of aramid, (e) internal abrasion in a nylon rope, and (f and g) axial compression in aramid rope showing kink bands and angular break.

property, and break can be avoided by working with an adequate safety factor. In long-term use, rope strength decreases, usually at a fairly low rate for a long period but then rapidly decreasing as more fibres break and transfer higher stresses to the remaining fibres. The decay from several forms of "fatigue" involves:

- *Hysteresis heating.* Internal losses and friction between fibres generates heat that weakens the fibres.
- *Creep rupture.* Under load, polymer fibres extend with time and eventually break. This is a particular problem with HMPE ropes, especially in warmer conditions. For example, Spectra 900 has been found to break after 182 days at 15% of break load. However other, HMPE yarns, such as Spectra 1000 and various forms of Dyneema, have improved resistance to creep.
- *Internal abrasion.* Cyclic loading leads to slippage between neighbouring fibres. The slip can be relative axial movement because of angle changes in rope extension or twisting, relative rotation or scissoring. The shear forces cause axial splitting, peeling of fibre surfaces and eventual fibre failure; see Figure 11.13e. Nylon, especially when wet, is particularly prone to this effect because the shear cracks are at angle and cross the fibre in the length of a few diameters. In polyester fibres, the cracks are almost parallel to the fibre, so the effect is much less severe. Suitable finishes reduce the problem.
- *External abrasion.* Passage over guides or other surface leads to surface wear. This is one reason for jacketing ropes.
- *Axial compression fatigue.* If rope components do not share tension evenly, some will go into axial compression when rope tension is low. Twisting causes more severe axial compression. The result is the development of kink bands within fibres, which eventually develop into cracks and breakage. Highly oriented fibres, such as aramids, are particularly prone to this form of fatigue, Figure 11.13f and g.

11.8   Conclusion

11.8.1   Future trends

The twentieth century was a time of great innovation in ropes, cordage, and nets that resulted from new fibres, new constructions, and advances in machinery. The situation is now more stable, and major innovations are unlikely in the mainstream cordage industry. Ropes are now made in as large a size as seems sensible. There is more opportunity for developments in the uses of ropes to be discussed in Volume 2, including the potential for fibre ropes competing in the markets now occupied by wire ropes.

Carbon fibres, which can be spun into yarns, have the potential to make ropes and cords with higher strength and axial stiffness. However, their brittleness makes them unsuitable for most uses of ropes, and any advantage over other high-performance fibres is probably not worth the effort. Carbon nanotubes or rolled up graphene would provide a step change in properties, which could be exploited if ways could be found to convert them into yarns. They would probably be most effective if consolidated with resin to form solid wires.

The twentieth century also saw the rope industry change in Europe and America from a larger number of smaller companies to a few major ropemakers. In China, the industry still ranges from farmers making agricultural cordage in their sheds to substantial companies that are successful in exporting to the Western consumer markets for everyday ropes. They are less advanced in manufacturing ropes for demanding engineering uses, but this is changing.

11.8.2   Sources of information

Himmelfarb, who was the US Navy ropemaker, provides much information on traditional ropes that is still relevant today in his book, *The Technology of Cordage Fibres and Ropes* (Himmelfarb, 1957). The modern equivalent is the *Handbook of Fibre Rope Technology* by McKenna et al. (2004). Another source of information, which resulted from intensive study as oil rigs moved to deeper water and fibre ropes had to be used instead of wire ropes, is *Deepwater Fibre Moorings: An Engineers' Design Guide* (Tension Technology International and Noble Denton, 1999).

A great deal of information is on the Internet. This ranges from general accounts in Wikipedia to much detailed data on cordage types, specifications, and properties on the websites of ropemakers. The websites of machinery companies provide descriptions of machines. Typical examples are www.bridon.com for ropes and www.roblon.com for machinery. A search on Google yields many more, although it is necessary to navigate through entries, such as *Ropemakers*, Bridport's Music Pub. More technical information is on the website of the consulting company: Tension Technology International Ltd: www.tensiontech.com.

References

Brandon, K., Bunting, H., 2004. A guide to net-making. http://duo.irational.org/red_net/a_ guide_to_net_making.pdf (accessed 02.06.15).
Cortland, 2014. Plasma® Synthetic Rope. www.cortlandcompany.com/plasma (accessed 20.11.14).
Denton, M.J., Daniels, P.N. (Eds.), 2002. Textile Terms and Definitions, eleventh ed., The Textile Institute, Manchester (a continually updated version of "TT&D" is now available on-line).
Gilbert, K.R., 1954. Rope making. In: Singer, C., Holmyard, E.J., Hall, A.R. (Eds.), A History of Technology, vol. 1. Clarendon Press, Oxford, UK, pp. 451–455.
Gillispie, C.C. (Ed.), 1950. A Diderot Pictorial Encycopedia of Trades and Industry, 458 plates selected from "L'Encyclopedaa od Denis Diderot". Dover Publications, New York.
Hearle, J.W.S., Lomas, B., Cookem, W.D., 1998. Atlas of Fibre Fracture and Damage to Textiles, second ed. Woodhead Publishing, Cambridge.
Himmelfarb, D., 1957. The Technology of Cordage Fibres and Ropes. Leonard Hill, London.
Layard, A.H., 1853. A Second Series of the Monuments of Nineveh Including Bas-Reliefs from the Palace of Sennacherib and Bronzes from the Ruins of Nimroud. John Murray, London.
McKenna, H.A., Hearle, J.W.S., O'Hear, N., 2004. Handbook of Fibre Rope Technology. Woodhead Publishing, Cambridge.
Tension Technology International and Noble Denton, 1999. Deepwater Fibre Mooings: An Engineers' Design Guide. Oilfields Publications, Ledbury.
Tyson, W., 1966. Rope: A History of the Hard Fibre Cordage Industry in the United Kingdom. Wheatlands Journals, London.

Index

Note: Page numbers followed by *f* indicate figures and *t* indicate tables.